9/1/78

Preface

Far too often, students with only a background of high-school algebra and trigonometry find the first semester of calculus a traumatic experience because they have to absorb too many concepts too rapidly. Ideally, a student entering a beginning course in calculus should have a basic understanding of the elementary functions, be able to draw a quick sketch of the graph of such a function, and also have had some experience with the notion of a limit. The purpose of this book is to bridge the gap between high-school mathematics and calculus in these areas. Specifically, it is intended as a precalculus course for a student who has completed a program through intermediate algebra and trigonometry, although the latter is not a prerequisite for the course.

The main theme throughout the book is the relationship of a function and its graph. Our effort is to teach the reader how to quickly analyze a simple function and to sketch, with a minimum of plotting points, a graph which illustrates the basic properties of the function. One aspect of this work is a consideration of such properties as domain, range, intercepts, symmetry, regions of increase or decrease, concavity, and asymptotes. Another important aspect is an understanding of how many functions can be generated from four elementary functions—the unit, identity, sine, and exponential functions—and how, correspondingly, their properties and graphs can be quickly obtained.

The notion of a limit is introduced in the beginning of the book in connection with an elementary treatment of sequences. This serves as a springboard to a later introduction of the notion of continuity of a function and the concept of a "slope function" or derivative of a function—concepts that are essential to a real understanding of the relationship between a function and its graph.

There are some things that we do *not* attempt to do in this book. We do not develop or emphasize the structure of the real number field or fields in general, and there will be no heavy emphasis on set theory or logic.

In the development of this text we owe much to many people. Particularly we are indebted to Larew M. Collister and George B. Pedrick for their critical evaluation of the manuscript, to Miss Pamela Conaghan of Scott, Foresman and Company for her cooperation and assistance in transforming the manuscript into this finished text, and to our wives for their encouragement and understanding during the creative process.

Charles J. A. Halberg, Jr.
John F. Devlin

Contents

1 Sequences and Limits

	Introduction	1
1.1	Sequences	1
1.2	The Limit of a Sequence	4
1.3	Absolute Value	7
1.4	Other Interpretations of $\lim_{n \to \infty} x_n = X$	11
1.5	Convergent and Bounded Sequences	13
1.6	Convergence of the Geometric Sequence (Optional)	16
1.7	Least Upper Bound, Greatest Lower Bound	18
1.8	Monotonic Sequences and Convergence	21
1.9	Algebra of Sequences	25
1.10	Some Properties of Limits	29
	Appendix	34

2 Functions

	Introduction	35
2.1	The Function Concept	35
2.2	Some Terminology and Notation	39
2.3	Graphs of Functions	42
2.4	Monotone Functions	51
2.5	Inverse Functions	57
2.6	Constant and Linear Functions	62
2.7	Parallel and Perpendicular Lines	73
2.8	The Distance from a Point to a Line	77
2.9	The Absolute Value Function	79
2.10	Continuity	82
2.11	The Limit of a Function	87
2.12	Algebra of Functions	90
2.13	Some Properties of the Limits of Functions	92
2.14	Symmetry	95
2.15	Symmetric Functions	100
2.16	The Graph of g^{-1} Is a Reflection of the Graph of g.	105

3 Quadratic Functions

	Introduction	110
3.1	The Quadratic Function $q(x) = x^2$	110

3.2	Multiplication of a Function by a Real Number	118
3.3	More about Quadratic Functions	123
3.4	Addition of a Constant to a Function	128
3.5	Composition of Functions	129
3.6	The General Quadratic Function	134
3.7	Analysis of the General Quadratic	139
3.8	The Quadratic Equation—A Graphical Interpretation	145
3.9	Tangent Lines	150
3.10	The Slope Function	154

4 Polynomial Functions

	Introduction	158
4.1	Polynomial Functions	158
4.2	Graphing Polynomial Functions	163
4.3	Continuity and the Intermediate Value Theorem	169
4.4	Slope Functions of Polynomial Functions	172
4.5	Some Properties of Slope Functions	183
4.6	Concavity	190
4.7	Basic Properties of Graphs of Polynomial Functions	196
4.8	The Remainder and Factor Theorems	202

5 Zeros of Polynomial Functions and Complex Numbers

	Introduction	207
5.1	Operations on Complex Numbers	208
5.2	Complex Conjugates and the Quotient of Complex Numbers	210
5.3	Complex Solutions of Quadratic Equations	213
5.4	The Fundamental Theorem of Algebra	215
5.5	Properties of Irreducible Quadratic Expressions	222
5.6	Zeros of Polynomials	227
5.7	Polynomials with Rational Coefficients	229
5.8	Isolating Zeros of Polynomials	234

6 Trigonometric Functions; Sine and Cosine

	Introduction	237
6.1	Measures of Angles	237
6.2	Sine and Cosine Functions Defined	246
6.3	Properties of the Sine Function	249
6.4	Properties of the Cosine Function	256
6.5	Some Identities Involving Cosine and Sine	261
6.6	The Slope Functions of the Sine and Cosine Functions	267

7 Other Trigonometric Functions

	Introduction	270
7.1	The Tangent Function	270
7.2	Derivation of the Slope Function of the Tangent Function (Optional)	278
7.3	The Cotangent Function	280
7.4	The Secant and Cosecant Functions	284
7.5	More Identities	291

8 Exponential and Logarithmic Functions

	Introduction	294
8.1	The Fundamental Law of Growth and Decay	294
8.2	Rational Exponents	301
8.3	More about Rational Exponents	307
8.4	The Function $\exp_a$ Defined on R	311
8.5	The Graph of $\exp_a$	313
8.6	The Slope Function of $\exp_a$ (Optional)	318
8.7	Concavity of the Graph of $\exp_a$ (Optional)	323
8.8	The Inverse of an Exponential Function	325
8.9	Logarithms to the Base a	328
8.10	Some Basic Properties of Logarithmic Functions	331
8.11	Further Properties of Logarithmic Functions	336
8.12	Computing with Common Logarithms	338
8.13	Logarithms to Different Bases	340

9 The Rational Functions

	Introduction	344
9.1	The Rational Functions	344
9.2	Graphs of Rational Functions	347
9.3	Derivatives of Products, Quotients and Composites of Functions	351
9.4	More about Graphs of Rational Functions	355
9.5	Horizontal Asymptotes	359
9.6	Vertical Asymptotes	364

10 Introduction to Polar Coordinates

	Introduction	374
10.1	Polar Coordinates	374
10.2	The Relationship of Polar and Rectangular Coordinates	376
10.3	Conic Sections in Polar Coordinates	379
10.4	Other Polar Curves (Optional)	382

Tables   384

Answers for Selected Exercises   387

Index   408

1
Sequences and Limits

Introduction

A fundamental concept in the branch of mathematics called analysis is that of a limit. In this chapter we shall endeavor to provide a natural introduction to this concept through an elementary study of sequences. From time to time we shall enlarge and amplify this introduction in connection with later topics in the book.

1.1 Sequences

A sequence of real numbers is a correspondence which relates to every natural number n an element a_n of the set of real numbers R. We will denote such a sequence by

$$\langle a_1, a_2, a_3, \ldots, a_n, \ldots \rangle, \quad \text{or simply by } \langle a_n \rangle,$$

where a_1 is the real number corresponding to 1, a_2 is the real number corresponding to 2, and so on. You are undoubtedly already familiar with many sequences.

Example 1 $\langle 1, 2, 4, \ldots, 2^{n-1}, \ldots \rangle$ or $\langle a_n \rangle = \langle 2^{n-1} \rangle$

Example 2 $\langle 1, 3, 5, \ldots, 2n-1, \ldots \rangle$ or $\langle a_n \rangle = \langle 2n-1 \rangle$

Example 3 $\left\langle 1, \frac{1}{2}, \frac{1}{3}, \ldots, \frac{1}{n}, \ldots \right\rangle$ or $\langle a_n \rangle = \left\langle \frac{1}{n} \right\rangle$

Example 1 is a special case of a **geometric sequence,** which is a sequence of the form

$$\langle a, ar, ar^2, \ldots, ar^{n-1}, \ldots \rangle \text{ or } \langle a_n \rangle = \langle ar^{n-1} \rangle,$$

where a and r are some constant real numbers. The number a is called the "first term" and r is called the "common ratio." Note that every term after the first is r times the term preceding it; that is, the ratio of any term to the one preceding it is r.

The second sequence given above is a special case of an **arithmetic sequence,** which is a sequence of the form

$$\langle a, a+b, a+2b, \ldots, a+(n-1)b, \ldots \rangle \text{ or } \langle a_n \rangle = \langle a+(n-1)b \rangle,$$

where a and b are constant real numbers. The number a is called the "first term" and b is called the "common difference." Note that every term after the first is b plus the term preceding it; that is, the difference of two consecutive terms is b.

Arithmetic and geometric sequences are special cases of **recursive sequences,** that is, sequences having some initial terms given, with successive terms determined by those preceding them through what is called a recursion formula. For example, the geometric sequence can be defined by

$$a_1 = a \text{ and } a_k = ra_{k-1}, \quad k = 2, 3, \ldots.$$

The arithmetic sequence can be defined by

$$a_1 = a \text{ and } a_k = a_{k-1} + b, \quad k = 2, 3, \ldots.$$

Example 4 A famous recursive sequence with a history dating back to the twelfth century is the Fibonacci sequence. The Fibonacci sequence is defined by

$$a_1 = 0, a_2 = 1, a_k = a_{k-1} + a_{k-2}, \quad k = 3, 4, 5, \ldots.$$

The first few terms of this sequence are

$$\langle 0, 1, 1, 2, 3, 5, 8, 13, 21, \ldots \rangle.$$

For a sequence to be well-defined there must be some rule which gives a_n for every positive integer n. It may not be possible to do this by a simple formula, as is shown by the sequence $\langle p_n \rangle$ where p_n is defined as the nth prime number. (There is a definite procedure for finding the nth prime number, but, as yet, there is no known simple formula for it.)

It is important to remember that one cannot determine from the first few terms of a sequence the rule that determines successive terms. For example, if you are told that the first four terms of a sequence are

$$\langle 0, 1, 2, 3, \ldots \rangle,$$

you might be tempted to jump to the conclusion that the next term is **4**. Note that the first four terms of each of the following sequences are also 0, 1, 2, 3, . . . but the next term is not 4 in either case.

$$\langle a_n \rangle : a_1 = 0,\ a_2 = 1,\ a_k = 2a_{k-1} - (a_{k-2})^2 \quad \text{for } k \geq 3$$
$$\langle b_n \rangle : b_n = n^4 - 10n^3 + 35n^2 - 49n + 23$$

1.1 EXERCISES

For each of Exercises 1–15, write the first six terms of the given sequence. Then indicate whether the sequence is an arithmetic sequence, a geometric sequence, or neither.

1. $\langle a_n \rangle = \left\langle \dfrac{1}{n^2} \right\rangle.$
2. $\langle a_n \rangle = \langle 2n + 1 \rangle.$
3. $\langle a_n \rangle = \langle (\tfrac{1}{2})^n \rangle.$
4. $\langle a_n \rangle = \langle (-1)^{n+1} \rangle.$
5. $\langle a_n \rangle = \left\langle \dfrac{2n+1}{3n} \right\rangle.$
6. $\langle a_n \rangle = \langle n^2 \rangle.$
7. $\langle a_n \rangle = \langle n(n-1) \rangle.$
8. $\langle a_n \rangle = \langle 3 \rangle.$
9. $\langle a_n \rangle = \langle n^2 - 6n + 8 \rangle.$
10. $\langle a_n \rangle = \left\langle 2 - \dfrac{7}{n} + \dfrac{4}{n^2} \right\rangle.$
11. $a_1 = 2,\ a_k = a_{k-1} - 5,\ k = 2, 3, 4, \ldots$
12. $a_1 = 10,\ a_k = 0.1 a_{k-1},\ k = 2, 3, 4, \ldots$
13. $a_1 = 1,\ a_k = a_{k-1} + k,\ k = 2, 3, 4, \ldots$
14. $a_1 = 1,\ a_k = k \cdot a_{k-1},\ k = 2, 3, 4, \ldots$
15. $a_1 = 1,\ a_2 = 2,\ a_k = a_{k-2} \cdot a_{k-1},\ k = 3, 4, 5, \ldots$

Find the missing terms in each of the following arithmetic sequences.

16. $\langle 14, —, 22, —, —, —, \ldots \rangle$
17. $\langle 4, —, —, 67, —, —, \ldots \rangle$
18. $\langle —, —, —, -5, -10, —, \ldots \rangle$
19. $\langle 4, —, —, —, —, 5, \ldots \rangle$

Find the missing terms in each of the following geometric sequences.

20. $\langle 2, 4, —, 16, —, —, \ldots \rangle$
21. $\langle 3, —, —, 81, —, —, \ldots \rangle$
22. $\langle -9, —, —, \tfrac{1}{3}, —, —, \ldots \rangle$
23. $\langle \sqrt{2}, —, —, 4, —, —, \ldots \rangle$
24. Find an x such that $3x + 5$, $5x$, and $6x - 1$ are successive terms in an arithmetic sequence.
25. Find an x such that x, $2x - 1$, and $4x$ are successive terms in a geometric sequence.
26. Show that the reciprocals of the terms in a geometric sequence also form a geometric sequence.
27. Show that the squares of the terms in a geometric sequence also form a geometric sequence.

1.2 The Limit of a Sequence

An important characteristic of a sequence is the behavior of its nth term as n gets larger and larger. Some of the standard ways of describing certain behaviors are introduced in the following examples.

Example 1 $\langle 1, 1, 1, 1, \ldots \rangle$ where $\langle a_n \rangle = \langle 1 \rangle$. In this sequence the nth term is 1 for every n. We call a sequence like this a **constant sequence.**

Example 2 $\langle \frac{\pi}{2}, \frac{2\pi}{3}, \frac{3\pi}{4}, \frac{4\pi}{5}, \ldots \rangle$ where $\langle b_n \rangle = \langle \frac{n\pi}{n+1} \rangle$. In this sequence b_n increases as n increases. That is, $b_{n+1} > b_n$ for all n. This is an example of an **increasing sequence.** Further, as n gets larger and larger, b_n gets closer and closer to π.

Example 3 $\langle \frac{1}{2}, \frac{2}{5}, \frac{3}{10}, \frac{4}{17}, \ldots \rangle$ where $\langle c_n \rangle = \langle \frac{n}{n^2+1} \rangle$. In this sequence $c_{n+1} < c_n$ for all n, so it is an example of a **decreasing sequence.** It is also true that c_n can be made to differ from zero by as small an amount as we desire provided we make n large enough. (Do you see that $c_n < \frac{1}{1000}$ if $n > 1000$ and that $c_n < \frac{1}{10,000}$ is $n > 10,000$?)

Example 4 $\langle 3, \frac{3}{2}, \frac{9}{4}, \frac{15}{8}, \ldots \rangle$ where $\langle d_n \rangle = \langle 2 + (-\frac{1}{2})^n \rangle$. In this case notice that $d_2 < d_1$, but $d_3 > d_2$, and so on. This is neither an increasing nor a decreasing sequence. However, as n increases, the values of d_n tend to "cluster" around the number 2. We can make the nth term arbitrarily close to 2 (that is, as close as we wish) by choosing n sufficiently large.

We characterize the behavior of these sequences by saying that:

1. "The limit of a_n as n approaches infinity is 1," which we write in symbols as $\lim_{n \to \infty} a_n = 1$ or $a_n \to 1$.

2. The limit of b_n as n approaches infinity is π, or $\lim_{n \to \infty} b_n = \pi$, or $b_n \to \pi$.

3. $\lim_{n \to \infty} c_n = 0$ or $c_n \to 0$.

4. $\lim_{n \to \infty} d_n = 2$, or $d_n \to 2$.

We wish to be very precise about this terminology. If $\langle x_n \rangle$ is a sequence of real numbers and X is a real number, then exactly what do we mean by saying "The limit of x_n as n approaches infinity is X, or $\lim_{n \to \infty} x_n = X$"?

Roughly we mean that x_n can be made to be as close to X as we want for *all* values of n that are *sufficiently large*. But what do we mean by 'x_n can be made to be as "close" to X as we want'? What do we mean by "sufficiently large"?

To say x_n is "close" to X has a geometric sound to it, and indeed when we use such terminology we are thinking of the representations of the real numbers x_n and X as points on a real number line. When we say x_n is close to X we mean the distance from the point on the real number line corresponding to x_n to the point corresponding to X is small. Let us, for the moment, represent this distance by $d(x_n, X)$. Then to say that we can "make x_n to be as close to X as we want" means that we can "make $d(x_n, X)$ as small as we want." By "making $d(x_n, X)$ as small as we want" we mean we can make it smaller than any positive number we might choose. For example, we can make $d(x_n, X) < \frac{1}{1000}$ or $d(x_n, X) < \frac{1}{10,000}$.

To be very concrete let us consider a specific example:

$$\langle x_n \rangle = \left\langle \frac{n-1}{n} \right\rangle.$$

The first few terms of the sequence $\langle x_n \rangle$ are:

$$0, \frac{1}{2}, \frac{2}{3}, \frac{3}{4}, \frac{4}{5}, \frac{5}{6}, \frac{6}{7}, \frac{7}{8}, \frac{8}{9}, \frac{9}{10}, \ldots$$

It appears that as n gets large, x_n gets very close to 1, that is, $\lim_{n \to \infty} x_n = 1$.

Note that

$$d(x_1, 1) = d(0, 1) = 1$$
$$d(x_2, 1) = d\left(\frac{1}{2}, 1\right) = \frac{1}{2}$$
$$\vdots$$
$$d(x_{10}, 1) = d\left(\frac{9}{10}, 1\right) = \frac{1}{10}$$
$$\vdots$$

and in general

$$d(x_n, 1) = d\left(\frac{n-1}{n}, 1\right) = 1 - \frac{n-1}{n} = \frac{1}{n}.$$

We see, then, that we can make $d(x_n, 1) < \frac{1}{1000}$, not for *all* n, *but for all* n *that are sufficiently large*. In this case $d(x_n, 1) < \frac{1}{1000}$ for all $n > 1000$, since $\frac{1}{n} < \frac{1}{1000}$ and $d(x_n, 1) = \frac{1}{n} < \frac{1}{1000}$. In the same way we see $d(x_n, 1) = \frac{1}{n}$ will be less than $\frac{1}{10,000}$ for all $n > 10,000$. In fact if ϵ is any positive number, $d(x_n, 1) = \dfrac{1}{n}$ will be less than ϵ for all $n > \dfrac{1}{\epsilon}$, since $n > \dfrac{1}{\epsilon}$ is equivalent to the statement $\dfrac{1}{n} < \epsilon$.

We are now ready to make a formal definition of the statement "the limit of x_n as n approaches infinity is X" or in symbols $\lim\limits_{n \to \infty} x_n = X$, or $x_n \to X$.

Definition 1.2a The statement $\lim\limits_{n \to \infty} x_n = X$ means that given any positive number ϵ, there is a positive integer N such that if $n > N$, then $d(x_n, X) < \epsilon$.

The order of the statement is very important. First it says given any number $\epsilon > 0$, *then* there is an integer $N > 0$ such that $d(x_n, X) < \epsilon$, when $n > N$. The statement *does not say* that there is an integer $N > 0$ such that if $n > N$ then $d(x_n, X) < \epsilon$ for any positive number ϵ. The point is that if any positive number ϵ is picked *then* there is an $N > 0$ such that $d(x_n, X) < \epsilon$ when $n > N$ *for that particular* ϵ. In general, the smaller ϵ is, the bigger N will be.

Let us consider the example of the sequence where $\langle x_n \rangle = \left\langle \dfrac{1}{n} \right\rangle$. If we pick $\epsilon = \dfrac{1}{1000}$, then $d(x_n, 0) = \dfrac{1}{n}$ will be less than ϵ if $n > 1000$. If we pick $\epsilon = \dfrac{1}{2000}$ then $d(x_n, 0) < \epsilon$ if $n > 2000$. If ϵ is an arbitrary positive number, then there must be an integer N such that $N\epsilon > 1$. (This is a consequence of a property of real numbers called the "Archimedean law": If a, b are positive real numbers, then there is an integer N such that $Na > b$). Since $N\epsilon > 1$, $\dfrac{1}{N} < \epsilon$ and if $n > N$ then $\dfrac{1}{n} < \dfrac{1}{N} < \epsilon$, and therefore $d(x_n, 0) = \dfrac{1}{n} < \epsilon$ if $n > N$.

1.2 EXERCISES

1. In Example 2 on page 4, how large does n have to be in order for it to be true that $\pi - b_n < \dfrac{\pi}{100}$? That $\pi - b_n < \dfrac{\pi}{1000}$?

2. In Example 4 on page 4, how large does n have to be in order that the difference between d_n and 2 be less than $\tfrac{1}{100}$? Less than $\tfrac{1}{1000}$?

For each of the sequences in Exercises 3–10, write out as many terms as you need in order to form a conjecture as to whether or not the sequence has a limit. If you think the sequence has a limit, try to guess its value.

3. $\left\langle \dfrac{4-n}{n^2} \right\rangle$

4. $\left\langle \dfrac{5n+1}{n} \right\rangle$

5. $\left\langle \dfrac{n^2-3}{n} \right\rangle$

6. $\left\langle \dfrac{12n+17}{3n+2} \right\rangle$

7. $\langle 2^n \rangle$

8. $\langle (-1)^n \rangle$

9. $\left\langle \dfrac{1}{n} - \dfrac{1}{n+1} \right\rangle$

10. $\left\langle \sqrt{\dfrac{9n+3}{n}} \right\rangle$

In each of Exercises 11–14, a sequence $\langle x_n \rangle$ and its limit X are given. For each given ϵ, determine the values of n for which $d(x_n, X) < \epsilon$.

11. $\left\langle \dfrac{1}{n^2} \right\rangle$; $X = 0$.
   a) $\epsilon = 0.25$ b) $\epsilon = 0.04$ c) $\epsilon = 0.01$ d) $\epsilon = 0.0001$

12. $\left\langle 2 - \dfrac{1}{10^n} \right\rangle$; $X = 2$.
   a) $\epsilon = \dfrac{1}{11}$ b) $\epsilon = \dfrac{1}{150}$ c) $\epsilon = \dfrac{1}{999}$ d) $\epsilon = \dfrac{1}{1001}$

13. $\langle 3 + (\tfrac{1}{2})^n \rangle$; $X = 3$.
   a) $\epsilon = \dfrac{1}{3}$ b) $\epsilon = \dfrac{1}{7}$ c) $\epsilon = 0.01$ d) $\epsilon = 0.005$

14. $\left\langle \dfrac{1}{\sqrt{n+1}} \right\rangle$; $X = 0$.
   a) $\epsilon = \dfrac{1}{5}$ b) $\epsilon = \dfrac{1}{10}$ c) $\epsilon = \dfrac{1}{100}$ d) $\epsilon = \dfrac{1}{1000}$

15. Consider the sequence $\langle b_n \rangle = \left\langle \dfrac{n\pi}{n+1} \right\rangle$ from Example 2 on page 4.
   a) Express b_{n+1} in terms of n.
   b) Show that $b_{n+1} - b_n = \dfrac{\pi}{(n+1)(n+2)}$.
   c) Explain how the result in part b shows that $\langle b_n \rangle$ is an increasing sequence.

1.3 Absolute Value

We regard distance as a nonnegative quantity and we have used the symbol $d(x, y)$ to denote the distance between the points on the real number line corresponding to the real numbers x and y. If we use the same symbols for the numbers and the points they represent, then if x lies to the right of y, $x - y$ is positive and $d(x, y) = x - y$. If x lies to the left of y, $y - x$ is positive and $d(x, y) = y - x$. If $x = y$, $d(x, y) = 0$. (See Figure 1-1.)

Figure 1-1

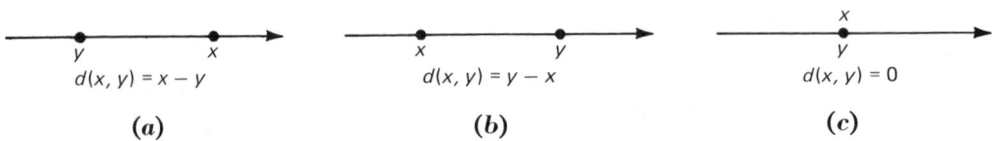

Combining these ideas we could have defined $d(x, y)$ as follows:

$$d(x, y) = \begin{cases} x - y \text{ if } x - y \geq 0 \\ y - x \text{ if } y - x > 0. \end{cases}$$

There is a standard symbol for $d(x, y)$ called the **absolute value** of $x - y$, denoted by $|x - y|$, and defined just as above:

If $x, y \in R$, then $|x - y| = \begin{cases} x - y \text{ if } x - y \geq 0 \\ y - x \text{ if } y - x > 0. \end{cases}$

By letting $y = 0$, we obtain the standard definition of the absolute value of a real number x:

$$|x| = \begin{cases} x \text{ if } x \geq 0 \\ -x \text{ if } -x > 0 \end{cases} \quad \text{(i.e. } x < 0\text{).}$$

Our approach to these definitions stresses the interpretation of $|x - y|$ as $d(x, y)$, and $|x|$ as $d(x, 0)$. (See Figure 1-2.)

Figure 1-2

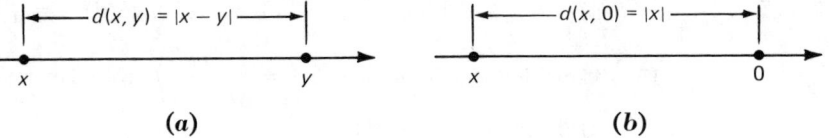

Certain properties of $d(x, y)$ may be illustrated geometrically. For example, $d(x, 0) = d(-x, 0)$, or $|x| = |-x|$, merely says x and $-x$ are the same distance from the origin. (See Figure 1-3.)

Figure 1-3

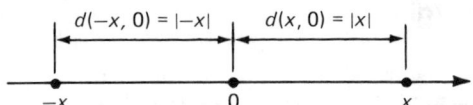

Another example is $d(x, y) \leq d(x, 0) + d(y, 0)$ which says that the distance between two points on the real number line cannot exceed the sum of their distances from 0. In fact it may be seen that $d(x, y) < d(x, 0) + d(y, 0)$ if x and y lie on the same side of 0 and $d(x, y) = d(x, 0) + d(y, 0)$ if x and y lie on opposite sides of 0. (See Figure 1-4.) This inequality given in absolute value notation is $|x - y| \leq |x| + |y|$.

Figure 1-4

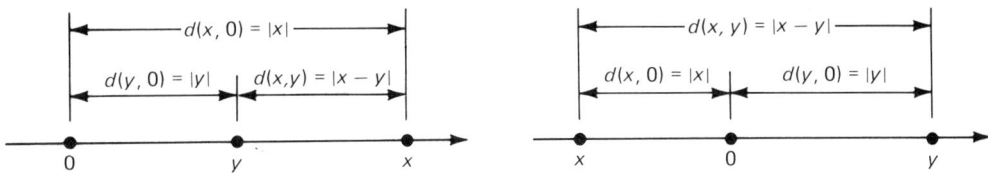

Another important inequality we will need is $||x| - |y|| \leq |x - y|$. If either x or y is zero, we have equality. If neither is 0, the inequality says that the positive difference of the distances of two points from the origin is never greater than the distance between them. As shown in Figure 1-5, $||x| - |y|| < |x - y|$ if the two points are on opposite sides of 0. If the two points are on the same side of the origin we see that the positive difference of their distances from the origin is equal to the distance between them; that is, $||x| - |y|| = |x - y|$.

Figure 1-5

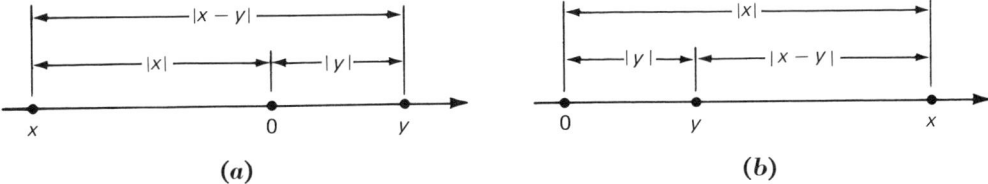

(a)   (b)

A list of the properties of $d(x, y) = |x - y|$ which we will want to use in future work is as follows:

1. $|x - y| \geq 0$, or $d(x, y) \geq 0$.
2. $|x - y| = 0$ if and only if $x = y$, or $d(x, y) = 0$ if and only if $x = y$.
3. $|x - y| = |y - x|$ or $d(x, y) = d(y, x)$.
4. $|x + y| \leq |x| + |y|$ or $d(x, -y) \leq d(x, 0) + d(y, 0)$.
5. $|x - y| \leq |x| + |y|$ or $d(x, y) \leq d(x, 0) + d(y, 0)$.
6. $||x| - |y|| \leq |x - y|$ or $|d(x, 0) - d(y, 0)| \leq d(x, y)$.
7. $|xy| = |x| |y|$ or $d(xy, 0) = d(x, 0) d(y, 0)$.

The equivalence of $d(x, y)$ and $|x - y|$ also provides us with a handy means of solving many conditions involving absolute value.

Example 1 To solve $|x - 4| < 5$, we use the fact that $|x - 4| = d(x, 4)$. Thus the solutions are those numbers x that are less than 5 units from 4. As Figure 1-6 shows, this set is the open interval $\{x \mid -1 < x < 9\}$, which we will also denote by $(-1, 9)$.

Example 2 To solve $|x| \leq 6$, remember that $|x| = d(x, 0)$. Thus the solutions are those numbers x whose distances from the origin are less than or equal to 6, namely the closed interval $\{x \mid -6 \leq x \leq 6\}$, denoted by $[-6, 6]$. See Figure 1-7.

Example 3 To solve $|x + 2| > 3$, first note that $|x + 2| = |x - (-2)| = d(x, -2)$. The solutions are those numbers x whose distance from -2 is greater than 3. This is the set $\{x \mid x < -5 \text{ or } x > 1\}$, which is the union of infinite intervals (or half lines) $(-\infty, -5) \cup (1, \infty)$. See Figure 1-8.

Figure 1-6

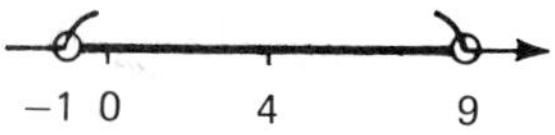

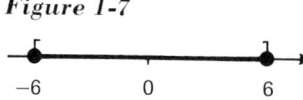

Figure 1-7

Figure 1-8

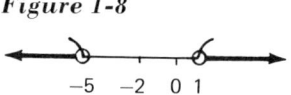

1.3 EXERCISES

For each of Exercises 1–6, describe the values of x for which the given statement is true.

1. $|x - 5| = x - 5$.
2. $|6 - x| = x - 6$.
3. $|x + 5| = -x - 5$.
4. $|x + 3| = 3 - x$.
5. $|x^2 - 4| = 4 - x^2$.
6. $|x^2 + 4| = 4 + x^2$.

Solve each of the following conditions.

7. $|x - 3| = 7$.
8. $|x + 1| = 6$.
9. $|3 - x| > 5$.
10. $|4 + x| < 6$.
11. $|x + 6| > 1$.
12. $|x - 9| < 4$.
13. $|x + 10| \leq 2$.
14. $|x - 2| \geq 3$.
15. $|x - 10| < \epsilon$.
16. $|x - a| < \epsilon$.

Describe the union and the intersection of each of the following pairs of intervals.

17. $[-10, 5)$ and $(-5, 10]$.
18. $(-\infty, 10]$ and $[-1, \infty)$.
19. $(-2, 3)$ and $(0, 5)$.
20. $[-2, 1]$ and $[1, 6]$.

21. Using the definition of the absolute value of a real number, show that $|x| = |-x|$ in each of the following cases.
    a) $x \geq 0$.   b) $x < 0$.
22. Draw diagrams illustrating that $|x + y| \leq |x| + |y|$ if
    a) x and y are on opposite sides of the origin.
    b) x and y are on the same side of the origin.
23. Beginning with the inequality $|a - b| \leq |a| + |b|$, let $a = x - y$ and $b = -y$ to show that $|x - y| \geq |x| - |y|$.
24. Use the definition of the absolute value of a real number to show that $|xy| = |x| \cdot |y|$ in each of the following cases.
    a) $x \geq 0$ and $y \geq 0$.
    b) $x \geq 0$ and $y < 0$.
    c) $x < 0$ and $y < 0$.

1.4 Other Interpretations of $\lim_{n \to \infty} x_n = X$

Using the ideas developed in Section 1.3, we can now consider alternative interpretations of the limit of a sequence. First, using the fact that $d(x_n, X) = |x_n - X|$, we can express the statement $\lim_{n \to \infty} x_n = X$ as follows:

> Given any positive number ϵ, there is a positive integer N such that if $n > N$, then $|x_n - X| < \epsilon$.

This interpretation enables us to prove a useful theorem.

Theorem 1.4a   *If $x_n \to X$, then $|x_n| \to |X|$.*

Proof   Since $x_n \to X$ we know that given $\epsilon > 0$, there is an N such that if $n > N$, then $|x_n - X| < \epsilon$. However, by property 6 in Section 1.3, $||x_n| - |X|| \leq |x_n - X|$. Hence, if $n > N$, $||x_n| - |X|| < \epsilon$. That is, $|x_n| \to |X|$.   □

The converse of this theorem is *not* true, as may be seen by considering the sequence $\langle (-1)^n \rangle$.

There is a useful geometric interpretation of the statement $\lim_{n \to \infty} x_n = X$.
As you know, there is a one-to-one correspondence between the real numbers R and the points on the real number line. If we think of the sequence $\langle x_n \rangle$ as a set of points on the real number line, we can interpret $\lim_{n \to \infty} x_n = X$ as follows:

$\lim_{n\to\infty} x_n = X$ *if and only if* for every open interval (a, b) containing X (that is, $a < X < b$), all but a finite number of the terms of $\langle x_n \rangle$ lie in (a, b).

Let us prove the "if" part. That is, let us assume that for any open interval (a, b) containing X, all but a finite number of the terms of $\langle x_n \rangle$ lie in (a, b), and show that $\lim_{n\to\infty} x_n = X$. Under the assumption above, it follows that if ϵ is any positive number, then all but a finite number of terms of $\langle x_n \rangle$ must lie in $(X - \epsilon, X + \epsilon)$, for this is clearly an open interval containing X. Now, any x_n in $(X - \epsilon, X + \epsilon)$ is within ϵ distance of X, so it follows that for all but a finite number of terms of $\langle x_n \rangle$, it is true that $d(x_n, X) < \epsilon$. Now suppose that x_N is the "last" of the finite set of terms in $\langle x_n \rangle$ that are *not* in $(X - \epsilon, X + \epsilon)$. It follows then that if $n > N$, then $d(x_n, X) < \epsilon$. Since ϵ is any positive number, this means that $\lim_{n\to\infty} x_n = X$.

We have now shown that by assuming

"If (a, b) is any open interval containing X, then all but a finite number of terms of $\langle x_n \rangle$ lie in (a, b)."

it follows that

"If ϵ is any positive number, then there is an integer N such that if $n > N$, then $d(x_n, X) < \epsilon$, or $\lim_{n\to\infty} x_n = X$."

We leave the proof of the "only if" part for the exercises.

1.4 EXERCISES

In each of Exercises 1–4, a sequence $\langle x_n \rangle$ and its limit X are given. For each given ϵ, determine the number of terms of the sequence that do not lie in $(X - \epsilon, X + \epsilon)$.

1. $\left\langle \dfrac{1}{\sqrt{n}} \right\rangle$; $X = 0$.

   a) $\epsilon = 0.5$ b) $\epsilon = 0.1$ c) $\epsilon = 0.01$ d) $\epsilon = 0.001$

2. $\left\langle \dfrac{128}{2^n} \right\rangle$; $X = 0$.

   a) $\epsilon = 0.1$ b) $\epsilon = 0.01$ c) $\epsilon = 0.001$ d) $\epsilon = 0.0001$

3. $\left\langle 2 + \dfrac{(-1)^n}{n} \right\rangle$; $X = 2$.

   a) $\epsilon = 0.1$ b) $\epsilon = 0.01$ c) $\epsilon = 0.001$ d) $\epsilon = 0.0001$

4. $\langle 5 + \dfrac{n}{n+1} \rangle$; $X = 6$.

   a) $\epsilon = 0.1$ b) $\epsilon = 0.01$ c) $\epsilon = 0.001$ d) $\epsilon = 0.0001$

In each of Exercises 5–8, find the smallest integer n that satisfies the given condition.

5. $\left| \dfrac{n+5}{n+3} - 1 \right| < 0.1$.

6. $\left| \dfrac{2n-11}{4n-7} - \dfrac{1}{2} \right| < 0.01$.

7. $\left| \dfrac{n^2}{4n^2 - 1} - \dfrac{1}{4} \right| < 0.001$.

8. $\left| \dfrac{5^n + 2}{5^{n+1} + 5} - \dfrac{1}{5} \right| < 0.0001$.

9. Suppose $\langle x_n \rangle$ and $\langle ax_n \rangle$ are sequences, $\lim_{n \to \infty} x_n = X$, and $a \in R$.

   a) Suppose $a \neq 0$, and let ϵ be an arbitrary positive real number. Show that there is an N such that if $n > N$, then $|x_n - X| < \dfrac{\epsilon}{|a|}$.

   $\left[\text{Hint: let } \epsilon' = \dfrac{\epsilon}{|a|}. \right]$

   b) Use this result to show that if $n > N$, then $|ax_n - aX| < \epsilon$. [Hint: $|x| \cdot |y| = |xy|$.]

   c) What does this result tell you about $\langle ax_n \rangle$?

   d) Now suppose $a = 0$. Show that $\lim_{n \to \infty} ax_n = aX$.

10. Prove that if $\lim_{n \to \infty} x_n = X$ and (a, b) is any open interval containing X, then all but a finite number of terms of $\langle x_n \rangle$ lie in (a, b).

11. Prove that a sequence $\langle x_n \rangle$ cannot have two limits X and Y. [Hint: let $\epsilon = \dfrac{|X - Y|}{2}$.]

1.5 Convergent and Bounded Sequences

It should be clear from the examples we have presented that not all sequences have a limit. If $\lim_{n \to \infty} x_n = X$, we say the sequence $\langle x_n \rangle$ has a limit X, or **converges to** X. If there is no real number X for which $\lim_{n \to \infty} x_n = X$, then we say the sequence $\langle x_n \rangle$ has no limit, or that it **diverges.** The sequence $\langle 2^n \rangle$ diverges, since 2^n gets arbitrarily large as n

approaches infinity. In fact it can be shown that all geometric sequences with common ratios with absolute value greater than 1 diverge, while those with common ratio less than one in absolute value converges to 0. (See page 16.)

Geometric sequences with common ratio r, $|r| > 1$, comprise a special subset of a class of sequences called *unbounded sequences*. We shall now introduce some terminology to explain what we mean by "bounded" and "unbounded" sequences.

Definition 1.5a Any set of real numbers S is **bounded above** if there is a real number U, called an "upper bound" for the set S, such that if s belongs to S, then $s \leq U$. If no such number U exists, we say the set is not bounded above or is **unbounded above.**

This definition of bounded above applies to all sets of real numbers. Since the terms of the sequences we are working with are sets of real numbers we can speak of sequences being bounded above or unbounded above.

Example 1 The sequence $\left\langle \dfrac{n}{n+1} \right\rangle$ is bounded above since $\dfrac{n}{n+1} < 1$ for every positive integer n.

Example 2 The sequence $\langle n \rangle$ is unbounded above because there is no real number which is greater than all positive integers.

A geometric interpretation of a sequence being bounded above is that there exists a point U on the real number line such that none of the points x_n lies to the right of U.

The definition of a set being bounded below is completely analogous to our definition of a set being bounded above.

Definition 1.5b A set of real numbers S is **bounded below** if there is a real number L, called a "lower bound" for the set S, such that if s belongs to S, then $s \geq L$.

Example 3 The sequence $\left\langle \dfrac{1}{n} \right\rangle$ is bounded below since $\dfrac{1}{n} > 0$ for every positive integer n.

Example 4 The sequence $\langle -n \rangle$ is not bounded below or is unbounded below since there is no real number which is less than every negative integer.

A geometric interpretation of a sequence being bounded below is that there is a point L on the real number line such that none of the points x_n lies to the left of L.

A sequence can be bounded below and unbounded above, as for example the sequence $\langle n^2 \rangle$ which has 1 for a lower bound and has no upper bound. The sequence $\langle -n + 2 \rangle$ is bounded above, (1 is an upper bound), but is not bounded below. The sequence $\left\langle \dfrac{n-1}{n} \right\rangle$ is both bounded above and bounded below having 1 for an upper bound and 0 for a lower bound.

Definition 1.5c A sequence $\langle x_n \rangle$ which is bounded above and bounded below is said to be **bounded**.

From our definitions and examples we see that any convergent sequence must be bounded and any unbounded sequence is divergent. The converses of these statements are not true. A bounded sequence need not be convergent and a divergent sequence may not be unbounded. The following example illustrates both of these statements.

Example 5 The sequence $\langle (-1)^n \rangle$ is clearly bounded, but it is not convergent and is therefore divergent. The sequence $\langle (-1)^n \rangle$ is not convergent since each of its terms is either 1 or -1. For the sequence to be convergent there would have to be a number X such that the distances from X to both 1 and -1 are both arbitrarily small. This is impossible.

1.5 EXERCISES

In each of Exercises 1–12, indicate whether the given sequence is (a) bounded, (b) bounded above only, (c) bounded below only, or (d) unbounded.

1. $\langle 2n \rangle$
2. $\langle 1 - 2n \rangle$
3. $\left\langle \dfrac{n}{2^n} \right\rangle$
4. $\left\langle n - \dfrac{1}{n} \right\rangle$
5. $\left\langle \dfrac{1}{n} - \dfrac{1}{n+1} \right\rangle$
6. $\left\langle \left(\dfrac{1}{n}\right)^n \right\rangle$
7. $\langle (-1)^n n \rangle$
8. $\left\langle \dfrac{(-1)^n}{n} \right\rangle$
9. $\left\langle \left(\dfrac{n+5}{n+2}\right)^2 \right\rangle$
10. $\left\langle \left(-\dfrac{1}{n}\right)^n \right\rangle$
11. $\left\langle \dfrac{3n}{n+3} - \dfrac{n+3}{2n} \right\rangle$
12. $\left\langle \dfrac{n^2}{2n+5} - \dfrac{n^2}{2n+1} \right\rangle$

Use Definitions 1.5a, 1.5b, and 1.5c to prove each of the following theorems,

13. If $\langle x_n \rangle$ is bounded above, then $\langle -x_n \rangle$ is bounded below.
14. If $\langle x_n \rangle$ is bounded, then $\langle -x_n \rangle$ is bounded.
15. If $\langle x_n \rangle$ and $\langle y_n \rangle$ are bounded, then $\{x_n\} \cup \{y_n\}$ is a bounded set.
16. If $\langle x_n \rangle$ is bounded above and $\langle y_n \rangle$ is bounded below, then $\{x_n\} \cap \{y_n\}$ is a bounded set.

1.6 Convergence of the Geometric Sequence (*Optional*)

We shall presently show that the geometric sequence $\langle ar^{n-1} \rangle$ converges to zero if $|r| < 1$ and diverges if $a \neq 0$ and $|r| > 1$. But first we shall establish an important inequality:

If $h > 0$, then $(1 + h)^n > 1 + nh$ for all $n > 1$.

When $n = 2$ we see that

$$(1 + h)^2 = 1 + 2h + h^2 > 1 + 2h$$

or $$(1 + h)^2 > 1 + 2h.$$

If we multiply both sides of this inequality by $(1 + h)$, we have

$$(1 + h)(1 + h)^2 > (1 + h)(1 + 2h) = 1 + 3h + 2h^2,$$

and thus

$$(1 + h)^3 > 1 + 3h.$$

Now if for any integer k, it is true that

$$(1 + h)^k > 1 + kh,$$

and we multiply both sides of the inequality by $(1 + h)$, we see that

$$(1 + h)(1 + h)^k > (1 + h)(1 + kh)$$

or $$(1 + h)^{k+1} > 1 + (k + 1)h + kh^2,$$

and therefore

$$(1 + h)^{k+1} > 1 + (k + 1)h.$$

That is, if the inequality is true for any integer k, then it is true for the next integer $(k + 1)$. Now we know $(1 + h)^n > 1 + nh$ is true for $n = 2$ and $n = 3$. By the property above, since it holds for $n = 3$ it holds for the next integer $n = 4$. Correspondingly, since it holds for $n = 4$, it holds for the next integer $n = 5$, and so on. We have thus established that the inequality

$$(1 + h)^n > 1 + nh, \quad h > 0$$

holds for all integers $n > 1$.

We will now consider the special case of the geometric sequence

$\langle ar^{n-1} \rangle$ where $a = 1$ and $|r| < 1$. Since $\langle r^n \rangle$ contains all the terms of $\langle r^{n-1} \rangle$ except the first term, $r^0 = 1$, it follows that if $r^n \to 0$, then $r^{n-1} \to 0$. We will now prove that $r^n \to 0$.

Let ϵ be an arbitrary positive number and consider $|r|^n$. By hypothesis $|r| < 1$ which can be stated as $|r| = \dfrac{1}{1+h}$ where $h > 0$. This implies that $|r|^n = \dfrac{1}{(1+h)^n}$. But we have shown that $(1+h)^n > 1 + nh$ and therefore

$$|r|^n = \frac{1}{(1+h)^n} < \frac{1}{1+nh} < \frac{1}{nh}.$$

From the Archimedean Law we know there is an integer N such that $N \cdot (h\epsilon) > 1$ or $\dfrac{1}{Nh} < \epsilon$. It now follows that for any $n \geq N$ we have $\dfrac{1}{nh} \leq \dfrac{1}{Nh} < \epsilon$ and therefore:

$$|r^n - 0| = |r|^n < \frac{1}{nh} < \frac{1}{Nh} < \epsilon.$$

This means $\lim\limits_{n \to \infty} r^n = 0$.

We now consider the more general geometric sequence $\langle ar^{n-1} \rangle$ where a is any nonzero real number. (If $a = 0$, $\langle ar^{n-1} \rangle = \langle 0 \rangle$, the zero sequence.) Instead of proving directly that if $|r| < 1$, then $\lim\limits_{n \to \infty} ar^n = 0$, we prove a lemma (helping theorem) that gives us this result as a consequence of the fact that $\lim\limits_{n \to \infty} r^n = 0$.

Lemma 1.6a *If* $\lim\limits_{n \to \infty} x_n = X$ *and* $a \in R$, *then* $\lim\limits_{n \to \infty} ax_n = aX$.

Proof The result is obvious in case $a = 0$, so we assume $a \neq 0$ and let ϵ be an arbitrary positive number. Now consider $|ax_n - aX| = |a| \, |x_n - X|$. Since $\lim\limits_{n \to \infty} x_n = X$, we can make $|x_n - X|$ as small as we wish for sufficiently large n. In particular there must be an N, such that for $n > N$, $|x_n - X| < \dfrac{\epsilon}{|a|}$. But this means that if $n > N$, then

$$|ax_n - aX| = |a| \, |x_n - X| < |a| \frac{\epsilon}{|a|} = \epsilon$$

or $\qquad |ax_n - aX| < \epsilon$ if $n > N$.

This is the statement that $\lim\limits_{n \to \infty} ax_n = aX$. $\square$

This lemma gives us immediately the result that the geometric sequence $\langle ar^{n-1} \rangle$ converges to zero if $|r| < 1$.

A direct application of the definition of a limit of a sequence shows that if $c \in R$, then $\lim_{n \to \infty} c = c$. That is, a constant sequence, $\langle c \rangle$, converges to c. From this fact we see that the geometric sequence $\langle ar^{n-1} \rangle$ converges to a if $r = 1$.

The proofs that $\langle ar^{n-1} \rangle$ diverges if $a \neq 0$ and $r = -1$ or $|r| > 1$ are left as exercises.

1.6 EXERCISES

1. Prove that if $a \neq 0$ and $r = -1$, then $\langle ar^{n-1} \rangle$ diverges.
2. Prove that if $a \neq 0$ and $|r| > 1$, then $\langle ar^{n-1} \rangle$ diverges.

1.7 Least Upper Bound, Greatest Lower Bound

A set of real numbers that is bounded above has more than one upper bound. In fact, any number that is greater than a given upper bound is also an upper bound. It seems reasonable, however, to expect that if a set of real numbers is bounded above, then there is an upper bound less than all the other upper bounds. Indeed this is a basic assumption we make about the real numbers, called the *Completeness Axiom*.

Axiom 1.7a (***Completeness Axiom***) *For every nonempty set of real numbers S that is bounded above, there is a unique number c having the following properties:*

1. *c is an upper bound for S.*
2. *If b is any other upper bound for S, then $c < b$.*

This unique number c is called **the least upper bound** *of S.*

This axiom should not be difficult to accept since it is merely a formal statement of an intuitive assumption that we have already made. Note that while the axiom assures us that there exists a least upper bound for every nonempty set of real numbers which is bounded above, it does not tell us how to determine this number for any given set.

If S is any finite set of real numbers such as $\{-\sqrt{2}, \frac{1}{2}, 7, -\frac{3}{5}, 1\}$ we can easily determine the least upper bound of the set. We do this by comparing all possible pairs of elements to see which one is the largest. The largest element—in this case 7—is the least upper bound of the set.

But what about a set like A which is the set of all elements of the sequence $\left\langle \dfrac{n-1}{n} \right\rangle = \left\langle 1 - \dfrac{1}{n} \right\rangle$? It is an infinite set, $\{0, \tfrac{1}{2}, \tfrac{2}{3}, \tfrac{3}{4}, \tfrac{5}{6}, \ldots\}$. It should be clear that any number of the set is less than 1 since $n - 1 < n$ and therefore $\dfrac{n-1}{n} < 1$. But is 1 the *least* upper bound of the set A? It certainly seems to be, but can we prove this? Before we can prove that 1 is the least upper bound for the set A we need some additional properties of the real numbers. These properties can be derived from the Completeness Axiom. The first of these, called the Archimedean Law, we have already mentioned.

Theorem 1.7b (***The Archimedean Law***) *If a and b are positive real numbers, then there is a positive integer n such that $na > b$.*

In physical terms we can think of this law in the following way. If we have two sticks, one of length a and one of length b, then either a is longer than b, or if we have enough sticks of length a and lay them end to end, their total length will exceed that of the stick of length b.

The Archimedean Law can be used to establish the property that between any two distinct real numbers there is a rational number (i.e. a quotient of integers). (See Exercise 11, p. 20.)

Theorem 1.7c *If a and b are real numbers and $a < b$, then there is a rational number r such that $a < r < b$.*

In the exercises you will use this last result to show that 1 is the least upper bound of the set of terms of the sequence $\left\langle \dfrac{n-1}{n} \right\rangle$. Note that in this case 1 is not in the set for which it is the least upper bound. The least upper bound for a set S may or may not belong to the set S.

The Completeness Axiom assures us that any nonempty set of real numbers bounded above has a least upper bound. By analogy it seems reasonable that any nonempty set of real numbers bounded below should have something called a *greatest lower bound*. This is indeed the case.

Theorem 1.7d *For every nonempty set of real numbers S that is bounded below, there is a unique number c having the following properties:*

1. *c is a lower bound for S.*
2. *If b is any other lower bound for S, then $c > b$.*

This unique number c is called **the greatest lower bound** *of S.*

The least upper bound of a set of real numbers has a property that other upper bounds of the set do not have, in that we can determine members of the set as close to the least upper bound as we wish. (The greatest lower bound has a similar property that distinguishes it from the other lower bounds.) To illustrate this property, consider the set $A = \{1 - \frac{1}{n} \mid n \text{ is a positive integer}\}$, which has the least upper bound 1. By choosing appropriate values of n, we can determine members of A that are within ϵ units (where $\epsilon > 0$) of 1, no matter how small ϵ may be. For example, if we are asked to determine the members of A that are within 0.1 units of 1, we merely choose $n > 10$. If we wish to determine the members of A that are within 0.05 units of 1, we select $n > 20$. In general, for this example, we can determine an integer N such that

$$1 - \left(1 - \frac{1}{n}\right) < \epsilon \quad \text{if } n > N,$$

no matter how small ϵ may be.

1.7 EXERCISES

Give the greatest lower bound and the least upper bound, if they exist, for each of the following sets. For each bound, indicate whether or not it is a member of the set.

1. $\left\langle \frac{(-1)^n}{n} \right\rangle$

2. $\left\langle n - \frac{1}{n} \right\rangle$

3. $\left\langle \frac{10n}{10^n} \right\rangle$

4. $\langle 5 + (-1)^n n \rangle$

5. $\langle 10(-\frac{1}{2})^n \rangle$

6. $\{x \mid x \in R \text{ and } 1 \leq |x| < 3\}$
7. $\{x \mid x \in R \text{ and } 1 < |x| \leq 3\}$
8. $\{x \mid x \in R \text{ and } x^2 < 2\}$
9. $\{x \mid x \in R \text{ and } x^2 > 2\}$
10. $\left\{ \frac{x^2 - 1}{x - 1} \mid x \in R \text{ and } x \neq 1 \right\}$

11. Suppose that a and b are real numbers and $a < b$.
    a) Give reasons to support the following statements.
       1. $b - a$ is a positive number.
       2. There exists an integer $n \geq 1$ such that $n(b - a) > 1$.
    b) Use statement 2 above to show that $d(nb, na) > 1$.
    c) How does part b show that there must be an integer m such that $na < m < nb$?
    d) Use the result of part c to show that there is a rational number r such that $a < r < b$.

12. Consider the sequence $\left\langle \frac{n-1}{n} \right\rangle$.

a) Show that 1 is an upper bound for this sequence. That is, show that $\frac{n-1}{n} < 1$ for all n.

b) Let p be any number less than 1. How do you know there is a rational number $\frac{m}{n}$ such that $p < \frac{m}{n} < 1$?

c) Show that $\frac{m}{n} \leq \frac{n-1}{n}$.

d) Use the results of parts b and c to show that p cannot be an upper bound for $\left\langle \frac{n-1}{n} \right\rangle$.

e) Use the results of parts a and d to show that 1 is the least upper bound for $\left\langle \frac{n-1}{n} \right\rangle$.

13. Consider the sequence $\langle (\frac{1}{2})^n \rangle$.
    a) Show that this is a decreasing sequence, that is, that $x_n > x_{n+1}$ for all n.
    b) Show that 0 is a lower bound of this sequence.
    c) Use the fact that $\{2^n\}$ has no upper bound to show that if p is any positive number, then p cannot be a lower bound of $\langle (\frac{1}{2})^n \rangle$.

1.8 Monotonic Sequences and Convergence

Sequences such as $\langle n \rangle$, $\langle n^2 + 1 \rangle$ and $\left\langle \frac{n}{n+1} \right\rangle$ all have the property that their terms increase in numerical value as n increases. They are examples of a class of sequences called *increasing sequences*. Sequences such as $\left\langle \frac{1}{n} \right\rangle$, $\left\langle \frac{1}{n^2+1} \right\rangle$ and $\left\langle \frac{n+1}{n} \right\rangle$ all have the property that their terms decrease as n increases. They are examples of a class of sequences called *decreasing sequences*.

Definition 1.8a The sequence $\langle x_n \rangle$ is called:

1. an **increasing sequence** if $x_{n+1} > x_n$ for all positive integers n.
2. a **decreasing sequence** if $x_{n+1} < x_n$ for all positive integers n.

The sequence $\left\langle n + \frac{1 + (-1)^n}{2} \right\rangle = \langle 1, 3, 3, 5, 5, 7, 7, \ldots \rangle$ is not an increasing sequence by our definition, but it has the property that its terms never decrease as n increases. It is an example of a *nondecreasing sequence.* Similarly, $\dfrac{1}{n + \dfrac{1+(-1)^n}{2}} = \langle 1, \tfrac{1}{3}, \tfrac{1}{3}, \tfrac{1}{5}, \tfrac{1}{5}, \tfrac{1}{7}, \tfrac{1}{7}, \ldots \rangle$ is an example of a *nonincreasing sequence.*

Definition 1.8b The sequence $\langle x_n \rangle$ is called:

1. a **nondecreasing sequence** if $x_{n+1} \geq x_n$ for all positive integers n.
2. a **nonincreasing sequence** if $x_{n+1} \leq x_n$ for all positive integers n.

Note that any increasing sequence is also a nondecreasing sequence and any decreasing sequence is also a nonincreasing sequence. Also note that any constant sequence is both a nonincreasing and a nondecreasing sequence.

The set of all nonincreasing and nondecreasing sequences is called the set of **monotonic sequences.** They are important because under certain boundedness conditions they are convergent sequences.

Note that whenever a monotonic sequence $\langle x_n \rangle$ has a reciprocal sequence, $\dfrac{1}{\langle x_n \rangle} = \left\langle \dfrac{1}{x_n} \right\rangle$, then the reciprocal sequence is also monotonic. For example the sequence of positive integers $\langle n \rangle$ is an increasing sequence while its reciprocal sequence $\left\langle \dfrac{1}{n} \right\rangle$ is a decreasing sequence.

We have seen that the sequence $\left\langle \dfrac{n-1}{n} \right\rangle$ is an increasing sequence with a least upper bound of 1 and that $\lim\limits_{n \to \infty} \dfrac{n-1}{n} = 1$.

From our study of geometric sequences we can conclude that $\langle (\tfrac{1}{2})^n \rangle$ is a decreasing sequence with a greatest lower bound of 0 and that $\lim\limits_{n \to \infty} (\tfrac{1}{2})^n = 0$.

The limits of both of these sequences could have been obtained by a simple application of the following theorem.

Theorem 1.8a *Any nondecreasing sequence bounded above converges to its least upper bound.*

Proof Let $\langle x_n \rangle$ be a nondecreasing sequence with a least upper bound l. Then given any positive number ϵ, there is an integer N, such that $l - x_N < \epsilon$. Since $x_n \geq x_N$ for any $n > N$, we see $l - x_n \leq l - x_N < \epsilon$ for any $n > N$. Since l is an upper bound for $\langle x_n \rangle$, $l - x_n \geq 0$ for all n.

We can conclude therefore that for all $n > N$, $|l - x_n| < \epsilon$, which means $\lim_{n \to \infty} x_n = l$. $\square$

This theorem applies directly to the sequence $\left\langle \dfrac{n-1}{n} \right\rangle$, since it is an increasing, and therefore nondecreasing, sequence and is bounded above.

To apply the theorem to the sequence $\langle (\frac{1}{2})^n \rangle$ we note that $-\langle (\frac{1}{2})^n \rangle = \langle -(\frac{1}{2})^n \rangle$ is an increasing sequence with a least upper bound 0. From our theorem we conclude $\lim_{n \to \infty} (-(\frac{1}{2})^n) = 0$. But by Lemma 1.6a, $\lim_{n \to \infty} ax_n = a \lim_{n \to \infty} x_n$ and therefore $-\lim_{n \to \infty} (\frac{1}{2})^n = \lim_{n \to \infty} (-(\frac{1}{2})^n) = 0$ or $\lim_{n \to \infty} (\frac{1}{2})^n = 0$.

Precisely the same device allows us to show that any nonincreasing sequence bounded below converges to its greatest lower bound. All one needs to observe is that if $\langle x_n \rangle$ is bounded below with a greatest lower bound a, then $\langle -x_n \rangle$ is bounded above with a least upper bound $-a$. This gives us the following corollary to our theorem.

Corollary 1.8b *Any nonincreasing sequence bounded below converges to its greatest lower bound.*

You will find that you can apply Theorem 1.8a and its corollary to many sequences. For example, consider the sequence $\left\langle \dfrac{1}{n!} \right\rangle$. Now, $n!$ (read "n factorial") is defined as follows:

$$1! = 1, \quad 2! = 1 \cdot 2, \quad 3! = 1 \cdot 2 \cdot 3$$

and in general,

$$n! = 1 \cdot 2 \cdot 3 \cdot \ldots \cdot n.$$

Thus, $n! < (n+1)!$, so $\dfrac{1}{n!} > \dfrac{1}{(n+1)!}$ for any positive integer n. This shows us that our sequence is a decreasing sequence. In the exercises you will show that 0 is the greatest lower bound for the sequence. With these facts we conclude from our corollary that $\lim_{n \to \infty} \dfrac{1}{n!} = 0$.

There are ways of using the convergence of known sequences to prove the convergence of other sequences. For example, suppose we know that $\lim_{n \to \infty} a_n = l$ and $\lim_{n \to \infty} c_n = l$ and that $a_n \leq b_n \leq c_n$ for all n. It would seem reasonable that as a_n and c_n both approach l, b_n, which is squeezed between them, must also approach l. This is indeed the case and it is not even necessary that $a_n \leq b_n \leq c_n$ for all n, but only for all sufficiently large n.

Theorem 1.8c *If* $\lim_{n\to\infty} a_n = \lim_{n\to\infty} c_n = l$ *and there is an integer* N *such that for all* $n > N$, $a_n \leq b_n \leq c_n$, *then* $\lim_{n\to\infty} b_n = l$.

Proof From our hypotheses we know there is an integer N_1 such that $n > N_1$ implies $l - \epsilon < a_n < l + \epsilon$, and an integer N_2 such that $n > N_2$ implies $l - \epsilon < c_n < l + \epsilon$. But then if $n > N^* = \max\{N, N_1, N_2\}$, we have $l - \epsilon < a_n \leq b_n \leq c_n < l + \epsilon$ or $|b_n - l| < \epsilon$ if $n > N^*$. Thus $\lim_{n\to\infty} b_n = l$. □

We can use this theorem to prove that $\langle \frac{1}{n^n} \rangle$ converges to 0. All we need to do is show that for all sufficiently large n, $\frac{1}{2^n} > \frac{1}{n^n} > 0$. Then, since the sequences $\langle \frac{1}{2^n} \rangle$ and $\langle 0 \rangle$ both converge to 0, by our theorem $\langle \frac{1}{n^n} \rangle$ converges to 0. (See Exercise 16, page 25.)

1.8 EXERCISES

Find the value of each of the following expressions.

1. $\dfrac{6!}{7!}$

2. $\dfrac{8!}{4!}$

3. $\dfrac{6!}{2!4!}$

4. $\dfrac{10!}{4!6!}$

Determine whether each of the following sequences is a nondecreasing sequence or a nonincreasing sequence.

5. $\langle n(n-1) \rangle$

6. $\langle \frac{1}{n} - \frac{1}{n+2} \rangle$

7. $\langle n^2 + n - 2 \rangle$

8. $\langle n - \frac{1}{n} \rangle$

9. $\langle \left(\frac{1}{n}\right)^n \rangle$

10. $\langle \sqrt{\frac{n^3+1}{n^3}} \rangle$ [Hint: If $0 < a < b$, then $\sqrt{a} < \sqrt{b}$.]

Simplify each of the following expressions.

11. $\dfrac{(n+3)!}{(n+2)!}$

12. $(n+1)!(n+2)$

13. $\dfrac{7!(n+2)!n}{6!n!}$

14. $\dfrac{3!(n+3)!}{(n+2)!}$

15. To show that the greatest lower bound of $\left\langle \frac{1}{n!} \right\rangle$ is 0,

    a) Show that 0 is a lower bound of the sequence.
    b) Show that for all $n \geq 4$, $\frac{1}{n!} < \frac{1}{2^n}$.
    c) Use this result to show that if p is any positive number, then p cannot be a lower bound for $\left\langle \frac{1}{n!} \right\rangle$. (See Exercise 13c on page 21.)

16. a) Show that for $n \geq 2$, $\frac{1}{n^n} \leq \frac{1}{2^n}$.
    b) Use Theorem 1.8c to prove that $\lim_{n \to \infty} \frac{1}{n^n} = 0$.

17. Suppose $x_n > 0$ for all n. Show that $\langle x_n \rangle$ is a nondecreasing sequence if and only if $\frac{x_n}{x_{n+1}} \leq 1$ for all n, and a nonincreasing sequence if and only if $\frac{x_n}{x_{n+1}} \geq 1$ for all n.

Use the theorem from Exercise 17 to determine whether each of the following sequences is nonincreasing or nondecreasing.

18. $\left\langle \frac{n!}{2^n} \right\rangle$

19. $\left\langle \frac{n}{(n+1)!} \right\rangle$

20. $\left\langle \frac{(n+1)!}{1 \cdot 3 \cdot 5 \cdots (2n+1)} \right\rangle$

21. $\left\langle \frac{2^{2n}(n!)^2}{(2n)!} \right\rangle$

1.9 Algebra of Sequences

If we are given two sequences $\langle a_n \rangle$ and $\langle b_n \rangle$ and consider the expressions $\langle a_n + b_n \rangle$ and $\langle a_n \cdot b_n \rangle$ we see that we have two new sequences.

Example 1 $\langle a_n \rangle = \langle n+1 \rangle$, $\langle b_n \rangle = \langle n \rangle$
$\langle a_n + b_n \rangle = \langle (n+1) + n \rangle = \langle 2n+1 \rangle$
$\langle a_n \cdot b_n \rangle = \langle (n+1) \cdot n \rangle = \langle n^2 + n \rangle$

It would seem very natural to call these new sequences the sum and product, respectively, of the sequences $\langle a_n \rangle$ and $\langle b_n \rangle$ and, indeed, we do.

We similarly define the difference and quotient of these sequences as $\langle a_n - b_n \rangle$ and $\left\langle \dfrac{a_n}{b_n} \right\rangle$, respectively, noting that for $\left\langle \dfrac{a_n}{b_n} \right\rangle$ to be defined, $b_n \neq 0$ for all n. If we now introduce the following notation,

$$\langle a_n \rangle + \langle b_n \rangle = \langle a_n + b_n \rangle$$
$$\langle a_n \rangle - \langle b_n \rangle = \langle a_n - b_n \rangle$$
$$\langle a_n \rangle \cdot \langle b_n \rangle = \langle a_n b_n \rangle$$
$$\langle a_n \rangle / \langle b_n \rangle = \langle a_n / b_n \rangle, \quad b_n \neq 0,$$

we have defined the basic algebraic operations for the set of all sequences.

Example 2 $\langle n \rangle + \left\langle \dfrac{1}{n} \right\rangle = \left\langle n + \dfrac{1}{n} \right\rangle = \left\langle \dfrac{n^2 + 1}{n} \right\rangle$

Example 3 $\left\langle \dfrac{1}{n} \right\rangle - \left\langle \dfrac{1}{n+1} \right\rangle = \left\langle \dfrac{1}{n} - \dfrac{1}{n+1} \right\rangle = \left\langle \dfrac{1}{n(n+1)} \right\rangle$

Example 4 $\langle n \rangle \cdot \left\langle \dfrac{1}{n+1} \right\rangle = \left\langle n \cdot \dfrac{1}{n+1} \right\rangle = \left\langle \dfrac{n}{n+1} \right\rangle$

Example 5 $\dfrac{\langle n-1 \rangle}{\langle n \rangle} = \left\langle \dfrac{n-1}{n} \right\rangle$

It is no accident that most of the properties of the algebraic operations on real numbers, such as commutativity and associativity of addition and multiplication, also hold for the corresponding operations on sequences. They are in a very real sense inherited properties. The following proof of the commutativity of multiplication of sequences illustrates this point.

By our definition, $\langle a_n \rangle \cdot \langle b_n \rangle = \langle a_n \cdot b_n \rangle$ and $\langle b_n \rangle \cdot \langle a_n \rangle = \langle b_n \cdot a_n \rangle$. By two sequences being equal we mean that the corresponding terms of the two sequences are equal. Since multiplication of real numbers is commutative, we have $a_n \cdot b_n = b_n \cdot a_n$ for every positive integer n, and therefore $\langle a_n \cdot b_n \rangle = \langle b_n \cdot a_n \rangle$. It now follows that

$$\langle a_n \rangle \cdot \langle b_n \rangle = \langle a_n \cdot b_n \rangle = \langle b_n \cdot a_n \rangle = \langle b_n \rangle \cdot \langle a_n \rangle,$$

or multiplication of sequences is commutative.

We observe that a rather common practice is to denote $\langle a \rangle \cdot \langle c_n \rangle = \langle a c_n \rangle$, where $\langle a \rangle$ is a constant sequence, by $a \langle c_n \rangle$.

Example 6 $\langle 2 \rangle \cdot \left\langle n + \dfrac{1}{2} \right\rangle = 2 \cdot \left\langle n + \dfrac{1}{2} \right\rangle = \langle 2n + 1 \rangle$

Example 7 $\left\langle \dfrac{n}{3n+6} \right\rangle = \left\langle \dfrac{1}{3} \cdot \dfrac{n}{n+2} \right\rangle = \left\langle \dfrac{1}{3} \right\rangle \left\langle \dfrac{n}{n+2} \right\rangle = \dfrac{1}{3} \left\langle \dfrac{n}{n+2} \right\rangle$

Example 8 $2\left\langle\dfrac{1}{n}\right\rangle + 3\left\langle\dfrac{1}{n^2}\right\rangle = \left\langle\dfrac{2}{n}\right\rangle + \left\langle\dfrac{3}{n^2}\right\rangle = \left\langle\dfrac{2n+3}{n^2}\right\rangle$

Example 9 $\alpha\langle a_n\rangle + \beta\langle b_n\rangle = \langle\alpha a_n\rangle + \langle\beta b_n\rangle = \langle\alpha a_n + \beta b_n\rangle$

We list some of the basic properties of addition and multiplication of sequences.

Commutativity of addition and multiplication

$$\langle a_n\rangle + \langle b_n\rangle = \langle b_n\rangle + \langle a_n\rangle$$
$$\langle a_n\rangle \cdot \langle b_n\rangle = \langle b_n\rangle \cdot \langle a_n\rangle$$

Associativity of addition and multiplication

$$\langle a_n\rangle + [\langle b_n\rangle + \langle c_n\rangle] = [\langle a_n\rangle + \langle b_n\rangle] + \langle c_n\rangle$$
$$\langle a_n\rangle \cdot [\langle b_n\rangle \cdot \langle c_n\rangle] = [\langle a_n\rangle \cdot \langle b_n\rangle] \cdot \langle c_n\rangle$$

Distributivity of multiplication over addition

$$\langle a_n\rangle \cdot [\langle b_n\rangle + \langle c_n\rangle] = \langle a_n\rangle \cdot \langle b_n\rangle + \langle a_n\rangle \cdot \langle c_n\rangle$$
$$[\langle b_n\rangle + \langle c_n\rangle] \cdot \langle a_n\rangle = \langle b_n\rangle \cdot \langle a_n\rangle + \langle c_n\rangle \cdot \langle a_n\rangle$$

By using these properties and the definitions of the algebraic operations we can often take a complicated sequence and express it in terms of simpler sequences:

Example 10 $\left\langle\dfrac{(n+2)^2}{n^2}\right\rangle = \left\langle\left(\dfrac{n+2}{n}\right)^2\right\rangle = \left\langle\left(1+\dfrac{2}{n}\right)^2\right\rangle = \left\langle 1+\dfrac{2}{n}\right\rangle\left\langle 1+\dfrac{2}{n}\right\rangle$

We leave the proofs of the various properties of these algebraic operations for the exercises. We take special note of a property that holds for multiplication of real numbers, that does not hold for multiplication of sequences. If a and b are real numbers and $ab = 0$, then at least one of the numbers a or b must be 0. However if $\langle a_n\rangle \cdot \langle b_n\rangle = \langle 0\rangle$, this *does not* imply that one of the sequences $\langle a_n\rangle$ or $\langle b_n\rangle$ must be the zero sequence, $\langle 0\rangle$. For example, if $a_n = 1 + (-1)^n$ and $b_n = 1 + (-1)^{n+1}$, then all the even terms of the sequence $\langle a_n\rangle$ are equal to 2 and the odd terms are zero, while the odd terms of $\langle b_n\rangle$ are equal to 2 and the even terms are zero. Thus neither $\langle a_n\rangle$ or $\langle b_n\rangle$ is the sequence $\langle 0\rangle$, but

$$\begin{aligned}\langle a_n\rangle \cdot \langle b_n\rangle &= \langle(1+(-1)^n)(1+(-1)^{n+1})\rangle \\ &= \langle 1+(-1)^n+(-1)^{n+1}+(-1)^{2n+1}\rangle \\ &= \langle 1+(-1)^n+(-1)^{n+1}+(-1)\rangle \\ &= \langle(-1)^n(1+(-1))\rangle \\ &= \langle 0\rangle.\end{aligned}$$

1.9 EXERCISES

Find $\langle a_n + b_n \rangle$ and $\langle a_n - b_n \rangle$ in each of the following exercises.

1. $\langle a_n \rangle = \langle n^4 - 5n^2 + 2n - 7 \rangle$; $\langle b_n \rangle = \langle 2n^3 - 2n^2 - 5n \rangle$.
2. $\langle a_n \rangle = \langle n^4 + 3n^3 - 7n + 11 \rangle$; $\langle b_n \rangle = \langle -3n^4 - n^3 + 6n^2 + 2n - 1 \rangle$.
3. $\langle a_n \rangle = \left\langle \dfrac{2n+1}{5n} \right\rangle$; $\langle b_n \rangle = \left\langle \dfrac{4n-2}{4n} \right\rangle$.
4. $\langle a_n \rangle = \left\langle \dfrac{5n}{2n+1} \right\rangle$; $\langle b_n \rangle = \left\langle \dfrac{3n}{3n-2} \right\rangle$.
5. $\langle a_n \rangle = \left\langle \dfrac{2n-5}{n^2+3n} \right\rangle$; $\langle b_n \rangle = \left\langle \dfrac{1}{n+3} \right\rangle$.
6. $\langle a_n \rangle = \left\langle \dfrac{n+2}{(n+1)^2} \right\rangle$; $\langle b_n \rangle = \left\langle \dfrac{3}{n+1} \right\rangle$.
7. $\langle a_n \rangle = \left\langle \dfrac{3}{n^2+5n+6} \right\rangle$; $\langle b_n \rangle = \left\langle \dfrac{2}{n^2+4n+4} \right\rangle$.
8. $\langle a_n \rangle = \left\langle \dfrac{2}{4n^2-1} \right\rangle$; $\langle b_n \rangle = \left\langle \dfrac{1}{2n^2+n} \right\rangle$.

Find $\langle a_n \cdot b_n \rangle$ in each of the following exercises.

9. $\left\langle \dfrac{(n+5)^2}{n^2} \right\rangle \cdot \left\langle \dfrac{4n^2}{4n+20} \right\rangle$.
10. $\left\langle \dfrac{4n-6}{4n+8} \right\rangle \cdot \left\langle \dfrac{6n+12}{4n^2-9} \right\rangle$.
11. $\left\langle \dfrac{5n+10}{15n} \right\rangle \cdot \left\langle \dfrac{n^2-4}{(n+2)^2} \right\rangle$.

Find $\langle a_n/b_n \rangle$ in each of the following exercises.

12. $\langle a_n \rangle = \langle n^3 + 27 \rangle$; $\langle b_n \rangle = \langle n + 3 \rangle$.
13. $\langle a_n \rangle = \left\langle \dfrac{7n-14}{4n+8} \right\rangle$; $\langle b_n \rangle = \left\langle \dfrac{1}{n^2+5n+6} \right\rangle$.
14. $\langle a_n \rangle = \left\langle \dfrac{16n^2-9}{4n+1} \right\rangle$; $\langle b_n \rangle = \langle 3 - 4n \rangle$.
15. Prove that addition of sequences is commutative.
16. Prove that multiplication of sequences is associative.
17. Prove that addition of sequences is associative.
18. Prove that multiplication of sequences is distributive over addition of sequences.
19. Prove that $\langle 1 \rangle \cdot \langle a_n \rangle = \langle a_n \rangle \cdot \langle 1 \rangle = \langle a_n \rangle$.
20. Prove that $\langle a_n \rangle + \langle 0 \rangle = \langle 0 \rangle + \langle a_n \rangle = \langle a_n \rangle$.
21. Prove that $\langle 0 \rangle \cdot \langle a_n \rangle = \langle a_n \rangle \cdot \langle 0 \rangle = \langle 0 \rangle$.

1.10 Some Properties of Limits

If $\langle a_n \rangle$ and $\langle b_n \rangle$ are convergent sequences, one might naturally wonder whether the sum, difference, product, and quotient of these sequences are convergent sequences. Let us consider a few examples and form some conjectures.

Example 1 Suppose $\langle a_n \rangle = \left\langle \dfrac{n-1}{n} \right\rangle$ and $\langle b_n \rangle = \left\langle \dfrac{n+2}{n} \right\rangle$.

In this case $\langle a_n + b_n \rangle = \left\langle \dfrac{n-1}{n} + \dfrac{n+2}{n} \right\rangle = \left\langle \dfrac{2n+1}{n} \right\rangle$

and $\langle a_n - b_n \rangle = \left\langle \dfrac{n-1}{n} - \dfrac{n+2}{n} \right\rangle = \left\langle \dfrac{-3}{n} \right\rangle$.

Now, $a_n \to 1$, $b_n \to 1$, while

$$a_n + b_n \to 2 = 1 + 1$$
$$a_n - b_n \to 0 = 1 - 1.$$

Example 2 Suppose $\langle a_n \rangle = \left\langle \dfrac{n^2+1}{n^2} \right\rangle$ and $\langle b_n \rangle = \left\langle \dfrac{2}{n} \right\rangle$.

In this instance

$\langle a_n + b_n \rangle = \left\langle \dfrac{n^2+1}{n^2} + \dfrac{2}{n} \right\rangle = \left\langle \dfrac{n^2+2n+1}{n^2} \right\rangle = \left\langle \left(\dfrac{n+1}{n}\right)^2 \right\rangle$

and $\langle a_n - b_n \rangle = \left\langle \dfrac{n^2+1}{n^2} - \dfrac{2}{n} \right\rangle = \left\langle \dfrac{n^2-2n+1}{n^2} \right\rangle = \left\langle \left(\dfrac{n-1}{n}\right)^2 \right\rangle$.

Thus, $a_n \to 1$, $b_n \to 0$, while

$$a_n + b_n \to 1 = 1 + 0$$
$$a_n - b_n \to 1 = 1 - 0.$$

Example 3 Suppose $\langle a_n \rangle = \left\langle \dfrac{n+1}{n} \right\rangle$ and $\langle b_n \rangle = \left\langle \dfrac{1-n}{n} \right\rangle$.

Here $\langle a_n + b_n \rangle = \left\langle \dfrac{n+1}{n} + \dfrac{1-n}{n} \right\rangle = \left\langle \dfrac{2}{n} \right\rangle$

and $\langle a_n - b_n \rangle = \left\langle \dfrac{n+1}{n} - \dfrac{1-n}{n} \right\rangle = \left\langle \dfrac{2n}{n} \right\rangle = \langle 2 \rangle$.

Thus, $a_n \to 1$, $b_n \to -1$, while

$$a_n + b_n \to 0 = 1 + (-1)$$
$$a_n - b_n \to 2 = 1 - (-1).$$

These three examples lead us to conjecture:

If $a_n \to A$ and $b_n \to B$, then $a_n + b_n \to A + B$ and
$$a_n - b_n \to A - B.$$

Now let us consider some products and quotients.

Example 4 Suppose $\langle a_n \rangle = \left\langle \dfrac{n-1}{n} \right\rangle$ and $b_n = \left\langle \dfrac{n+2}{n} \right\rangle$.

In this case $\langle a_n b_n \rangle = \left\langle \dfrac{n-1}{n} \cdot \dfrac{n+2}{n} \right\rangle = \left\langle \dfrac{n^2 + n - 2}{n^2} \right\rangle$

and $\left\langle \dfrac{a_n}{b_n} \right\rangle = \left\langle \dfrac{n-1}{n} \Big/ \dfrac{n+2}{n} \right\rangle = \left\langle \dfrac{n-1}{n+2} \right\rangle.$

Thus, $a_n \to 1$, $b_n \to 1$, while

$$a_n b_n \to 1 = 1 \cdot 1$$
$$\frac{a_n}{b_n} \to 1 = \frac{1}{1}.$$

Example 5 Suppose $\langle a_n \rangle = \left\langle \dfrac{n^2 + 1}{n^2} \right\rangle$ and $\langle b_n \rangle = \left\langle \dfrac{2}{n} \right\rangle$.

Here $\langle a_n b_n \rangle = \left\langle \dfrac{n^2+1}{n^2} \cdot \dfrac{2}{n} \right\rangle = \left\langle \dfrac{2n^2 + 2}{n^3} \right\rangle$

and $\left\langle \dfrac{a_n}{b_n} \right\rangle = \left\langle \dfrac{n^2+1}{n^2} \Big/ \dfrac{2}{n} \right\rangle = \left\langle \dfrac{n^2+1}{2n} \right\rangle.$

Thus, $a_n \to 1$, $b_n \to 0$, while

$$a_n b_n \to 0 = 1 \cdot 0$$
$$\frac{a_n}{b_n} \text{ diverges.}$$

Here we might conjecture that:

If $a_n \to A$ and $b_n \to B$, then $a_n b_n \to A \cdot B$ and $\dfrac{a_n}{b_n} \to \dfrac{A}{B}$, provided in the latter case that $B \neq 0$.

After a moment's reflection we might add in the case of $\dfrac{a_n}{b_n}$ the condition that $b_n \neq 0$ for any n, since if $b_n = 0$, then $\dfrac{a_n}{b_n}$ is not defined.

All of our conjectures are true.

Theorem 1.10a *If $\langle a_n \rangle$ and $\langle b_n \rangle$ are convergent sequences, with limits A and B respectively, then*

1. *$\langle a_n \rangle + \langle b_n \rangle$ is convergent with limit $A + B$*
2. *$\langle a_n \rangle - \langle b_n \rangle$ is convergent with limit $A - B$*
3. *$\langle a_n \rangle \cdot \langle b_n \rangle$ is convergent with limit $A \cdot B$*
4. *$\dfrac{\langle a_n \rangle}{\langle b_n \rangle}$ is convergent with limit $\dfrac{A}{B}$, provided $\dfrac{a_n}{b_n}$ is defined for all n and $B \neq 0$.*

The results of this theorem can also be stated as follows:

1. The limit of the sum of two convergent sequences is the sum of their limits.
2. The limit of the difference of two convergent sequences is the difference of their limits.
3. The limit of the product of two convergent sequences is the product of their limits.
4. If the quotient of two convergent sequences is defined, then the limit of their quotient is equal to the quotient of the limits, provided the limit of the denominator is not zero.

We take note that it is possible to have the sum, difference, product or quotient of two divergent sequences be a convergent sequence.

Example 6 The sequences $\langle n \rangle$ and $\langle 2 - n \rangle$ are both divergent, but their sum $\langle n + (2 - n) \rangle = \langle 2 \rangle$ is convergent, and their quotient $\dfrac{\langle n \rangle}{\langle 2 - n \rangle} = \left\langle -1 + \dfrac{2}{2 - n} \right\rangle$ is convergent.

Example 7 Earlier we noted that the product of the two divergent sequences $\langle 1 + (-1)^n \rangle$ and $\langle 1 + (-1)^{n+1} \rangle$ is the convergent sequence $\langle 0 \rangle$. Note that $\langle (-1)^n \rangle$ is a divergent sequence, but $\langle (-1)^n \rangle \cdot \langle (-1)^n \rangle$ is convergent.

The proofs of the first three conclusions of Theorem 1.10a are not too difficult and are left as exercises. The proof that $\dfrac{a_n}{b_n} \to \dfrac{A}{B}$ is not so easy. We can, however, use the third conclusion to simplify our proof of the fourth conclusion. For if we can show that $\dfrac{1}{b_n} \to \dfrac{1}{B}$, then by the third conclusion $\dfrac{a_n}{b_n} = a_n \cdot \dfrac{1}{b_n} \to A \cdot \dfrac{1}{B} = \dfrac{A}{B}$. This proof will appear in the appendix to this chapter, page 34.

1.10 EXERCISES

In each of the following exercises, find the limits of $\langle a_n + b_n \rangle$, $\langle a_n - b_n \rangle$, $\langle a_n \cdot b_n \rangle$, and $\left\langle \dfrac{a_n}{b_n} \right\rangle$, if they exist.

1. $\langle a_n \rangle = \langle n + 3 \rangle$; $\langle b_n \rangle = \langle n^3 + 27 \rangle$.
2. $\langle a_n \rangle = \langle 2n + 5 \rangle$; $\langle b_n \rangle = \langle 3n - 1 \rangle$.
3. $\langle a_n \rangle = \left\langle \dfrac{2n + 1}{5n} \right\rangle$; $\langle b_n \rangle = \left\langle \dfrac{4n - 2}{4n} \right\rangle$.
4. $\langle a_n \rangle = \left\langle \dfrac{5n}{2n + 1} \right\rangle$; $\langle b_n \rangle = \left\langle \dfrac{3n}{3n - 2} \right\rangle$.
5. $\langle a_n \rangle = \left\langle \dfrac{2n - 5}{n^2 + 3n} \right\rangle$; $\langle b_n \rangle = \left\langle \dfrac{1}{n + 3} \right\rangle$.
6. $\langle a_n \rangle = \left\langle \dfrac{n + 2}{(n + 1)^2} \right\rangle$; $\langle b_n \rangle = \left\langle \dfrac{3}{n + 1} \right\rangle$.
7. $\langle a_n \rangle = \left\langle \dfrac{3}{n^2 + 5n + 6} \right\rangle$; $\langle b_n \rangle = \left\langle \dfrac{2}{n^2 + 4n + 4} \right\rangle$.
8. $\langle a_n \rangle = \left\langle \dfrac{2}{4n^2 - 1} \right\rangle$; $\langle b_n \rangle = \left\langle \dfrac{1}{2n^2 + n} \right\rangle$.
9. $\langle a_n \rangle = \left\langle \dfrac{(n + 5)^2}{n^2} \right\rangle$; $\langle b_n \rangle = \left\langle \dfrac{4n^2}{4n + 20} \right\rangle$.
10. $\langle a_n \rangle = \left\langle \dfrac{4n - 6}{4n + 8} \right\rangle$; $\langle b_n \rangle = \left\langle \dfrac{6n + 12}{4n^2 - 9} \right\rangle$.
11. $\langle a_n \rangle = \left\langle \dfrac{5n + 10}{15n} \right\rangle$; $\langle b_n \rangle = \left\langle \dfrac{n^2 - 4}{(n + 2)^2} \right\rangle$.
12. $\langle a_n \rangle = \left\langle \dfrac{16n^2 - 9}{4n + 1} \right\rangle$; $\langle b_n \rangle = \langle 3 - 4n \rangle$.

13. Prove that if $\langle a_n \rangle$ converges to A and $\langle b_n \rangle$ converges to B, then
    a) $\langle a_n + b_n \rangle$ converges to $A + B$.
    b) $\langle a_n - b_n \rangle$ converges to $A - B$.

14. Suppose that $\langle a_n \rangle$ and $\langle b_n \rangle$ are convergent sequences with limits A and B respectively.
    a) Show that $|a_n b_n - AB| \leq |b_n| \cdot |a_n - A| + |A| \cdot |b_n - B|$.
    [Hint: $|a_n b_n - AB| = |a_n b_n - Ab_n + Ab_n - AB|$].
    b) Show that there is an N_1, such that $|b_n - B| < 1$ if $n > N_1$. Show that this means $|b_n| < |B| + 1$ if $n > N_1$.
    c) Given $\epsilon > 0$, show that there is an N_2 such that
    $$|a_n - A| < \dfrac{\epsilon}{2(|B| + 1)} \quad \text{if } n > N_2.$$

d) If $A \neq 0$, show that there is an N_3 such that

$$|b_n - B| < \frac{\epsilon}{2|A|} \quad \text{if } n > N_3.$$

e) Let $N^* = \max \{N_1, N_2, N_3\}$. Use the results of parts a–d to show that $|a_n b_n - AB| < \epsilon$ if $n > N^*$.

f) Note that if $A = 0$, then $|a_n b_n - AB| \leq |b_n| \cdot |a_n - A|$. Use all your findings to prove that $\langle a_n b_n \rangle$ is a convergent sequence with limit AB.

Appendix to Chapter 1

Theorem 1.10b *If $b_n \to b$, $b_n \neq 0$ for any n, and $b \neq 0$ then $\dfrac{1}{b_n} \to \dfrac{1}{b}$.*

Proof First, since $b \neq 0$, $|b| = \alpha > 0$ and therefore $|b| - \dfrac{\alpha}{2} > 0$. Now we know from Theorem 1.4a that since $b_n \to b$, $|b_n| \to |b|$. Thus there is an N_1 such that if $n > N_1$, then $||b_n| - |b|| < \dfrac{\alpha}{2}$, and hence $|b| - \dfrac{\alpha}{2} < |b_n|$ and $\dfrac{1}{|b_n|} < \dfrac{1}{|b| - \dfrac{\alpha}{2}}$ if $n > N_1$.

Second, keeping this conclusion in mind, let us take an arbitrary positive number ϵ and consider which values of n make $\left|\dfrac{1}{b_n} - \dfrac{1}{b}\right| < \epsilon$. We observe that $\left|\dfrac{1}{b_n} - \dfrac{1}{b}\right| = \left|\dfrac{b - b_n}{b_n b}\right| = \dfrac{|b_n - b|}{|b||b_n|}$. We also know then, that if $n > N_1$, it must follow that

$$\left|\dfrac{1}{b_n} - \dfrac{1}{b}\right| = \dfrac{|b_n - b|}{|b||b_n|} < \dfrac{|b_n - b|}{|b|\left(|b| - \dfrac{\alpha}{2}\right)},$$

since $\dfrac{1}{|b_n|} < \dfrac{1}{|b| - \dfrac{\alpha}{2}}$ if $n > N_1$. Now since $b_n \to b$ we can make $|b_n - b|$ smaller than any positive number for all sufficiently large values of n. In particular we can make $|b_n - b| < \epsilon\left[|b|\left(|b| - \dfrac{\alpha}{2}\right)\right]$ for all $n > N_2$, where N_2 is a sufficiently large integer. But then if $N = \max\{N_1, N_2\}$ so that $n > N$ implies $n > N_1$ and $n > N_2$, we see that if $n > N$,

$$\left|\dfrac{1}{b_n} - \dfrac{1}{b}\right| < \dfrac{|b_n - b|}{|b|\left(|b| - \dfrac{\alpha}{2}\right)} = |b_n - b| \cdot \dfrac{1}{|b|\left(|b| - \dfrac{\alpha}{2}\right)}$$

and $\qquad |b_n - b| < \epsilon\left[|b|\left(|b| - \dfrac{\alpha}{2}\right)\right].$

From these two facts we see that if $n > N$, then

$$\left|\dfrac{1}{b_n} - \dfrac{1}{b}\right| < \epsilon, \qquad \text{that is,} \qquad \dfrac{1}{b_n} \to \dfrac{1}{b}. \quad \square$$

2
Functions

Introduction

The development of mathematics has depended quite often on the extension or generalization of a concept. The development of the concept of number, from the positive integers to the set of all integers, then to rational numbers and then again to the set of real numbers R, illustrates this fact. In each of these generalizations the earlier concept was preserved as a special case of the extended concept. The concept of a sequence as a correspondence relating the positive integers to some subset of the real numbers has some fairly obvious generalizations. One of these, of course, is the notion of a correspondence between two (possibly the same) subsets of R. More general than this is the notion of a correspondence between two sets of things.

Everyone is familiar with, and uses, the notion of a correspondence between two sets of things. To every item in a cart full of groceries, there corresponds a specific price. To each of the rocks in a pile, there corresponds a weight in pounds. To each individual in a group of people, there corresponds a special name. These are examples of correspondences between two sets: A set of grocery items and a set of prices, a set of rocks and a set of weights, a set of people and a set of names.

2.1 The Function Concept

Perhaps the easiest, and most natural way to think of a **function** is as a special kind of correspondence between two sets. When we think of a function in this way, we call one of the sets the **domain** of the function the other set the **range** of the function. Each of these sets consists of distinct elements, that is, no two elements are equal. To each element in the set called the domain, there corresponds *precisely* one element in the set called the range. Each element in the range corresponds to at least one element in the domain, but there is no limit to how many elements in the domain that it may correspond.

Example 1		*Example 2*		*Example 3*		
			Rational	Real		Real
Rock	Weight	Integer	No.	No.		No.
1	$\to$ 3.0 lbs	1	$\to$ 1	-1	$\to$	1
2	$\to$ 4.1 lbs	2	$\to \frac{1}{2}$	$\sqrt{2}$	$\to$	2
3	$\to$ 1.5 lbs	3	$\to \frac{1}{3}$	π	$\to$	π^2
4	$\to$ 3.0 lbs	4	$\to \frac{1}{4}$	.		.
5	$\to$ 1.5 lbs	.	.	.		.
6	$\to$ 3.0 lbs	.	.	.		.
		n	$\to \frac{1}{n}$	x	$\to$	x^2
		.	.	.		.
		.	.	.		.
		.	.	.		.

Notice that when we display our correspondence, or function, as in these three examples, we have a list of pairs. The first element, or **first component,** of the pair is from the domain of the function. The second element, or **second component,** is from the range of the function. We are, thus, exhibiting the functions as sets of ordered pairs. As you recall, a pair of things is said to be an **ordered pair** when one of them is designated as the first element, or first component, of the pair and the other as the second. The standard symbol for an ordered pair is (a, b). This notation indicates that a is the first component and b the second component.

In Example 1, we might have simplified our listing by just exhibiting the set of ordered pairs in this way:

$$\{(1, 3.0), (2, 4.1), (3, 1.5), (4, 3.0), (5, 1.5), (6, 3.0)\}$$

Although the notion of a function can be defined and viewed in various ways, it has become a more or less standard practice to define a function formally as follows:

Definition 2.1a A **function** is a set of ordered pairs such that no two distinct members of the set have the same first component.

When a function is considered as a set of ordered pairs, the ordered pairs can be regarded as defining a rule of correspondence between the set of first components of the pairs and the set of second components. On the other hand, when a function is viewed as being a correspondence between two sets, to each distinct element of the first set, there corresponds precisely one element of the second set. The rule of correspondence then determines the components of the ordered pairs of the function. The difference between these two conceptions of a function is just a difference of viewpoint. The following example illustrates each point of view.

Example 4 Suppose that an automobile travels along a highway for a period of three hours at a constant speed of 50 miles per hour. In one hour's time, the car will travel 50 miles; in $1\frac{1}{2}$ hours' time, the car will travel 75 miles; and in $\frac{1}{2}$ hour's time, the car will travel 25 miles. In fact, since we are assuming a constant speed of 50 miles per hour, corresponding to any time t, from 0 to 3 hours, there is a distance $50t$, expressed in miles, which the car will travel. This correspondence may be described by the conditions $d = 50t$, $0 \leq t \leq 3$.

For this example, the conditions $d = 50t$, $0 \leq t \leq 3$, are a convenient way of describing a function as a correspondence which relates to numerical values of time numerical values of distance. It can, however, also be convenient to think of the function as a set of ordered pairs of the form $(t, 50t)$, where t is any real number.

Example 4 exhibits a function whose domain and range are infinite rather than finite sets. Most of the functions we shall consider will be of this type.

Since we have not placed any restriction on what kinds of objects may be components of the ordered pairs of a function, the definition is extremely general. For example, a set of ordered pairs with squares as first components and circles as second components is a function, provided that the same square does not occur as the first component of two different ordered pairs. Even if we restrict ourselves to functions that are sets of ordered pairs of real numbers—and such functions will be our main interest—the definition is so general that it includes as functions a very wide range of sets. But it is precisely this generality that will allow us to study a great variety of mathematical topics from the point of view of functions and to see a unity about them that would otherwise not be apparent.

For example, in Chapter 1 you studied sequences. Given a sequence $\langle a_n \rangle$, we have corresponding to each positive integer a number, or an object a_n. Thus, it is convenient to think of any sequence as a function whose domain is the positive integers. In fact, this is often the way that *sequence* is defined.

Example 5 The sequence $\langle n^2 \rangle$, which equals

$$\langle 1, 4, 9, \ldots, n^2, \ldots \rangle,$$

can be thought of as the correspondence which relates every positive integer to its square, *or* as the set of ordered pairs (n, n^2), where n is a positive integer.

The general sequence $\langle a_n \rangle$ can be thought of as a correspondence for which to every positive integer n, there corresponds a_n, *or* as the set of ordered pairs (n, a_n) where n is a positive integer.

To conclude these introductory remarks, we emphasize that a function may be viewed as a special kind of correspondence between two sets or as a special kind of set of ordered pairs. Thus, although we have chosen to define a function *formally* as a special kind of set of ordered pairs, we will not restrict ourselves to this language in all of our discussions. In certain situations the idea of a function as a correspondence will be useful; in others the idea of a function as a set of ordered pairs will be useful. We will use whichever point of view seems appropriate to the specific situation.

2.1 EXERCISES

In each of the Exercises 1–6 a verbal statement is given that defines a correspondence between two sets. Consider some of the ordered pairs (x, y) determined by this correspondence, and state whether or not the correspondence is a function. If the correspondence is not a function, explain why not.

1. In a leap year, the month y contains exactly x days.
2. y is a United States senator from the state of x.
3. y is the letter of the alphabet that follows the vowel x.
4. The day of the week x contains y hours.
5. y is the remainder obtained when the integer x is divided by 3.
6. y is a major league baseball team from the city of x.
7. List all the ordered pairs in the function defined in Exercise 3.
8. List the ordered pairs in the function defined in Exercise 4.

In each of the Exercises 9–12 the form of the ordered pairs in a set of ordered pairs of real numbers is given. Name some of the ordered pairs in the set and state whether or not the set is a function. If the set is not a function, explain why not.

9. $(x, x - 3)$
10. (n, n^3)
11. $(C, \frac{9}{5}C + 32)$
12. $(t^2, t - 3)$

13. Mr. and Mrs. Jones invited Mr. and Mrs. Williams and Mr. and Mrs. Thomas to dinner. The figure to the right shows the arrangement of chairs at the table. One function that is a correspondence between the set of people and the set of chairs consists of the following ordered pairs: (Mr. Jones, A), (Mrs. Jones, B), (Mr. Williams, C), (Mrs. Williams, D), (Mr. Thomas, E), (Mrs. Thomas, F). Use this information in answering the following exercises.

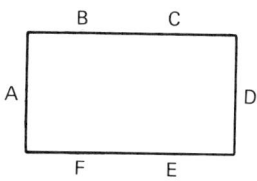

a) If Mr. Jones is to occupy chair A and Mrs. Jones chair D, there are four functions involving people and chairs that meet the requirement that no two persons of the same sex be seated in adjacent chairs. List the ordered pairs in each of these four functions.

b) If, in addition to the requirements in part a, it is further required that no two spouses be seated in adjacent chairs, there are only two functions that will fulfill all the requirements. Which of the functions you tabulated for part a meet the requirements of part b as well?

For each of Exercises 14–19, write five ordered pairs (n, a_n) determined by the given sequence.

14. $\langle 2 \rangle$

15. $\langle 6n - 5 \rangle$

16. $\left\langle \dfrac{10}{n} \right\rangle$

17. $\langle n^2 + 3n \rangle$

18. $\langle n - n! \rangle$

19. $\langle n + |-n| \rangle$

2.2 Some Terminology and Notation

As we have seen, functions can be sets of ordered pairs of many things. In this book, we will mainly be concerned with functions which are ordered pairs of *real numbers*. Another way of saying this is that we will mainly be concerned with functions with domains and ranges which are subsets of R.

There are some fairly standard symbols for denoting functions. Generally speaking, lower-case English letters, such as f, g, or h, are used to represent functions, with f occurring most frequently. Upper-case letters are sometimes used, however, such as F and G. On occasion, letters of the Greek alphabet, such as Γ and Ψ, are used to designate functions. Any letter from any alphabet could be used, of course, as well as any symbol you might want to invent. The point to keep in mind when using a symbol like f to represent a function is that this symbol stands for a set of ordered pairs, no two of which have the same first component. A list of a few more examples of functions follows.

Example 1 $f = \{(x, y) \mid y = x^2, x \in I\}$. This notation is read: f is the set of ordered pairs (x, y) such that $y = x^2$ and x is an integer. The domain of f, $D(f) = I$, and the range of f, $R(f)$ is the set of all squares of integers. Note that f is a sequence.

Example 2 $g = \{(x, y) \mid y = x^2, x \in R\}$. Thus, g is the set of ordered pairs (x, y) such that $y = x^2$ and x is a real number. The domain of g, $D(g) = R$, and the range of g, $R(g)$, is the set of all squares of real numbers. Do you see why $R(g)$ is equal to the set of all nonnegative members of R?

The function f in Example 1 is a subset of the function g in Example 2. We can say that g is an **extension** of f or f is a **restriction** of g. Is g a sequence?

Example 3 $h = \{(x, y) \mid y = x + 1, x \in R\}$. The function h is the set of all ordered pairs (x, y) such that $y = x + 1$ and x is a real number. The domain of h, $D(h)$, is R. Do you see that $R(h)$ is also R? Note also that $h = \{(x, x + 1) \mid x \in R\}$.

If f is a function with domain A and range B, we often express this relationship in the following way:

$$f: A \to B \quad \text{or} \quad A \xrightarrow{f} B.$$

Each of these symbols is read as "f is a function on A onto B." More generally, it is possible to talk about a function on A *into* B, which means that A is the domain of f and the range of f is a subset of B. For example, the function

$$f = \{(x, y) \mid x \in I \text{ and } y = 2x\}$$

is a function on I, the set of integers, *into* R, the set of all real numbers, because the range of this function is a proper subset of R. Function f is a function on I *onto* the set of all even integers because the range of f is equal to the set of even integers. When we state that f is a function on A onto B, we will always mean that A is the domain of f and B is the range of f.

If f is a function and (x, y) is an element of f, we will write $y = f(x)$, which is read as "y equals f of x." Thus, the notation $y = f(x)$ is just another way of indicating that $(x, y) \in f$. If we think of a function as determining a correspondence between the elements of its domain and the elements of its range, then $y = f(x)$ should be interpreted as meaning that y is the element in the range of f corresponding to the element x in the domain of f. We could use the notation,

$$f = \{(x, f(x)) \mid x \text{ is in the domain of } f\},$$

which provides a rather circular statement about f, but does stress the meaning of the notation $f(x)$.

Historically, it has been a common practice to think of functions in "kinetic" terms. This is particularly true when a function is serving as a mathematical model of a physical event.

Example 4 If a ball is dropped from the top of a tall building, it falls a distance of d feet in t seconds, where d can be determined approximately by the formula $d = 16t^2$. If the total time the ball falls before it reaches the ground is 4 seconds, then the function

$$f = \{(t, 16t^2) \mid 0 \leq t \leq 4\}$$

serves as a mathematical model for this physical occurrence. For any ordered pair of real numbers belonging to f, the first component is the time in seconds and the second component is the distance in feet that the ball has fallen in this amount of time.

In Example 4 it is very natural to think of the first component of the ordered pairs $(t, f(t))$ as varying and the second as varying also, but dependent upon the value of the first component according to the formula $f(t) = 16t^2$. For this reason, if we denote the ordered pairs of a function by $(t, f(t))$, then t is called the **independent variable** and $f(t)$ is called the **dependent variable.** It is common practice to say that the dependent variable is "a function of" the independent variable, and we shall on occasion use this terminology. We must remember, however, that the dependent variable is not actually a function itself; the function is, of course, the set of all ordered pairs $(t, f(t))$.

The notation $f(x) = y$ is also useful for listing specific elements in the range of a function. For example, suppose that we are given the function

$$f = \{(x, y) \mid x \in I \text{ and } y = x^2\}.$$

Since $f(x) = y = x^2$, we see that $f(0) = 0$, $f(1) = 1$, $f(4) = f(-4) = 16$, and so on.

You will find in many textbooks statements like "Consider the function $f(x) = x^2$, $1 \leq x \leq 2$, ..." It is more precise to say "Consider the function f *defined by* $f(x) = x^2$, . . . ," since f denotes the function and $f(x)$ denotes an element in the range of the function. We will follow the practice of speaking of "the function f defined by . . ." long enough to establish firmly this distinction. Later, however, for the sake of brevity, we will become less precise and more informal and use phrases like "the function $f(x) = x^2$."

Another method of designating a function, which is both precise and brief and which we will use occasionally, is the following:

$$f : f(x) = x^2, \quad 0 \leq x \leq 1.$$

This sentence is read: "The function f is defined by $f(x) = x^2$, where the domain of f is the set of elements x such that $0 \leq x \leq 1$."

At times, an equation that defines a function f will be given with no mention of a domain for the function. In such cases, we will assume that the domain consists of all the real numbers, x, for which the equation produces an associated real number, $f(x)$. (This assumption is sometimes called the *principle of maximal domain*.) For example, consider the function $f: y = \sqrt{x - 3}$. Since there is no real number that is the square root of $x - 3$ if $x - 3$ is negative, it follows that the domain of f consists of those real numbers for which $x - 3 \geq 0$. That is, the domain is $\{x \mid x \geq 3\}$, which is called the set of *meaningful replacements* for x.

Once we know the domain of this function, we can determine its range. Since $y = \sqrt{x - 3}$ and $x - 3 \geq 0$, the range must consist of those real numbers that are the principal square roots of nonnegative real numbers. Hence, the range is $\{y \mid y \geq 0\}$.

2.2 EXERCISES

1. Given $f = \{(x, f(x)) \mid f(x) = 6 - x^2\}$, determine the value of each of the following.
   a) $f(3)$ b) $f(-3)$ c) $f(\sqrt{2})$ d) $f(2a)$ e) $f(3a - 1)$
2. Given $h: h(x) = 3x^2 - 2x + 1$, determine each of the following.
   a) $h(-1)$ b) $h(\frac{1}{2})$ c) $h(-\sqrt{3})$ d) $h(a - 1)$ e) $h(2a + 1)$
3. Given $F(x) = x^3$, determine each of the following.
   a) $F(-2)$ b) $F(\frac{2}{3})$ c) $F(-\frac{1}{3})$ d) $F(a^2)$ e) $F(1 - a)$
4. Given $\Psi = \{(x, y) \mid 3x + 2y = 7\}$, determine each of the following.
   a) $\Psi(-1.2)$ b) $\Psi(\frac{5}{2})$ c) $\Psi(-3\frac{1}{3})$ d) $\Psi(-a)$ e) $\Psi(\Psi(a))$

For each function below, describe the domain and the range in set notation.

5. $\{(x, y) \mid 3x + 4y = 12\}$.
6. $\{(x, y) \mid y^3 = x\}$.
7. $\{(x, y) \mid xy = 18\}$.
8. $g: g(x) = \dfrac{1}{\sqrt{x}}$.
9. $f: f(x) = \sqrt{x}$.
10. $h: h(x) = \sqrt{25 - x^2}$.
11. $F: F(x) = 4\sqrt{2x^2 - 18}$.
12. $G: G(x) = \dfrac{x - 1}{x}$.

You know that the domain of a sequence is the set of positive integers. Describe the range of each of the following sequences.

13. $\langle -n \rangle$
14. $\langle n + 2 \rangle$
15. $\langle \sqrt{n^2} \rangle$
16. $\langle 2n \rangle$
17. $\langle 2n - 1 \rangle$
18. $\langle 5n \rangle$

2.3 Graphs of Functions

As we have said, a function determines a correspondence that relates each member of the domain of the function to a unique member of its range. It is sometimes helpful to think of this correspondence as "transferring" elements of the domain into elements of the range. In this interpretation, one usually thinks of a point in the domain as moving, or being mapped, onto a point in the range. When we use a figure of this kind to represent a correspondence defined by a function, we use words in describing it that are suggested by the figure. Instead of referring to a function on A into B, we speak of a *mapping* of A into B and of *points* in A and B, instead of elements in A and B. We will also speak of points in B as being *image points* of points in A.

For example, a particular function that maps the set of real numbers onto the set of positive real numbers might be represented as in Figure 2–1.

Figure 2-1

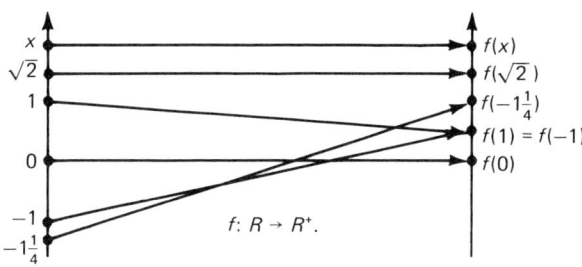

When we wish to emphasize the fact that a function is a mapping, we will sometimes use the notation

$$x \to f(x).$$

The notation $x \to x^2$ is one way of describing the function $f = \{(x, x^2) \mid x \in R\}$ as a mapping of R into R.

It should be clear that if we try to represent the mapping $x \to x^2$ graphically, as above, we can only indicate a few representative points of the mapping. In fact, if we could draw all of the arrows representing the mapping $x \to x^2$, the space between the two vertical lines that lies above the mapping of 0 into 0 would be entirely covered! (It is an interesting, but somewhat difficult exercise, to see which points, between the vertical lines, would not be covered.)

Another means of graphically representing a mapping is with a rectangular coordinate system. A rectangular coordinate system is established in the plane by choosing two perpendicular lines in the plane, called **coordinate axes,** and denoting their intersection, called the **origin,** by O. A real-number scale with origin at 0 is established on each of the axes, and the unit of length is assumed to be the same on both axes. Figure 2–2 on page 44 illustrates the coordinate axes as they are conventionally drawn. Note that the horizontal axis is labeled as the x-axis and the vertical axis as the y-axis. The arrowheads indicate the positive directions for the number scales on the x- and y-axes.

Figure 2-2

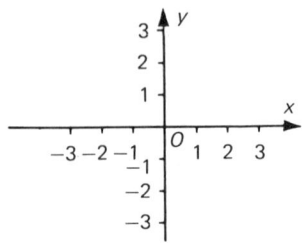

Let P be an arbitrary point in the plane, as shown in Figure 2–3. If a line is drawn through P perpendicular to the x-axis, this line will intersect the x-axis in a point corresponding to the real number a in the number scale on the x-axis. Similarly, a line through P perpendicular to the y-axis intersects the y-axis in a point corresponding to the real number b on the number scale of the y-axis. The components a and b of the ordered pair of real numbers (a, b) are then said to be the **coordinates** of the point P. The number a is called the x-coordinate (or sometimes the *abscissa*) of the point P and b is called the y-coordinate (or sometimes the *ordinate*) of the point P.

Figure 2-3

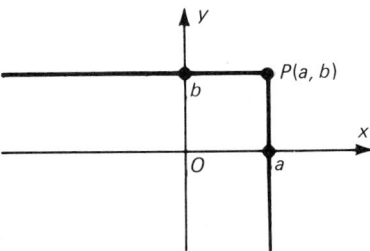

By the procedure just described, an ordered pair of real numbers can be made to correspond to each point in the plane. Similarly, given any ordered pair of real numbers (a, b), we can draw a line perpendicular to the x-axis that passes through the point in the x-axis corresponding to the real number a, and a line perpendicular to the y-axis that passes through the point in the y-axis corresponding to the real number b. These lines will intersect in a point P having coordinates (a, b).

Hence, the introduction of rectangular coordinates sets up a one-to-one correspondence between ordered pairs of real numbers and points in the plane. As noted above, the plane with such a rectangular coordinate system is often called the **Cartesian plane,** and it provides us with a geometric model of the set of all ordered pairs of real numbers.

The portion of the x-axis to the right of the origin, where each point has a positive abscissa, is the positive x-axis; and the portion of the x-axis to the left of the origin, where each point has a negative abscissa, is the negative x-axis. Similarly, the portion of the y-axis above the origin, where each point has a positive ordinate, is the positive y-axis; and the portion of the y-axis below the origin, where each point has a negative ordinate, is the negative y-axis.

The coordinate axes separate the plane into four distinct regions, called **quadrants.** The points above the x-axis and to the right of the y-axis constitute the first quadrant. Both the coordinates of any point in the first quadrant are positive, which is indicated in Figure 2–4 by the symbol $(+, +)$. The other quadrants and the signs of the coordinates of points in these quadrants are also indicated in the figure.

Figure 2-4

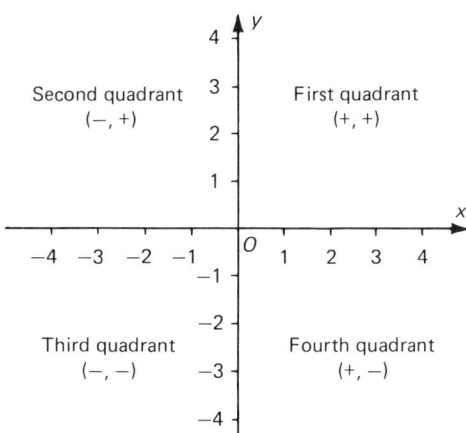

We may, on occasion, refer to the "upper half-plane," meaning all points above the x-axis, or to the "lower half-plane," which consists of all points below the x-axis. Similarly, references to the "right half-plane" and the "left half-plane" would mean all points to the right of the y-axis and all points to the left of the y-axis, respectively.

A vertical line that intersects the x-axis in a point corresponding to the real number a has the property that each point in this line has a for its x-coordinate. Hence, we say that this vertical line is determined by the condition $x = a$. Similarly, a horizontal line that intersects the y-axis in the point corresponding to the real number b has the property that each of its points has b for its y-coordinate. Therefore, we say that this horizontal line is determined by the condition $y = b$. Thus, the point in the plane with coordinates (a, b) is the intersection of the two lines determined by the conditions $x = a$ and $y = b$.

Of course, a point in the plane and the ordered pair of real numbers that are its coordinates are not the same thing. We will, however, adopt the convenient practice of referring to "the point (a, b)." It is also true that the vertical line whose points all have x-coordinate a and the condition $x = a$ are not the same thing. But again, we will often be somewhat imprecise and speak of the "line $x = a$" instead of "the line in the coordinate plane determined by the condition $x = a$."

As we stated earlier, we will be dealing almost exclusively with functions which are sets of ordered pairs of real numbers. Such functions are technically referred to as *real-valued functions of a single real variable*. They are called real-valued functions because their ranges are subsets of R. They are functions of a single real variable because the elements of their domains are single real numbers. (If the domain elements of a function are ordered pairs, the function is called a function of two variables; if the domain elements are ordered triples, the function is called a function of three variables; and so on.) Unless we specifically state otherwise, whenever we refer to a function, we will mean a real-valued function of a single real variable.

An important and useful property of such functions is that they can be represented by graphs in the Cartesian plane. We will emphasize this representation and always consider the graph of any function that we are investigating in detail. Functions are usually defined by first giving a rule of correspondence and then stipulating the domain of the function. From this definition, it is usually possible to make a rough sketch of a portion of the graph of the function in the Cartesian plane. We will shortly be considering a number of tests that will greatly aid us in sketching graphs. These tests will help us determine certain characteristics of the graph of the given function, such as where the graph is rising or falling as we move to the right or to the left, where the high and low points on the graph are located if any such exist, and whether or not the graph has certain symmetries.

It will not usually be our goal to produce a detailed and accurate graph of a function, but, rather, a useful sketch showing its basic characteristics. We hope that you will eventually become skillful in sketching graphs and that, when you are given a function for study, you will begin by attempting to sketch its graph.

Since no two distinct elements of a function have the same first component, we can think of a graph of a function as a set of points in the plane which satisfies the condition that no two points have the same abscissa, or x-coordinate. The graphs in Figure 2–5 satisfy this condition and, hence, are graphs of functions.

The graphs in Figure 2–6 are not graphs of functions because, in each graph, at least two distinct points have the same x-coordinate.

The functions corresponding to the graphs shown in Figure 2–5 consist of the sets of ordered pairs of numbers that are the coordinates of the points in each of the graphs. Notice that no vertical line in the plane would intersect any of the graphs of functions in more than one point. Remember that every point in a vertical line in the Cartesian plane has the same abscissa, or x-coordinate. Therefore, if a vertical line were to intersect the graph of a function in more than one point, then these points of intersection would have to have the same x-coordinate. This is impos-

Figure 2-5

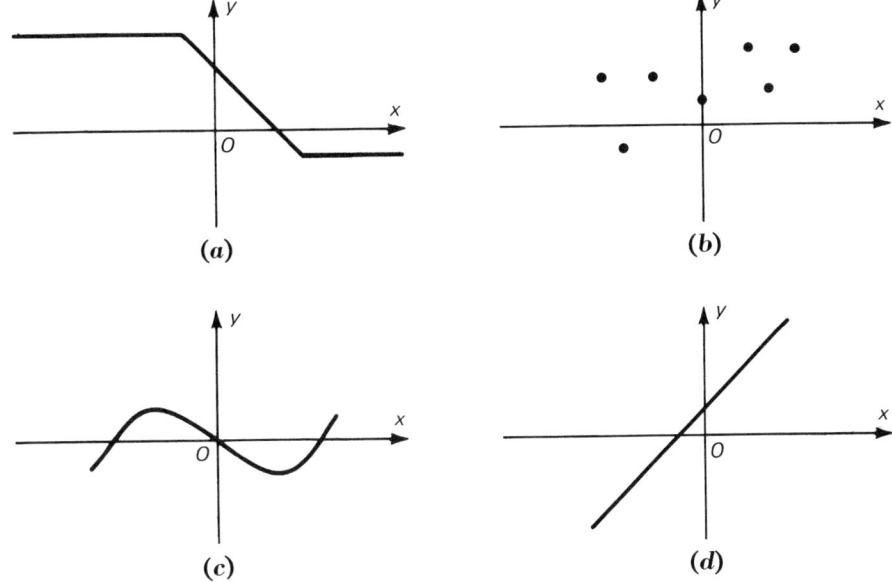

Figure 2-6

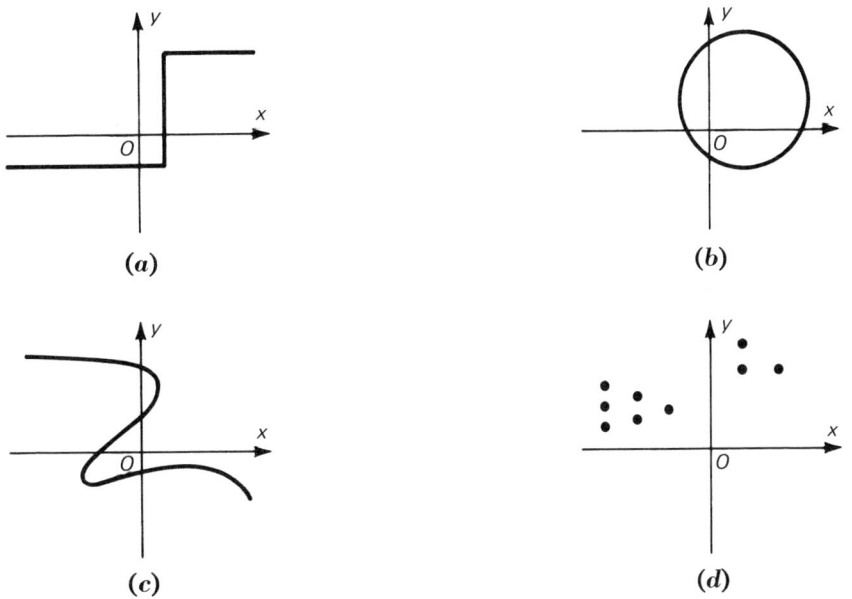

sible, of course, because a function is a set of ordered pairs in which no two pairs have the same first component.

We will now discuss a method of graphing the domain and the range of a function. The projection on a line l of a set of points S in the same plane as l is a subset of l. This subset consists of the points of l that are obtained by drawing through each point of S the line that is perpendicular to l.

Thus, the projection of the graph of a function on the x-axis would be the set of points on the x-axis obtained by drawing a vertical line through each point of the graph. The x-coordinates of the points contained in the projection on the x-axis make up the set of real numbers that form the domain of the function. (See Figure 2–7.)

Similarly, if one draws a horizontal line through each point in the graph of a function, the set of points that are the intersections of these lines with the y-axis is the projection of the graph on the y-axis. The set of ordinates, or y-coordinates, of this set of points on the y-axis is the range of the function. (See Figure 2–7.)

Figure 2-7

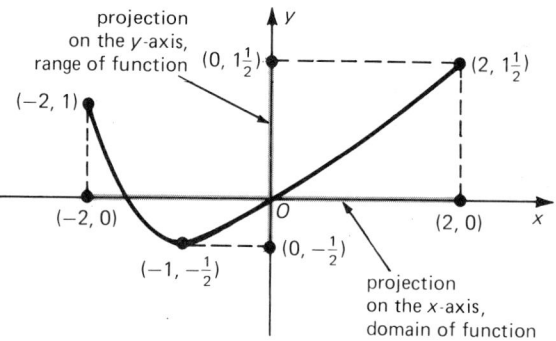

Since there is a one-to-one correspondence between the x-axis and the real-number line (that is, the point $(x, 0)$ on the x-axis corresponds to the point x on the real line and, conversely, point x on the real line corresponds to $(x, 0)$ on the x-axis), we see that the projection of the graph of a function on the x-axis is a "replica" of the graph of its domain. This means that the points of the projection on the x-axis correspond to the points in the graph of the domain on the real-number line. Because of this correspondence, we will take the liberty of referring to the projection of a function f on the x-axis as the graph of the domain of f.

By analogy, there is a one-to-one correspondence between the y-axis and the real-number line, with point $(0, y)$ and point y corresponding to each other. Hence, the projection of the graph of a function on the y-axis is a replica of the graph of the range of the function, and we will refer to this projection on the y-axis as the graph of the range of the function.

In Figure 2–7, the domain of the function represented by the graph is $\{x \mid -2 \leq x \leq 2\}$, and the graph of the domain is the closed interval on the x-axis with endpoints $(-2, 0)$ and $(2, 0)$. Similarly, the range of the function is $\{y \mid -\frac{1}{2} \leq y \leq 1\frac{1}{2}\}$, and the graph of the range is the closed interval on the y-axis with endpoints $(0, -\frac{1}{2})$ and $(0, 1\frac{1}{2})$.

We take special note that the graph of the domain of a sequence is the set of points on the x-axis which correspond to the positive integers, namely $(1, 0), (2, 0), (3, 0), \ldots, (n, 0), \ldots$, and the graph of the range of a sequence is set of points on the y-axis, which may be a finite or infinite set; $(0, a_1), (0, a_2), (0, a_3), \ldots, (0, a_n), \ldots$. This is illustrated in Figure 2–8.

Figure 2-8

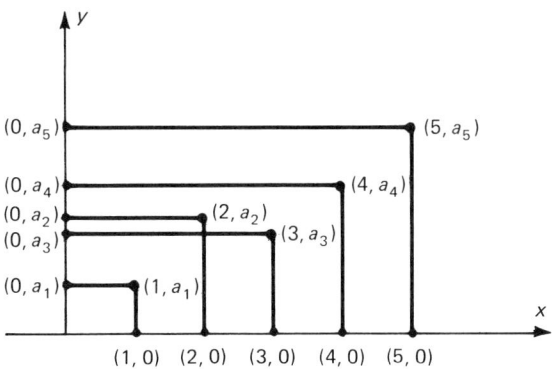

2.3 EXERCISES

1. The following figure illustrates the function g, which maps each integer onto its triple.

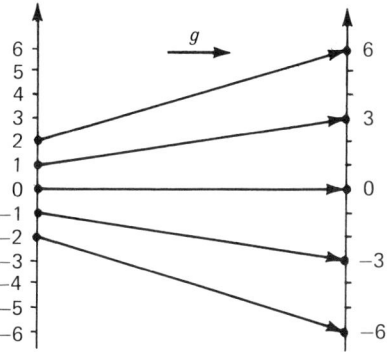

a) What is $g(0)$? $g(1)$? $g(-1)$?
b) What is $g(-2)$? $g(2)$? $g(g(3))$?
c) Is g a mapping of I *onto* I? Explain your answer.
d) Would the function $G : x \to 3x$ map R onto R? Explain your answer.

For each of the functions defined below, draw a mapping (similar to the one in Exercise 1) to illustrate the function.

2. $y = x + 1$.
3. $x \to 2x - 1$.
4. $x \to -x$.
5. $y = x^2$.
6. $x \to x^2 + x$.
7. $y = |x|$.

For each of Exercises 8–11, indicate whether or not the given graph is the graph of a function. Then for each function describe the domain and range.

8.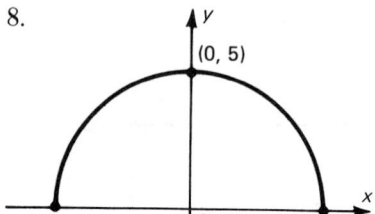

9.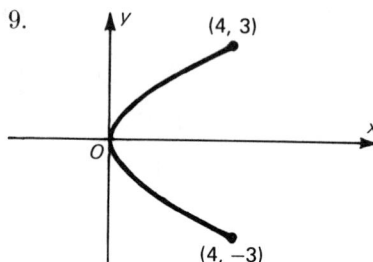

10.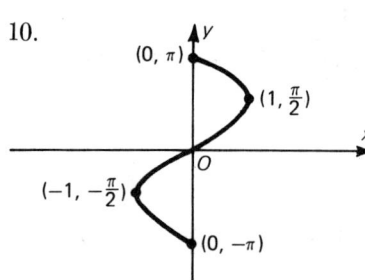

11.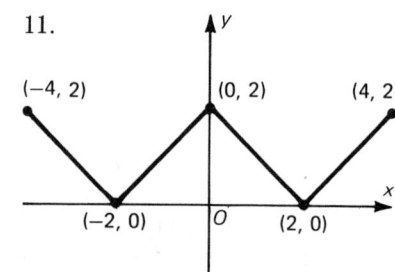

In each of Exercises 12–17, two equations that determine a pair of parallel lines are given. For each exercise, (a) when possible, graph the lines in the Cartesian plane, and (b) determine the distance between the lines.

12. $x = 2; x = -1$.
13. $x = -1; x = -4$.
14. $y = -2; y = 6$.
15. $y = 4; y = d$.
16. $x = a; x = b$.
17. $y = c; y = d$.

In each of Exercises 18–23, an equation for a line and the coordinates of a point are given. For each exercise, (a) when possible, graph the line and the point in the Cartesian plane, and (b) determine the distance between the line and the point.

18. $x = 3$; $(7, 2)$.
19. $y = 1$; $(-3, 4)$.
20. $x = -5$; $(-1, 4)$.
21. $y = -6$; $(3, -1)$.
22. $x = a$; (c, d).
23. $y = b$; (c, d).

24. What is the x-coordinate of any point in the y-axis? What is the y-coordinate of any point in the x-axis? What are the coordinates of the origin?

2.4 Monotone Functions

By recalling our discussion of monotone sequences in Section 1.8, we can graphically interpret the behavior of an increasing sequence. Note that the graph of an increasing sequence has the following property, (P): If a vertical and horizontal line intersect in a point of the graph of an increasing sequence, then all points of the graph to the right of the vertical line lie above the horizontal line. In fact we could have defined an increasing sequence as any sequence whose graph has property P.

Example 1 The graph of the increasing sequence $\langle \sqrt{n} \rangle$ in Figure 2–9 has property P.

Figure 2-9

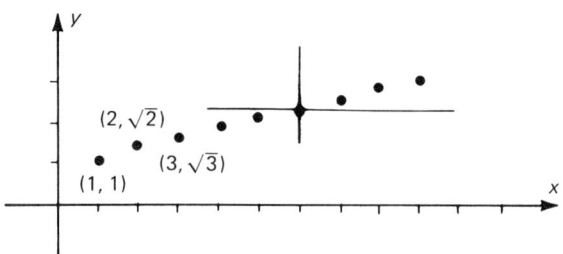

We can now extend the notion of "increasing" to functions which are not sequences—by defining an increasing function as one whose graph has property P. Note that there are properties analogous to P which characterize the other types of monotonic sequences, namely the nondecreasing, decreasing and nonincreasing sequences. Thus, by using these analogous properties, as we did property P, we can extend the notion of "monotone" from sequences to functions. If you interpret the following definitions graphically you will realize that this is precisely what we have done.

Definition 2.4a A function f is said to be an **increasing function** provided that, for every two points x_1 and x_2 in the domain of f, $x_2 > x_1$ implies that $f(x_2) > f(x_1)$. If, for every two points x_1 and x_2 in the domain of f, $x_2 > x_1$ implies $f(x_2) \geq f(x_1)$, then f is said to be a **nondecreasing function.**

Take note that every increasing function is a nondecreasing function, but that the converse is not true. Graphs of two nondecreasing functions are shown in Figure 2–10. The graph in (b), but not the one in (a), is the graph of an increasing function. This means that, given two points in the graph of the function, the one further to the right is always the higher of the two. In informal terms, we say that increasing functions have graphs that rise as they go to the right.

Figure 2-10

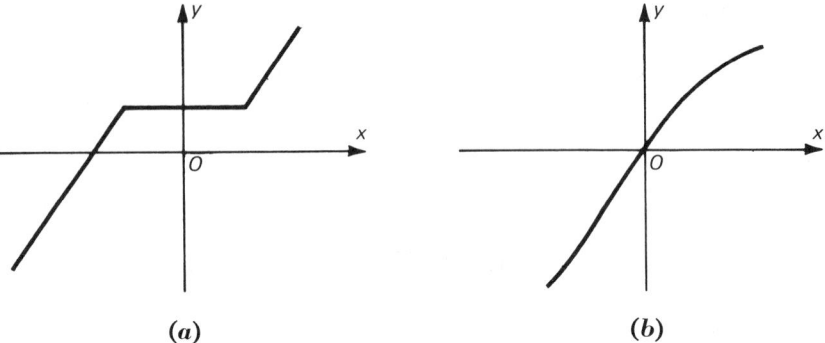

(a) (b)

We now give a similar definition for functions whose graphs are falling.

Definition 2.4b A function is said to be a **decreasing function** provided that, for every two points x_1 and x_2 in the domain of f, $x_2 > x_1$ implies that $f(x_2) < f(x_1)$. If for every two points in the domain of f, $x_2 > x_1$ implies that $f(x_2) \leq f(x_1)$, then f is said to be a **nonincreasing function.**

Graphs of two nonincreasing functions are shown in Figure 2-11. The graph in (b), but not the one in (a), is the graph of a decreasing function. This means that, given any two points of the graph of the function, the one further to the right is the lower of the two. Obviously, a decreasing function is a special type of nonincreasing function. In informal terms, we may say that decreasing functions have graphs that fall as they go to the right.

Figure 2-11

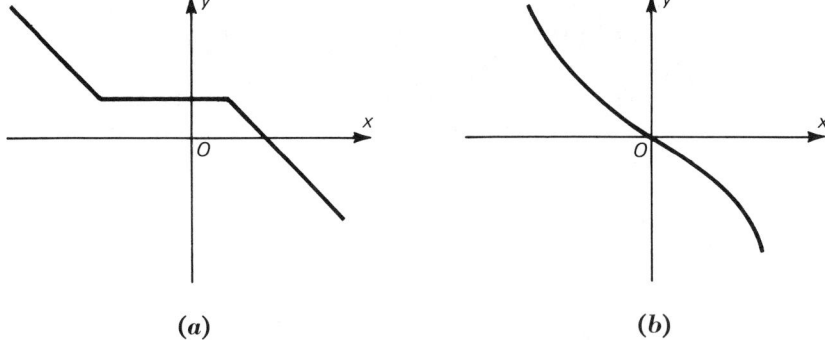

(a) (b)

Any subset of a function is also a function. This follows immediately from the fact that no two ordered pairs belonging to a subset of a function can have the same first component, because no two ordered pairs contained in the function itself have the same first component. A subset of a function is sometimes called a **restriction** of the function. The following example illustrates that, although it may not be possible to classify a function as either an increasing or a decreasing function, some of its restrictions may be increasing functions and some may be decreasing functions.

Figure 2-12

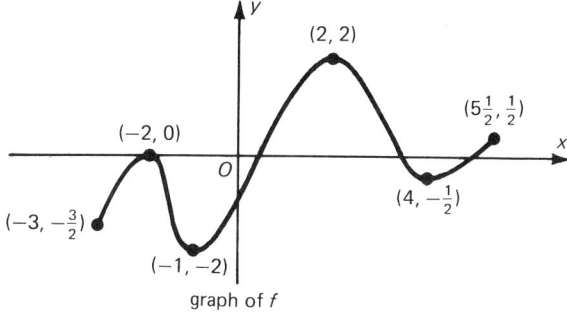

graph of f

Example 2 Let f be a function having the graph shown in Figure 2-12. How can you tell immediately that this is a graph of a function? A glance at the graph tells us that f is neither an increasing nor a decreasing function because its graph sometimes rises and sometimes falls as you go to the right. However, the restriction of f defined by

$$\hat{f} = \{(x, y) \mid (x, y) \, \epsilon f \text{ and } -1 \leq x \leq 2\}$$

is an increasing function, as shown in Figure 2-13(a). Similarly, the restriction of f defined by

$$\hat{\hat{f}} = \{(x, y) \mid (x, y) \, \epsilon f \text{ and } -2 \leq x \leq -1\}$$

is a decreasing function. The graph in Figure 2-13(b) is the graph of this restriction of function f.

Figure 2-13

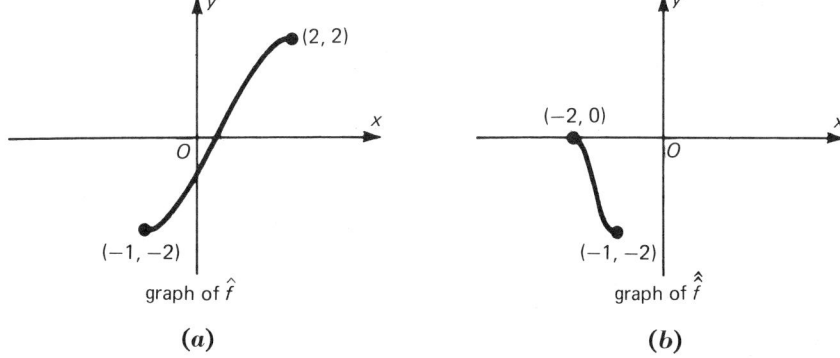

(a)   (b)

We can describe the properties of these particular restrictions of function f by saying that f is an increasing function on the interval $-1 \leq x \leq 2$ and f is a decreasing function on the interval $-2 \leq x \leq -1$. Over what other intervals is this particular function increasing or decreasing?

It is sometimes possible to determine intervals over which a function f is increasing or decreasing algebraically, by considering the difference $f(x_2) - f(x_1)$, where x_1 and x_2 are in the domain of f and $x_2 > x_1$.

Example 3 The function defined by $f(x) = x^2$ is an increasing function for all x such that $x \geq 0$. To see this, we assume that $x_2 > x_1$ and consider the difference

$$f(x_2) - f(x_1) = x_2{}^2 - x_1{}^2 = (x_2 - x_1)(x_2 + x_1).$$

Now $x_2 - x_1$ is positive because $x_2 > x_1$; furthermore, since x_1 is nonnegative and x_2 is positive, $x_1 + x_2$ is positive. Therefore, $(x_2 - x_1)(x_2 + x_1)$, which is equal to $f(x_2) - f(x_1)$, is also positive. Hence, $f(x_2) > f(x_1)$.

We have shown that, if $x_2 > x_1$, then $f(x_2) > f(x_1)$, which means that function f is an increasing function. If we compute the values of $f(x)$ when x is replaced by $0, \frac{1}{2}, 1, \frac{3}{2}$, and 2, we see that the points having coordinates $(0, 0)$, $(\frac{1}{2}, \frac{1}{4})$, $(1, 1)$, $(\frac{3}{2}, \frac{9}{4})$, and $(2, 4)$ all belong to the graph of this function. But how can we be sure that the graph of f does not look like the one in Figure 2–14? The answer is simple. The graph shown is obviously not the graph of an increasing function because it falls over certain intervals. This contradicts the fact already established that f is an increasing function. This fact is not enough, however, to show us that the graph could not look like the one in Figure 2–15.

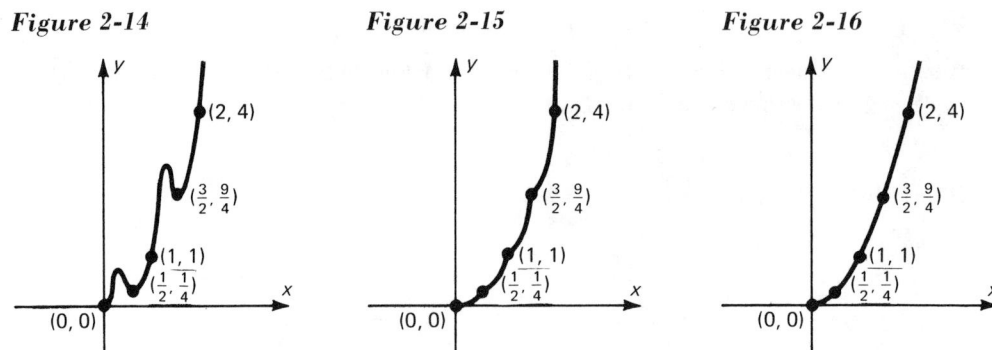

Figure 2-14   ***Figure 2-15***   ***Figure 2-16***

The graph of f actually looks like the one in Figure 2–16, although we have not yet established this.

If we consider the function defined by $g(x) = x^2$, with $x \leq 0$, we could give an analogous argument to show that g is a decreasing function.

The function q, defined by $q(x) = x^2$ for all $x \in R$, is neither an increasing nor a decreasing function. This is the case because, as you can readily ascertain, $q = f \cup g$ and the graph of q is the union of the graphs of f and g. We can say, however, that q is a decreasing function for $x \leq 0$ and is an increasing function for $x \geq 0$. Most of the functions that we will consider from now on will be the unions of functions that are either increasing or decreasing functions over intervals or half-lines.

As an interesting comment, we observe that the function f defined by $f(x) = a$, where a is a constant, is both a nonincreasing function and a nondecreasing function. This is so because, for every two points x_1 and x_2 in the domain of f, if $x_1 < x_2$, then $f(x_1) \leq f(x_2)$ and $f(x_1) \geq f(x_2)$ [that is, $f(x_1) = a = f(x_2)$].

Increasing or decreasing functions have a certain property that will prove to be very valuable in later considerations. This property is that there is a one-to-one correspondence between the domain and the range of such a function. Not only is every point in the domain associated with precisely one point in the range (as is the case for all functions), but every point in the range has only one point in the domain associated with it. This result is important enough to state as a theorem.

Theorem 2.4a *An increasing or decreasing function determines a one-to-one correspondence between its domain and its range. In this correspondence, each x in the domain of f corresponds to precisely one y in the range of f, namely $y = f(x)$, and for each y in the range of f, there is precisely one x in the domain of f such that $f(x) = y$.*

The proof of this theorem is not difficult, so we omit it.

2.4 EXERCISES

For each function in Exercises 1–4, a) define a restriction that is an increasing function, and b) define a restriction that is a decreasing function.

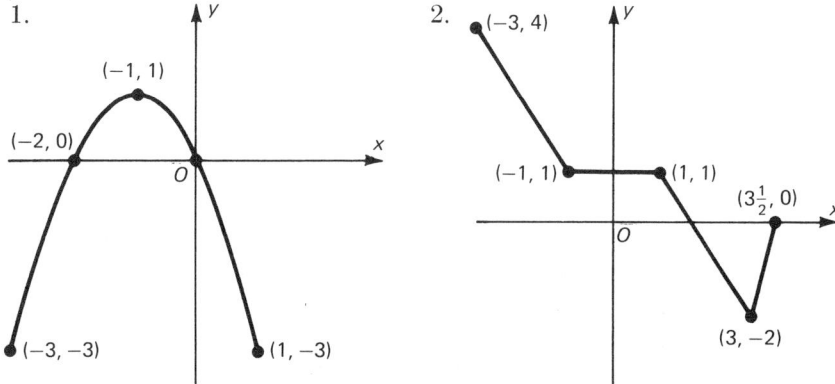

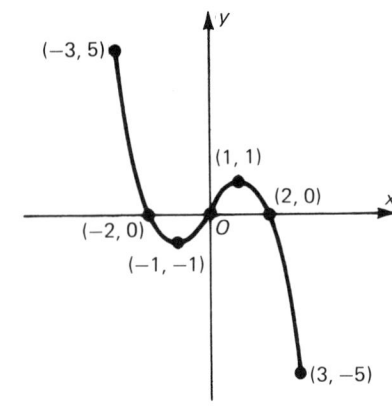

5. Draw a graph of a function g with all of the following properties:
   a) The domain of g is $\{x \mid -5 \leq x \leq 5\}$.
   b) The range of g is $\{y \mid 0 \leq y \leq 6\}$.
   c) $g(-1) = g(1) = 3$.
   d) g is a nonincreasing function.

In Exercises 6–10, indicate whether each of the following statements about function g of Exercise 5 is true or false, and explain your answer.

6. $g(-5) = 0$.
7. $g(0) < 3$.
8. If $x \geq 0$, then $g(x) \leq 3$.
9. If $x \leq 0$, then $g(x) \geq 3$.
10. If $0 < k < 6$, then there is exactly one number c which satisfies the conditions that $-5 < c < 5$ and $g(c) = k$.
11. Draw a graph of a function h that has all of the following properties:
    a) The domain of h is $\{x \mid -3 \leq x \leq 3\}$.
    b) The range of h is $\{y \mid -1 \leq y \leq 4\}$.
    c) $h(0) = 0$.
    d) h is an increasing function.

In Exercises 12–16, indicate whether each of the following statements about function h in Exercise 11 is true or false, and explain your answer.

12. $h(-3) = -1$.
13. If $h(c) = 4$, then $c = 3$.
14. If $x < 0$, then $h(x) < 0$.
15. If $x > 0$, then $h(x) > 0$.

16. If $-1 < k < 4$, then there is exactly one number c which satisfies the conditions that $-3 < c < 3$ and $f(c) = k$.
17. Prove that the function $f:f(x) = 2x - 3$ is an increasing function.
18. Prove that any function of the form $f:f(x) = mx + b$, $m < 0$ is a decreasing function.
19. Prove Theorem 2.4a for the case of an increasing function. [Hint: assume that f is an increasing function and that (a, c) and (b, c) belong to f, where $a \neq b$. Show that the assumption $a < b$ leads to a contradiction and that the assumption $a > b$ leads to a contradiction.

2.5 Inverse Functions

Often in discussing functions, we have illustrated them as mappings from the real-number line into the real-number line, as in the following example.

Example 1 In Figure 2-17 the arrow beginning at 1 and ending at 2 symbolizes that $(1, 2)$ is a member of f. Similarly, it is indicated that $(0, 0)$, $(-1, 0)$, $(2, -2)$, and (a, b) belong to f. Generally, we cannot show all the arrows, particularly if the function is an infinite set of ordered pairs, and we therefore limit ourselves to a representative set.

Figure 2-17

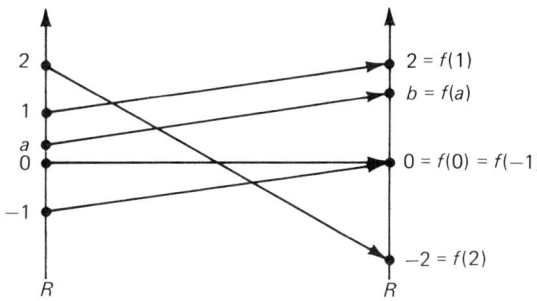

In such an illustration of a function, no more than one arrow can begin at any point in the first number line. However, any number of arrows *may* lead to a given point in the second number line.

Example 2 An extreme example is afforded by the function

$$f:f(x) = 1, \quad \text{for } x \in R.$$

In any illustration of this function as a mapping, all of the arrows must lead to the number 1 in the second line, as in Figure 2-18 on page 58.

Figure 2-18

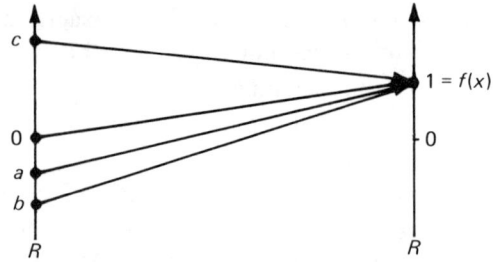

On the other hand, there are many functions whose illustrations as mappings would never have more than one arrow leading to the same point in the second line.

Example 3 For example, the function

$$g:g(x) = x + 1, \quad \text{for } x \in R,$$

is a function having this property. (See Figure 2–19.) Such functions are, of course, one-to-one functions. You should be able to think of many other examples of one-to-one functions.

Figure 2-19

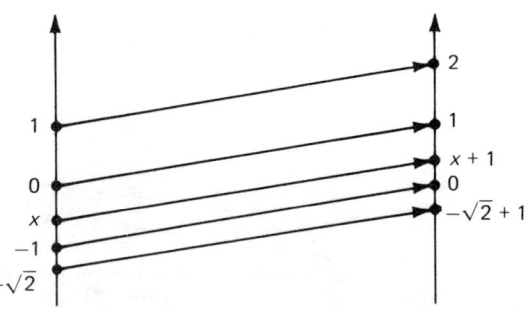

If we reverse all the arrows in Figure 2–19 illustrating g as a mapping, we have a new mapping, as shown in Figure 2–20. Since no two arrows led to the same point before we reversed them, no two arrows in this new

Figure 2-20

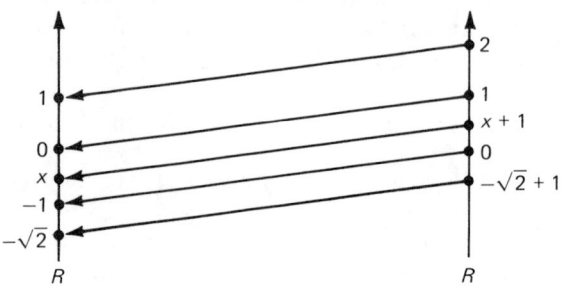

mapping start at the same point. Therefore, this new mapping is the representation of a new function, which we will call function h.

In the illustration of g, an arrow starting at a led to b; in the illustration of h, the arrow starting at b leads to a. That is, if $(a, b) \in g$, then $(b, a) \in h$. From this, we see that the domain of h is the range of g and the range of h is the domain of g.

If we reverse the arrows in some mappings, the resulting mapping is not the representation of a function. Figures 2–17 and 2–18 illustrate this fact.

With this preliminary discussion in mind, we are ready to give a definition of an *inverse function*.

Definition 2.5a If g is a function and
$$\{(a, b) \mid (b, a) \in g\}$$
is also a function, we say it is the **inverse function** of g and denote it by g^{-1}.

The symbol g^{-1} should not be confused with the algebraic symbol sometimes used for the reciprocal of a number. The symbol g^{-1} is read as "g inverse" or "the inverse of g." We will use $\frac{1}{g}$ to represent the reciprocal of the function g.

Not all functions have inverse functions. You should be able to explain why the function f, defined by $f(x) = 1$, for $x \in R$, for example, does not have an inverse.

When a function g does have an inverse function g^{-1}, then the domain of g^{-1} is the range of g, and the range of g^{-1} is the domain of g.

The use of the word "inverse" is very descriptive. If a function g has an inverse and g takes a into b, then g^{-1} does the inverse operation—it takes b into a. That is, g^{-1} undoes what g has done. But, further, if g^{-1} takes b into a, then g takes a into b. Thus, g undoes what g^{-1} does. These remarks lead us to the following theorems.

Theorem 2.5a *If the function g has an inverse function g^{-1}, then g is the inverse function of g^{-1}. In symbols, $(g^{-1})^{-1} = g$.*

The proof of this theorem is an easy consequence of Definition 2.5a and is left as an exercise.

Theorem 2.5b *If the function g has an inverse, then*
  1. *$g^{-1}(g(x)) = x$ for all x in the domain of g, and*
  2. *$g(g^{-1}(x)) = x$ for all x in the range of g.*

Proof *Part I* If x is in the domain of g, then $(x, g(x)) \in g$ and, by the definition of an inverse of a function, this means that $(g(x), x) \in g^{-1}$. But, by the definition of a function, $(g(x), x) \in g^{-1}$ means that $g^{-1}(g(x)) = x$.

Part II If x is in the range of g, then x is in the domain of g^{-1}, and, thus, $(x, g^{-1}(x)) \in g^{-1}$. This last statement implies that $(g^{-1}(x), x) \in (g^{-1})^{-1}$. But, by Theorem 2.6a, $(g^{-1})^{-1} = g$, so $(g^{-1}(x), x) \in g$. Finally, $(g^{-1}(x), x) \in g$ means that $g(g^{-1}(x)) = x$. □

By the definition of the inverse of a function, if it exists, it is unique. In other words, if a function g has inverse g^{-1}, then g^{-1} is the only function having the range of g as its domain that satisfies properties 1 and 2 of Theorem 2.5b.

Any function f that has an inverse must have the following properties and every function f having these properties has an inverse:

1. Every point in the domain of f maps onto precisely one point in the range of f.

2. Every point in the range of f is the image of precisely one point in the domain of f.

Thus, *a function has an inverse if and only if it is a one-to-one function.* (The shorthand notation 1:1 function is sometimes used for designating a one-to-one function.) Of course, the first condition stated above is satisfied by all functions. Satisfying the second condition is what distinguishes 1:1 functions from functions in general. An equivalent way of stating these properties is to say that a 1:1 function (or a function that has an inverse) is a set of ordered pairs such that (1) no two ordered pairs of the set have the same first component, and (2) no two ordered pairs of the set have the same second component.

Interpreted in terms of the graph of the function, the first requirement for a 1:1 function means that no vertical line intersects the graph of the function in more than one point, and the second requirement means that no horizontal line intersects the graph in more than one point.

In Theorem 2.4a on page 55, we established that an increasing or decreasing function is a one-to-one function. The following theorem is an immediate consequence.

Theorem 2.5c *Every increasing or decreasing function has an inverse function.*

Not every function that has an inverse is an increasing or decreasing function, however. Use the graph given in Figure 2–21 to justify this statement.

Figure 2-21

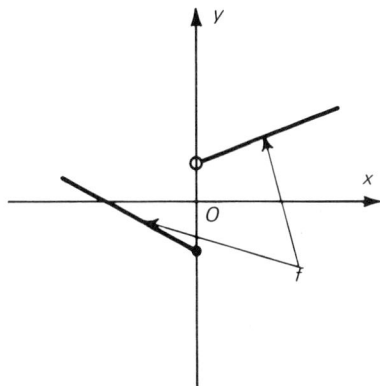

2.5 EXERCISES

For each of Exercises 1–4, state whether the given function has an inverse and tabulate the inverse if it exists. If the function does not have an inverse, explain why this is so.

1. $f = \{(-2, 3), (-1, 1), (0, -1), (1, 1), (2, 0)\}$.
2. $g = \{(0, 5), (1, 4), (2, 3), (3, 2), (4, 1), (5, 0)\}$.
3. $h = \{(-3, -3), (-2, -1), (0, 0), (1, 2), (3, 3)\}$.
4. $F = \{(0, 5), (3, 5), (5, 5), (7, 5)\}$.

In each of Exercises 5–10, a relationship is given which defines a function. If the function has an inverse, state a relationship that defines the inverse. If the function does not have an inverse, explain why this is so.

5. The second component of each ordered pair is 5 less than the first component. The domain is R.
6. The second component of each ordered pair is twice the first component. The domain is R.
7. The second component of each ordered pair is 1. The domain is R.
8. The second component of each ordered pair is the absolute value of the first component. The domain is R.
9. The second component of each ordered pair is the square of the first component. The domain is R.
10. The second component of each ordered pair is the principal square root of the first component. The domain is $\{x \mid x \in R \text{ and } x \geq 0\}$.

The property that $g(g^{-1}(x)) = x$ [Theorem 2.5b] may be used to find an equation that determines the inverse of a function g. For example, if g is defined by $g(x) = 3x + 1$, then by Theorem 2.5b, $3(g^{-1}(x)) + 1 = x$. Solving for $g^{-1}(x)$, we find that $g^{-1}(x) = \dfrac{x - 1}{3}$. Since the range of g is R, the domain of g^{-1} is R. For each of Exercises 11–16, use the technique just outlined to *find an equation that defines the inverse of the given function. Then describe the domain and range of this inverse. We will use the notation* D_f *for* $D(f)$.

11. $f : f(x) = 2 - 7x;\ D_f = R$.
12. $g : y = \dfrac{x-5}{4};\ D_g = R$.
13. $h : x \to \sqrt{x^2 - 3};\ D_h = \{x \mid x \in R \text{ and } x \geq \sqrt{3}\}$.
14. $F : F(x) = \sqrt{x} + 5;\ D_F = \{x \mid x \in R \text{ and } x \geq 0\}$.
15. $G : y = \dfrac{1}{2+x};\ D_G = \{x \mid x \in R \text{ and } x < -2\}$.
16. $H : x \to 4x^2 + 2;\ D_H = \{x \mid x \in R \text{ and } x \geq 0\}$.
17. Prove that, if the function g has an inverse function g^{-1}, then g is the inverse of g^{-1}. [In symbols, prove that $(g^{-1})^{-1} = g$.]

2.6 Constant and Linear Functions

When we considered the Cartesian plane, we saw that a line parallel to the y-axis is determined by the condition $x = a$, where a is a fixed real number. Such a line is the graph of a set of ordered pairs of the form

$$\{(a, y) \mid y \in R\}.$$

It is obvious that this set is not a function, since it violates the special condition required of a set of ordered pairs that is a function; namely, that no two members of the set may have the same first component. In fact, *every* ordered pair belonging to this set has the same first component a.

Another kind of line discussed in the text is one that is parallel to the x-axis. A line of this type is determined by the condition that all of its points have the same ordinate, that is, by the condition $y = a$, where a is a real number. Such a line is the graph of a set of ordered pairs of numbers of the form

$$\{(x, a) \mid x \in R\}.$$

This set is a function since no two of its elements have the same first component. Such a function may be defined as follows:

$$f : f(x) = a, \quad \text{for } x \in R.$$

The very descriptive name for such a function is a *constant function* because the second component of every element of the function is the constant a. Note that the constant function $f(x) = a$ is just an extension of the constant sequence $\langle a \rangle$. The x-axis is the graph of the constant function defined by

$$f(x) = 0, \quad \text{for } x \in R.$$

What can we now say about all the other straight lines in the plane, those that are not parallel to either the x- or the y-axis? Since they are not parallel to either of the axes, they must slant up to the right or down to the right. To develop a way of describing the slant of such a line, we will first consider nonvertical *segments* in the plane. The number that tells whether a nonvertical segment slants up or down and also how much it slants is called the *slope* of the segment, as indicated in the following definition.

Definition 2.6a The **slope** of a nonvertical segment with endpoints (x_1, y_1) and (x_2, y_2) is the number m given by

$$m = \frac{y_2 - y_1}{x_2 - x_1}.$$

Notice that, if (x_1, y_1) and (x_2, y_2) are the endpoints of a vertical segment, then $x_1 = x_2$. Hence, the slope is not defined for vertical segments because division by 0 is not defined.

Example 1 Segment $\overline{AB}$, shown in Figure 2-22, has slope $\dfrac{4 - (-2)}{-7 - 1} = -\dfrac{3}{4}$.

Figure 2-22

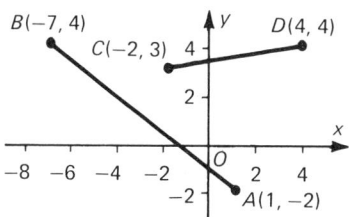

Example 2 Segment $\overline{CD}$, shown in Figure 2-22, has slope $\dfrac{4 - 3}{4 - (-2)} = \dfrac{1}{6}$.

It does not make any difference which of the endpoints you designate as (x_1, y_1) and which as (x_2, y_2), since

$$\frac{y_2 - y_1}{x_2 - x_1} = \frac{-(y_2 - y_1)}{-(x_2 - x_1)} = \frac{y_1 - y_2}{x_1 - x_2}.$$

It is simply customary to write the quotient as it is given in Definition 2.6a.

We can show without too much difficulty that any segment contained in a given nonvertical line has the same slope as any other segment in that line. The slope of any segment in a horizontal line is clearly 0, because all the y-coordinates are the same and, therefore,

$$\frac{y_2 - y_1}{x_2 - x_1} = \frac{0}{x_2 - x_1} = 0.$$

If the given nonvertical line is not horizontal, then any two segments in that line are the hypotenuses of two similar right triangles, as

Figure 2-23

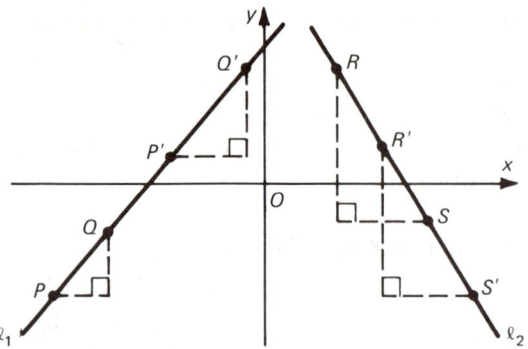

shown in Figure 2–23. Segments $\overline{PQ}$ and $\overline{P'Q'}$ in l_1 are the hypotenuses of two similar right triangles, and segments $\overline{RS}$ and $\overline{R'S'}$ in l_2 are the hypotenuses of similar right triangles. Using Definition 2.6a together with a familiar theorem from geometry concerning the ratios of corresponding sides of similar triangles, you should be able to prove that $\overline{PQ}$ and $\overline{P'Q'}$, for example, have the same slope.

We considered an arbitrary nonvertical and nonhorizontal line and showed that any two arbitrary segments in this line have the same slope. Since we also know that any segment in a horizontal line has slope 0, it follows that: Any two segments in the same nonvertical line have the same slope.

Thus, we can talk about the "slope of a line," meaning by this the slope of any segment of the line. If we think of a line as being traced out by a point moving across the plane from left to right, then the slope of the line is the quotient of the change in the ordinate and the increase in the abscissa over any given interval. If we choose the interval so that the increase in the abscissa is one unit as illustrated in Figure 2–24, then m, the slope of the line, is

$$\frac{y_2 - y_1}{(x_1 + 1) - x_1} = \frac{y_2 - y_1}{1} = y_2 - y_1.$$

Thus, we can think of the slope of a line as the change in the ordinate of a point moving along the line which corresponds to an increase of its

Figure 2-24

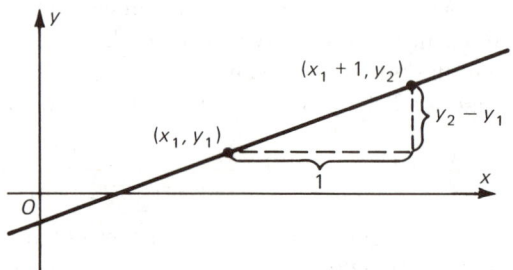

abscissa by one unit. Some specific examples are given in the following illustrations.

As illustrated in Figures 2-25 and 2-26, if a line slants up to the right, then a point moving up this line will have its ordinate increasing as its abscissa increases, and thus the slope of the line is positive; if a line slants down to the right, then a point moving down this line will have the value of its ordinate decreasing as its abscissa increases, and the slope of the line is negative.

Figure 2-25

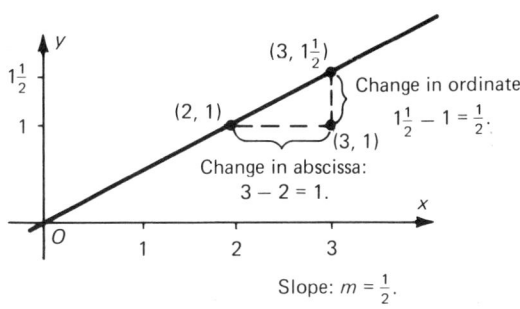

Figure 2-26

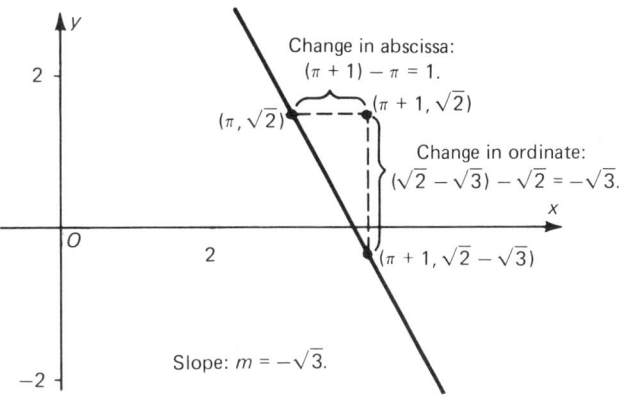

A little reflection should convince you that, given a point (a, b), it is possible to construct exactly one line containing (a, b) having a given real number for its slope. For example, suppose that you want to construct a line containing the point $(2, 1)$ with a slope of 3. For a line to have a slope of 3, the ordinate of a point in the line must increase by 3 when the abscissa increases by 1. Thus the point $(2 + 1, 1 + 3) = (3, 4)$ must also be in the line. The two points, $(2, 1)$ and $(3, 4)$, determine the line, as shown in Figure 2-27 on page 66.

2.6 CONSTANT AND LINEAR FUNCTIONS

Figure 2-27

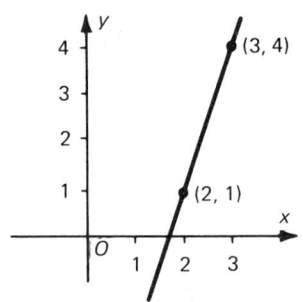

In general, to construct the line containing (a, b) with slope m, we do the following:

1. Locate point (a, b).
2. Locate point $(a + 1, b + m)$.
3. Draw the line determined by (a, b) and $(a + 1, b + m)$.

2.6a EXERCISES

For each of Exercises 1–6, find the slope of the line determined by the given points.

1. $(5, 5)$, $(14, -4)$.
2. $(-2, 3)$, $(5, 10)$.
3. $(6, 0)$, $(10, -4\sqrt{3})$.
4. $(0, -3)$, $(-5\sqrt{3}, 2)$.
5. $(6, 0)$, $(-4, 10\sqrt{3})$.
6. $(a, ma + b)$, $(c, mc + b)$.

For each of Exercises 7–12, use the idea of the slope of a line to determine whether the three points are collinear.

7. $(3, 2)$, $(7, -4)$, $(-1, 8)$.
8. $(-1, -1)$, $(0, 3)$, $(3, 7)$.
9. $(6, 6)$, $(1, -1)$, $(-4, -8)$.
10. (a, b), $\left(\dfrac{a + c}{2}, \dfrac{b + d}{2}\right)$, (c, d).
11. (a, b), $\left(\dfrac{2a + c}{3}, \dfrac{2b + d}{3}\right)$, (c, d).
12. (a, b), $\left(\dfrac{a + 2c}{3}, \dfrac{b + 2d}{3}\right)$, (c, d).

In each of Exercises 13–18, draw a graph of the line that passes through the given point with the given slope.

13. $(-2, -5)$, $m = 2$.
14. $(-4, 8)$, $m = -\frac{2}{3}$.

15. $(0, 0)$, $m = -\frac{1}{2}$.
16. $(-6, 0)$, $m = \frac{1}{2}$.
17. $(-7, 5)$, $m = -1$.
18. $(0, 5)$, $m = \frac{3}{4}$.
19. Prove that any two segments belonging to the same nonvertical line have the same slope.

The greater the slope of a line, the greater is the acute angle α that the line makes with all horizontal lines (see Figure 2–28(a)). The greater the absolute value of a negative slope, the smaller is the obtuse angle β that the line makes with all horizontal lines (see Figure 2–28(b)).

Figure 2-28

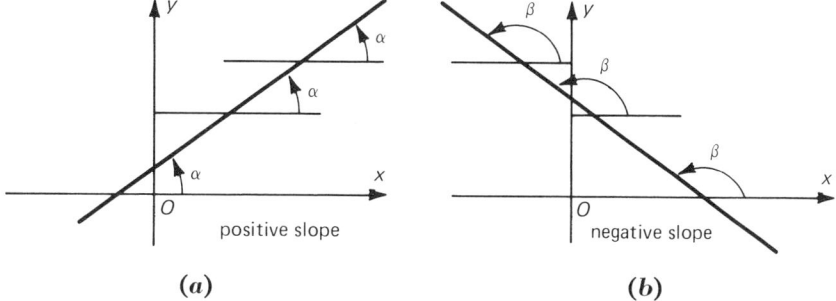

(a) positive slope   (b) negative slope

It is clear that a line with positive slope will form an acute angle with every horizontal line and that the measure of this angle will be greater than 0° and less than 90°. Similarly, the obtuse angle formed by a line with negative slope and a horizontal line will have a measure greater than 90° and less than 180°.

When two lines intersect at a point P, four angles are formed which are congruent in pairs. Among the four rays with endpoints at P, there are just two rays whose direction from P is positive (that is, to the right, if the ray is horizontal, and upward, if the ray is not horizontal). We then define **the angle of intersection** of the two lines to be the angle that is the union of the two positive rays. In each of Figures 2–29(a) and (b),

Figure 2-29

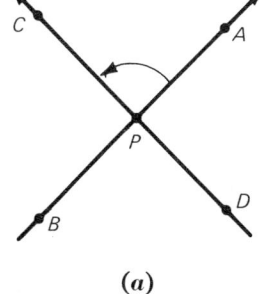 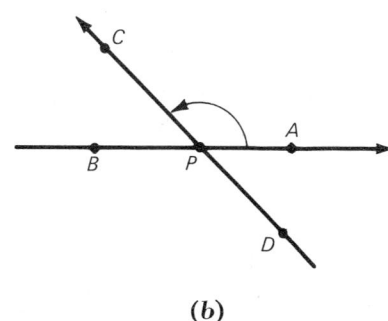

(a)   (b)

2.6 CONSTANT AND LINEAR FUNCTIONS

the angle of intersection of the two lines is ∢APC. We now define the **inclination** of a nonhorizontal line in the plane to be the measure of its angle of intersection with the x-axis.

Since the angle of intersection of a line l with the x-axis is congruent to the angle of intersection of l with any horizontal line, if we know two points through which l passes, we can determine the inclination of l. For example, in Figure 2-30, l passes through $A(3, -1)$ and $B(7, -5)$.

Figure 2-30

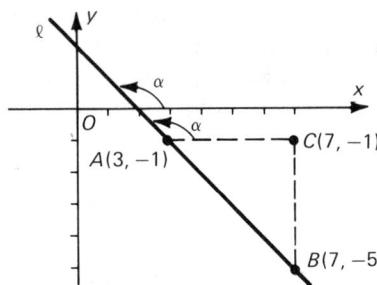

Since $d(C, B) = d(A, C) = 4$ and $\triangle ABC$ is a right triangle, it follows that

$$\tan \sphericalangle BAC = \frac{d(C, B)}{d(A, C)} = \frac{4}{4} = 1.$$

Therefore, $\sphericalangle BAC = 45°$, and hence, $\alpha = 135°$.

2.6b EXERCISES

In each of Exercises 1–6, indicate whether the inclination α of line l is in the interval $0° < \alpha < 90°$ or in the interval $90° < \alpha < 180°$.

1. l slants up to the right.
2. l slants down to the right.
3. l intersects the negative x-axis and the positive y-axis.
4. l intersects the positive x-axis and the positive y-axis.
5. The slope of l is negative.
6. The slope of l is positive.

In each of Exercises 7–12, draw a graph of the line that passes through the given point with the given inclination.

7. $(4, 8)$, $\alpha = 45°$.
8. $(0, 0)$, $\alpha = 135°$.
9. $(-5, 0)$, $\alpha = 120°$.
10. $(0, -4)$, $\alpha = 60°$.
11. $(1, 0)$, $\alpha = 150°$.
12. $(2, 3)$, $\alpha = 90°$.

In each of Exercises 13–18, find, to the nearest degree, the inclination of the line which passes through the given points.

13. $(4, -4)$, $(13, 5)$.
14. $(-3, -2)$, $(4, -9)$.
15. $(5, 0)$, $(9, -4\sqrt{3})$.
16. $(-3, 5)$, $(7, 10)$.
17. $(-2, -4)$, $(2, 8)$.
18. $(1, 1)$, $(11, -3)$.
19. Prove that if l is a line with nonzero slope m and inclination α, then $\tan \alpha = m$ when $m > 0$ and $\tan(180 - \alpha) = -m$ when $m < 0$.
20. If a line which slants up to the right has an inclination in the interval $0° < \alpha < 90°$, a line which slants down to the right has an inclination in the interval $90° < \alpha < 180°$, and a vertical line has inclination $\alpha = 90°$, what values might one assign to the inclination of a horizontal line?

Since no two distinct lines intersect in more than one point, it is clear that any nonvertical line is the graph of a function. We already know that every horizontal line is the graph of a function f that can be described by the condition $f(x) = a$, $a \in R$. Can a line that is nonvertical and also nonhorizontal, that is, a line with nonzero slope, be defined by some similar condition? The answer is yes. We shall now show what that condition is.

From your work in geometry, you should be able to justify the statement that any line with nonzero slope must intersect both the x- and the y-axes. (As a matter of fact, such a line must intersect every vertical line in precisely one point and every horizontal line in precisely one point.)

The point at which a nonhorizontal line intersects the x-axis is called its **x-intercept**. The point at which a nonvertical line intersects the y-axis is called its **y-intercept**. (Some authors define the x- and y-intercepts as the abscissa and ordinate, respectively, of the points we call the x-intercept and y-intercept.) Thus, a line with nonzero slope has an x-intercept, $(a, 0)$, and a y-intercept, $(0, b)$. (See Figure 2–31.)

Figure 2-31

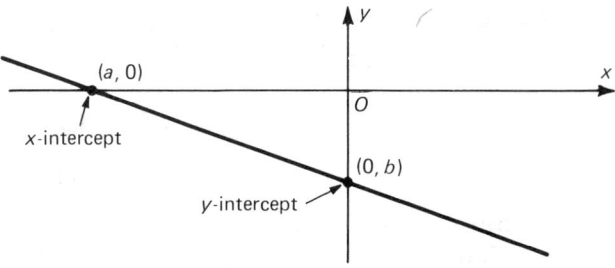

A line with nonzero slope has distinct intercepts unless it passes through the origin, in which case it has $(0, 0)$ as both its x- and its y-intercept.

Since a line with nonzero slope has precisely one intersection with each vertical line, it is the graph of a function having for its domain the set of all real numbers. Figure 2-32 shows a particular line l with slope $m \neq 0$, which is the graph of some function f.

Figure 2-32

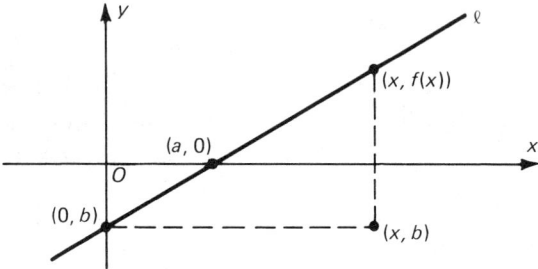

We would like to determine an equation that will give the value of $f(x)$ for each real number x. An arbitrary point in the line has coordinates $(x, f(x))$, since we have observed that l is the graph of the function f. If $(x, f(x))$ is a point different from the y-intercept $(0, b)$, then we know that

$$\frac{f(x) - b}{x - 0} = m.$$

This condition is, of course, equivalent to

$$f(x) = mx + b.$$

From this we see that an arbitrary point $(x, f(x))$ in l, including the point $(0, b)$, has coordinates $(x, mx + b)$. Since l intersects every vertical line, it follows that, for every real number x, there must be a point on the line having x as its abscissa. Therefore, the domain of function f is R. Thus, we are able to conclude that line l is the graph of a function f defined by

$$f : f(x) = mx + b, \quad \text{for } x \in R \text{ and } m \neq 0.$$

Such a function is called a *linear function*. As we noted earlier, a function whose graph is a horizontal line—that is, a line with slope zero—is called a constant function. The following definition makes this distinction clear.

Definition 2.6b If m and b are real numbers, the function defined by

$$f : f(x) = mx + b, \quad \text{for } x \in R,$$

is a **linear function** if $m \neq 0$. If $m = 0$, the function is a **constant function**.

Just as the constant function is an extension of the constant sequence, the linear function is an extension of the arithmetic sequence.

We have just shown that every line with nonzero slope is the graph of a linear function. We can also show that, conversely, the graph of any linear function is a line with nonzero slope. Given a linear function f defined by $f(x) = mx + b$, for $x \in R$, we merely observe that all of the coordinates of the points in its graph are of the form $(x, mx + b)$. But these are just the coordinates of the points in the line through $(0, b)$ with slope m, as we demonstrated earlier.

This means then that there is a one-to-one correspondence between the set of all lines in the plane with nonzero slope and the set of all linear functions. In other words, corresponding to each line l with nonzero slope, there is precisely one linear function f, and corresponding to each linear function f, there is precisely one line l with nonzero slope. Similarly, it can be shown that there is a one-to-one correspondence between the set of all constant functions and the set of all horizontal lines in the plane.

Every linear function f is defined by a linear equation of the form $f(x) = mx + b$. We will refer to this as the equation of the line l, where l is the graph of f.

Often we will be given some information about a line that is the graph of a linear function and, from this information, want to determine its linear equation. Some examples of how this might be done are given next.

Example 3 Suppose that we are told that a given line has slope -2 and y-intercept $(0, \sqrt{2})$. Then we can immediately write down its linear equation

$$f(x) = (-2)x + \sqrt{2} = -2x + \sqrt{2}.$$

Example 4 We might be told that the slope of a line is 3 and that it passes through the point $(1, 2)$. In this case, we could proceed as follows. Because $m = 3$, we know that the linear equation of the line is of the form $f(x) = 3x + b$. Now we must determine the value of b. The fact that $(1, 2)$ is a point in the line means that $f(1) = 2$. But, substituting 1 for x in $f(x) = 3x + b$, we have $f(1) = 3 + b$. Since $f(1) = 2$, this means that $3 + b = 2$, or $b = -1$. Thus, the equation for the line with slope 3 that contains point $(1, 2)$ is $f(x) = 3x - 1$.

Example 5 As yet another example, suppose that we are told that a line passes through the points $(1, 2)$ and $(-3, 4)$. First of all, the equation of the line is of the form $f(x) = mx + b$; the fact that $(1, 2)$ is in the line means that $f(1) = 2$; and the fact that $(-3, 4)$ is in the line means that $f(-3) = 4$. Therefore, we have

$$2 = f(1) = m \cdot 1 + b,$$
and
$$4 = f(-3) = m \cdot (-3) + b.$$

These equations are equivalent to

$$2 - m = b,$$
and
$$4 + 3m = b.$$

Hence, $2 - m = 4 + 3m$, or $4m = -2$. Thus, $m = -\frac{1}{2}$. Since $m = -\frac{1}{2}$, we see, from the equation $2 - m = b$, that $b = 2\frac{1}{2}$. Therefore, the equation of the line containing points $(1, 2)$ and $(-3, 4)$ is $f(x) = -\frac{1}{2}x + 2\frac{1}{2}$.

Notice that an alternative approach for this example would be to determine the slope of the line from the slope of the segment joining the given points. We could then proceed as in Example 4.

Examples 3, 4, and 5 demonstrate that it is easy to determine the linear equation of a line with nonzero slope, provided you know the slope and the y-intercept, or the slope and a point in the line, or two points contained in the line.

2.6c EXERCISES

In each of Exercises 1–8, write an equation for the line l which satisfies the given conditions. Then draw a graph of the line.

1. l has slope $\frac{2}{3}$ and y-intercept $(0, -5)$.
2. l has slope -2 and y-intercept $(0, 3)$.
3. l passes through $(4, 8)$ with slope $\frac{3}{4}$.
4. l passes through $(-3, -7)$ with slope $-\frac{1}{2}$.
5. l passes through $(-3, -2)$ and $(5, 6)$.
6. l passes through $(4, 2)$ and $(7, -6)$.
7. l has intercepts $(6, 0)$ and $(0, -4)$.
8. l has intercepts $(2, 0)$ and $(0, 3)$.

In each of Exercises 9–16, an equation of a line l is given. For each exercise, draw a graph of the line l.

9. $x + 5y = 0$.
10. $y = -\frac{5}{3}x + 9$.
11. $y - 2 = -4(x + 1)$.
12. $\dfrac{x}{3} + \dfrac{y}{5} = 1$.
13. $x + 2y - 8 = 0$.
14. $y = 3x - 2$.
15. $y + 3 = \frac{2}{3}(x - 2)$.
16. $\dfrac{x}{4} - \dfrac{y}{7} = 1$.

17. Prove that a linear function is an increasing function if the slope of its graph is greater than 0 and a decreasing function if the slope of its graph is less than 0.
18. Prove that if the domain of a constant function is restricted to the positive integers, then the function is a constant sequence.
19. Prove that if the domain of a linear function is restricted to the positive integers, then the function is an arithmetic sequence with common difference m.

2.7 Parallel and Perpendicular Lines

From geometry, you know that a pair of lines in the plane, l_1 and l_2, either intersect in a single point or are parallel. We would now like to see what the implications of this fact are with respect to the concept of slope.

If we consider the lines l_1 and l_2 in terms of their linear equations, $f_1(x) = m_1 x + b_1$ and $f_2(x) = m_2 x + b_2$, we see that the intersection of the two lines requires that there be a value of x such that $f_1(x) = f_2(x)$. When this is true, $(x, f_1(x)) = (x, f_2(x))$ is a common point of the lines l_1 and l_2. Since $f_1(x) = m_1 x + b_1$ and $f_2(x) = m_2 x + b_2$, the requirement that $f_1(x) = f_2(x)$ may be written as $m_1 x + b_1 = m_2 x + b_2$, or,

(1) $$(m_1 - m_2)x = b_2 - b_1.$$

There is a unique solution to this equation, namely,

$$x = \frac{b_2 - b_1}{m_1 - m_2},$$

if and only if $m_1 \neq m_2$; that is, if and only if l_1 and l_2 have different slopes. In this case, l_1 and l_2 intersect at precisely one point with coordinates

$$\left(\frac{b_2 - b_1}{m_1 - m_2},\ m_1 \left[\frac{b_2 - b_1}{m_1 - m_2}\right] + b_1\right) = \left(\frac{b_2 - b_1}{m_1 - m_2},\ \frac{m_1 b_2 - m_2 b_1}{m_1 - m_2}\right).$$

The preceding formula for the coordinates of the point of intersection is rather complicated and is presented only for the sake of completeness. In practice, it is not worthwhile to memorize this formula and use it to obtain the coordinates of the point of intersection. Instead, it is better to remember the method used in deriving the formula and to apply this method to each specific case.

Example 1 Let l_1 be the linear function defined by $f_1(x) = 3x + 5$ and let l_2 be the linear function defined by $f_2(x) = -2x - 1$. To determine the point of intersection of these two lines, we note that, when $f_1(x) = f_2(x)$, $3x + 5 = -2x - 1$, or $(3 - (-2))x = -1 - 5$. Hence, $x = \dfrac{-1 - 5}{3 + 2} = -\dfrac{6}{5}$. Substituting $x = -\frac{6}{5}$ in $f_1(x) = 3x + 5$, we have $f_1\left(-\frac{6}{5}\right) = 3\left(-\frac{6}{5}\right) + 5 = \frac{7}{5}$. Therefore, as Figure 2-33 shows, the intersection point of l_1 and l_2 is $\left(-\frac{6}{5}, \frac{7}{5}\right)$.

Figure 2-33

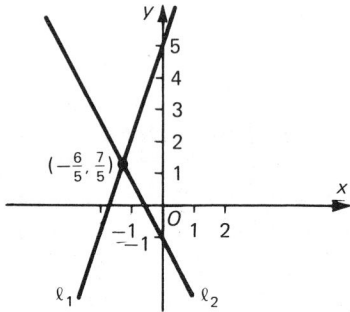

The important fact that should be remembered from the preceding discussion is: *Two lines in the plane having different slopes intersect in a single point, and therefore, are not parallel.*

If $m_1 = m_2$, then, of course, it must be the case that $b_1 \neq b_2$; for, otherwise, l_1 and l_2 would be the same line. But notice that if $m_1 = m_2$ and $b_1 \neq b_2$, then there can be no solution to equation (1) on page 73. That is, there is no value of x such that $m_1 x + b_1 = m_2 x + b_2$, which means that the lines l_1 and l_2 do not intersect and, therefore, are parallel. Thus: *Two lines in the plane having the same slope are parallel.* (See Figure 2-34.)

Figure 2-34

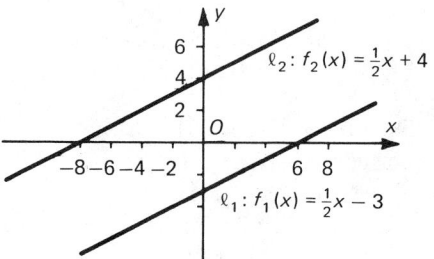

Now let us investigate the relationship between the slopes of two perpendicular lines. Let l_1 and l_2, with nonzero slopes m_1 and m_2, respectively, be two perpendicular lines that intersect at a point C with coordinates (a, b). The line $x = a + 1$ intersects l_1 and l_2 at the points A and B with coordinates $(a + 1, b + m_1)$ and $(a + 1, b + m_2)$, respectively. (See Figure 2-35.)

Figure 2-35

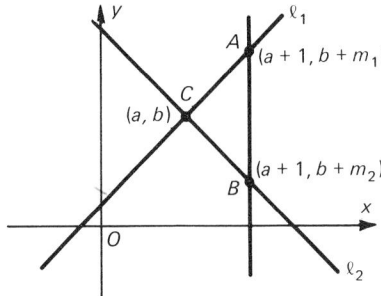

Because l_1 and l_2 are perpendicular, we know that $\triangle ABC$ is a right triangle. We shall let $d(A, B)$ denote the distance from A to B. Now $d(A, B) = |(b + m_1) - (b + m_2)| = |m_1 - m_2|$. Further,

$$d(C, B) = \sqrt{(a + 1 - a)^2 + (b + m_2 - b)^2} = \sqrt{1 + m_2^2},$$

and

$$d(C, A) = \sqrt{(a + 1 - a)^2 + (b + m_1 - b)^2} = \sqrt{1 + m_1^2}.$$

Therefore,

$$(m_1 - m_2)^2 = (1 + m_2^2) + (1 + m_1^2).$$

This last equation simplifies to $m_1 m_2 = -1$. Thus, if two nonvertical, nonhorizontal lines are perpendicular, then the product of their slopes is -1.

Now let us consider the converse of this statement. Suppose that l_1 and l_2 are two intersecting lines with nonzero slopes m_1 and m_2, respectively, and that $m_1 m_2 = -1$. Does it follow that l_1 and l_2 are perpendicular? If the lines intersect at a point C with coordinates (a, b), then the line $x = a + 1$ intersects l_1 and l_2 at points A and B with coordinates $(a + 1, b + m_1)$ and $(a + 1, b + m_2)$, respectively, as shown in Figure 2–35. Our task now is to show that $\triangle ABC$ is a right triangle, which will show that l_1 and l_2 are perpendicular.

First of all,

$$(d(A, B))^2 = (m_1 - m_2)^2 = m_1^2 - 2m_1 m_2 + m_2^2 = m_1^2 + m_2^2 + 2.$$

(The last step in this chain of equalities follows from the hypothesis that $m_1 m_2 = -1$.) Further,

$$(d(C, B))^2 + (d(C, A))^2 = (1 + m_2^2) + (1 + m_1^2) = m_1^2 + m_2^2 + 2.$$

Therefore, $(d(A, B))^2 = (d(C, B))^2 + (d(C, A))^2$, and $\triangle ABC$ is a right triangle. Hence, l_1 is perpendicular to l_2.

We may now combine the results of these two discussions into a single statement: Two nonvertical, nonhorizontal lines are perpendicular if and only if the product of their slopes is -1.

2.7 EXERCISES

In each of Exercises 1–6, find the equation of the line through the given point that is parallel to the given line. Then draw a graph of the two lines.

1. $(5, 5)$, $y = \frac{1}{2}x + 8$.
2. $(-6, -4)$, $\dfrac{x}{12} + \dfrac{y}{12} = 1$.
3. $(0, 0)$, $x + 2y - 8 = 0$.
4. $(3, -4)$, $5x + 3y - 30 = 0$.
5. $(-2, 0)$, $4x - y - 12 = 0$.
6. $(-7, -1)$, $2x + y - 6 = 0$.
7. a) The endpoints of one side of a triangle are located at $(0, 0)$ and $(12, 0)$. The slopes of the other two sides are $\frac{3}{4}$ and $-\frac{3}{2}$. Determine the equations of the lines that contain the other sides and the coordinates of the third vertex if the third vertex is in the first quadrant.
   b) Find the coordinates of the third vertex if it is in the fourth quadrant.
8. a) If $A(0, 9)$, $B(4, -1)$, and $C(13, 5)$ are three vertices of parallelogram $ABCD$, determine the equations of the lines that contain sides $\overline{AD}$ and $\overline{CD}$.
   b) Find the coordinates of vertex D.
9. Prove that the segments joining the midpoints of the sides of quadrilateral $ABCD$ whose vertices are $A(0, 0)$, $B(2a, 0)$, $C(2b, 2c)$, and $D(2d, 2e)$ form a parallelogram.
10. Prove that the segments joining the midpoints of the sides of rectangle $ABCD$ whose vertices are $A(0, 0)$, $B(2a, 0)$, $C(2a, 2b)$, and $D(0, 2b)$ form a rhombus.
11. Let the vertices of a triangle be $(0, 0)$, $(8, 0)$, and $(6, 8)$.
    a) Determine the equations of the lines that contain the medians of the triangle.
    b) Prove that the medians intersect in a common point (Hint: Find the coordinates of the point of intersection of two of the lines and show that this point is also on the third line.)
12. Generalize the results of Exercise 11 by using the coordinates $(0, 0)$, $(2a, 0)$, and $(2b, 2c)$ for the vertices of the triangle.

In each of Exercises 13–18, find the equation of the line through the given point which is perpendicular to the given line. Then draw a graph of the two lines.

13. $(5, 5)$, $y = \frac{1}{2}x + 8$.
14. $(-2, 5)$, $3x - 4y - 24 = 0$.

15. $(-6, -4)$, $\dfrac{x}{12} + \dfrac{y}{12} = 1$.
16. $(0, 0)$, $x + 2y - 8 = 0$.
17. $(-2, 0)$, $4x - y - 12 = 0$.
18. $(-7, -1)$, $2x + y - 6 = 0$.
19. a) Determine the equations of the lines that contain the altitudes of the triangle whose vertices are $(-6, 0)$, $(6, 0)$, and $(2, 8)$.
    b) Prove that these lines intersect in a common point.
20. Generalize the results of Exercise 19 by using the coordinates $(-a, 0)$, $(a, 0)$, and (b, c).

2.8 The Distance from a Point to a Line

We make much use of the formula for the distance between two points. Another extremely useful formula is that for the *perpendicular* distance from a point P to a nonvertical line l in the plane. If P is (x_0, y_0) and l is the graph of $f: f(x) = mx + b$, $x \in R$, then the perpendicular distance from P to l, $d(P, l)$, is given by:

$$d(P, l) = \frac{|mx_0 + b - y_0|}{\sqrt{1 + m^2}}.$$

There are many proofs of this formula, but the following uses only very elementary techniques.

Figure 2-36

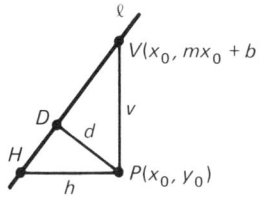

Consider a line l and a point P, not on l, as shown in Figure 2-36. The horizontal, vertical, and perpendicular line segments $\overline{HP}$, $\overline{VP}$, and $\overline{DP}$, respectively, have lengths h, v, and d. These lengths are the horizontal, vertical, and perpendicular distances, respectively, of the point P from the line l. Since $\angle PDV$ and $\angle HPV$ are both right angles, it is easy to see that $\triangle PDV$ is similar to $\triangle HPV$. We leave as an exercise the problem of showing that

$$d = \frac{1}{\sqrt{m^2 + 1}} v,$$

where $m = \dfrac{v}{h}$ is the slope of the line l. Hence, the perpendicular distance d of P from l is a constant multiple of the vertical distance v of P from l, where the constant factor is $\dfrac{1}{\sqrt{m^2 + 1}}$. Since the vertical distance v is clearly $|mx_0 + b - y_0|$, we have our formula.

It is easy to see that the formula holds even if l is a horizontal line, $m = 0$, or if P is on l, $y_0 = mx_0 + b$. If l is a vertical line $x = a$ then $d(P, l) = |a - x_0|$.

It should also be clear that the proof does not depend on the figure where l has positive slope and P lies below l. If the slope of l were negative, or P were above l, the figure would be different, but the reasoning the same.

2.8 EXERCISES

In each of Exercises 1–6, find the distance from the graph of the given function to the given point.

1. $f : f(x) = \frac{4}{3}x - 5$; $(3, -4)$.
2. $g : g(x) = \frac{5}{12}x - 10$; $(4, 8)$.
3. $h : y = \frac{1}{2}x$; $(-5, 10)$.
4. $F : y = x - 10$; $(-5, 5)$.
5. $G : 2x + 3y = 18$; $(-2, -7)$.
6. $H : 3x - 4y = 18$; $(-2, 7)$.

In each of Exercises 7 and 8, the graphs of the three given functions contain the sides of a triangle. For each triangle find the coordinates of its vertices and the lengths of its three altitudes.

7. $f : f(x) = \frac{1}{2}x - 5$. $g : g(x) = -\frac{1}{3}x + \frac{10}{3}$. $h : h(x) = 3x - 10$.
8. $F : x - 2y = 4$. $G : 5x + 4y = 48$. $H : 4x - y + 12 = 0$.

In each of Exercises 9 and 10, find the distance between the parallel lines that are the graphs of the given functions.

9. $f : f(x) = 2x + 5$, $g : g(x) = 2x - 8$.
10. $F : 2x - 3y = 18$, $G : 2x - 3y = 6$.
11. Show that the perpendicular distance from the origin to the graph of
$F : F(x) = mx + b$ is $\dfrac{|b|}{\sqrt{1 + m^2}}$.
12. Show that for the situation described in Figure 2–36,
$$d = \frac{1}{\sqrt{1 + m^2}} v,$$
where m is the slope of line l.
13. Show that the distance from the line $Ax + By + C = 0$ to the point (x_0, y_0) is given by $\dfrac{|Ax_0 + By_0 + C|}{\sqrt{A^2 + B^2}}$.

2.9 The Absolute Value Function

A very important and useful function is the function defined as the union of two restrictions of linear functions (remember that a restriction of a function g is a subset of g) as follows:

$$f(x) = \begin{cases} x & \text{if } x \geq 0, \\ -x & \text{if } x < 0. \end{cases}$$

It should be clear that f is the union of a restriction of g and a restriction of h, where $g(x) = -x$, $x \in R$ and $h(x) = x$, $x \in R$. The graph of this function is shown in Figure 2-37.

Figure 2-37

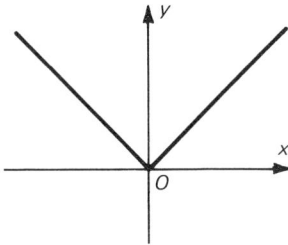

This function is the *absolute-value* function and could have been more simply defined by $f(x) = |x|$. We recall for you the definition of the absolute value of a real number.

Definition 2.9a If x is a real number, the **absolute value** of x is denoted by $|x|$, and is defined by

$$|x| = \begin{cases} x & \text{if } x \geq 0, \\ -x & \text{if } x < 0. \end{cases}$$

As you know, the absolute value of a number is never less than 0, and the only nonpositive absolute value of a number is the absolute value of 0, $|0| = 0$. Thus, the absolute-value function maps R onto the set consisting of all positive real numbers and zero; that is, the function has the set of all nonnegative real numbers as its range.

The preceding observations can be made from the graph of the absolute-value function given in Figure 2-37. It is clear that all ordinates of points on the graph are nonnegative, which means that $|x|$ is not negative. The graph also shows that the only point with 0 for its ordinate also has 0 for its abscissa; that is, $|0| = 0$.

Let us consider again the graph of the absolute value function. Since the triangle formed by the points $(0, 0)$, $(a, 0)$, and $(a, |a|)$ is an isosceles right triangle (see Figure 2–38), the distance from $(a, 0)$ to the origin is

Figure 2-38

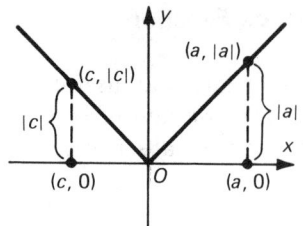

the same as the distance from $(a, 0)$ to $(a, |a|)$. From the distance formula, we know that the distance from $(a, 0)$ to $(0, 0)$ is as follows:

$$d((a, 0), (0, 0)) = \sqrt{(a - 0)^2 + (0 - 0)^2} = \sqrt{a^2}.$$

Since both $|a|$ and $\sqrt{a^2}$ represent the distance from $(a, 0)$ to the origin, we conclude that

$$|a| = \sqrt{a^2}.$$

In fact, $\sqrt{x^2}$ is an alternative way of defining $|x|$, for any real number x.

A certain amount of confusion often exists about what is meant by the symbol $\sqrt{c}$, where c is a positive number. The symbol $\sqrt{c}$ means the *nonnegative number* a such that $a^2 = c$. It is true that if $a^2 > 0$, then there are exactly two numbers whose squares are equal to a^2, namely a and $-a$. For example, the two numbers whose squares are equal to $(-2)^2$ are -2 and $-(-2) = 2$. However, when we write the symbol $\sqrt{a^2}$, we always mean the positive one of the two numbers a and $-a$, provided $a \neq 0$. If a is positive, this number is a, and if a is negative, this number is $-a$, which is positive. In other words, you should always keep in mind that

$$\sqrt{x^2} = |x|.$$

We previously noted that the distance from $(a, 0)$ to $(0, 0)$ is $|a|$. This is a special case of the fact that the distance from $(a, 0)$ to $(b, 0)$ is $|a - b|$. This follows from the fact that $d((a, 0), (b, 0)) = \sqrt{(b - a)^2} = \sqrt{(a - b)^2} = |a - b|$. Since $|a - b| = |b - a|$, we see that we can express the distance from $(a, 0)$ to $(b, 0)$ by either $|a - b|$ or $|b - a|$.

There are several important properties of the absolute-value function that we shall have occasion to use in some of our later work. They correspond to properties of the absolute value of a real number, which we now recall for you:

$$|ab| = |a|\,|b|$$
$$\left|\frac{a}{b}\right| = \frac{|a|}{|b|}$$
$$|a + b| \leq |a| + |b|$$
$$||a| - |b|| \leq |a - b| \leq |a| + |b|.$$

2.9 EXERCISES

State the values of x for which each of the following is true.

1. $\sqrt{(x-2)^2} = 2 - x$.
2. $\sqrt{(x+3)^2} = x + 3$.
3. $|(4-x)^2| = -(x-4)^2$.
4. $|(x-1)^3| = (1-x)^3$.

Use the definition of $|x|$ to prove each of the following.

5. If $x \in R$ then $|x^2| = |x|^2$.
6. If $x \in R$ then $-|x| \leq x \leq |x|$.
7. If $x \in R$ and $x \geq 0$, then $-x \leq |x| \leq x$.
8. Suppose $|x - y| < 1$. Show that $|x| < |y| + 1$.

Use the property that $|a| \cdot |b| = |ab|$ to solve each of the following inequalities.

9. $|\frac{2}{3}x| < 12$.
10. $|-0.02x| \geq 1.78$.
11. $|-5x| > 15$.
12. $|0.5x| \leq 4$.
13. a) Solve the inequality $|y| < 8$.
    b) Let $y = x + 4$. Use the solutions of part a to find the solutions of $|x + 4| < 8$.

Use the technique suggested in Exercise 13 to solve each of the following inequalities.

14. $|2x + 3| \leq 1$.
15. $|5x - 2| > 3$.
16. Consider the function $f: f(x) = 2x + |2 - x|$.
    a) Show that $f = g \cup h$, where $g: g(x) = x + 2$, $x \in R$ and $x < 2$, and $h: h(x) = 3x - 2$, $x \in R$ and $x \geq 2$.
    b) Use the results of part a to draw the graph of f.

Use the technique suggested by Exercise 16 to draw the graphs of the following functions.

17. $f: f(x) = x + |x + 1|$.
18. $f: f(x) = |x + 4| - x$.

2.10 Continuity

Once we know two points on the graph of a constant or a linear function, we can use a straightedge to draw as much of the graph of that function as the dimensions of our paper or blackboard allow. There are no "breaks" or "jumps" in the graph of such a function. Many other functions have graphs which, although not "straight" lines, also have no breaks or jumps and can be drawn in a continuous motion without lifting the pencil. Other functions, like function g illustrated in Figure 2-39, have graphs which do have breaks or jumps.

Figure 2-39

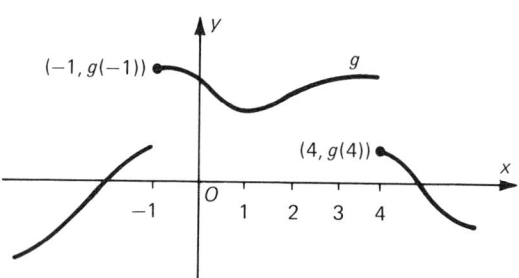

The everyday meaning of the word "continuous" as "unbroken" or "uninterrupted" suggested its use as a technical word in mathematics to describe a property which a function may or may not have at a given value of x. There are two requirements for a function to be continuous at c: point $(c, f(c))$ must be on the graph, and there must be no breaks in the graph near that point.

By this informal definition, the function f, whose graph is shown in Figure 2-40, is continuous at 3 and is discontinuous at 1. From an inspection of the graph of f, we see that the behavior of $f(x)$ for values of x near 1 is obviously different from the behavior of $f(x)$ for values of x near 3. How can we describe this difference in behavior?

Figure 2-40

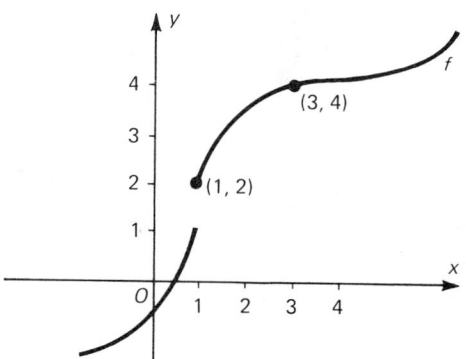

Let us draw a vertical and a horizontal line through point (3, 4), as shown in Figure 2-41.

Figure 2-41

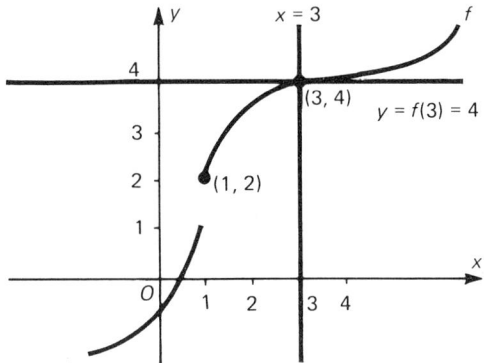

The set of all points $(x, f(x))$ in the graph of f which are less than one unit away from the horizontal line through $(3, f(3))$ must satisfy the condition $|f(x) - f(3)| < 1$. These points lie in the horizontal strip of the plane that is bounded by the two horizontal lines which are, respectively, one unit above and one unit below the horizontal line through $(3, f(3))$. This strip, indicated in Figure 2-42, is symmetric with respect to the horizontal line through $(3, f(3))$.

Figure 2-42

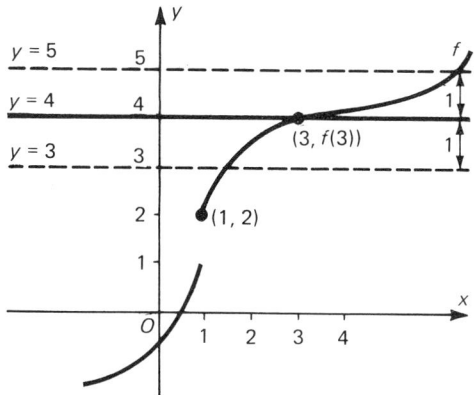

If we now consider the vertical line through $(3, f(3))$, as shown in Figure 2-43, it is clear that *all* of the points of the graph which are *sufficiently close* to this vertical line lie inside this horizontal strip. For example, we see that all of the points $(x, f(x))$ of the graph that are less than $\frac{1}{2}$ unit from the vertical line through $(3, f(3))$ lie inside the horizontal strip. All of those points $(x, f(x))$ of the graph which are less than $\frac{1}{2}$ unit from the vertical line through $(3, f(3))$ satisfy the condition that

$|x - 3| < \frac{1}{2}$, and these points are within the symmetric vertical strip bounded by the vertical lines $x = 2\frac{1}{2}$ and $x = 3\frac{1}{2}$. In other words, all those points of the graph of f which satisfy the condition that $|x - 3| < \frac{1}{2}$ also satisfy the condition that $|f(x) - f(3)| < 1$.

Figure 2-43

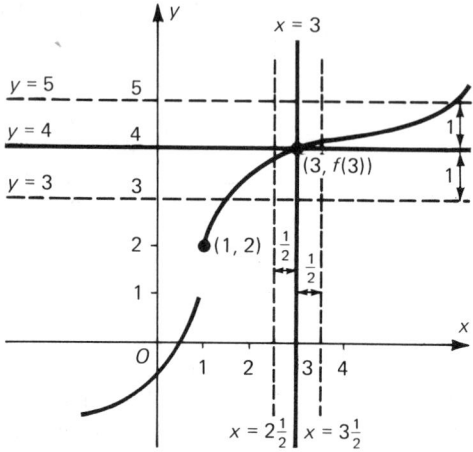

Now suppose that we construct narrower symmetric strips about the horizontal line through $(3, f(3))$. Some experimentation should convince you that you can construct as narrow a horizontal strip as you please that is symmetric with respect to $y = 3$ and you will still be able to construct a vertical strip symmetric with respect to $x = 3$ so that, if a point of the graph is within the vertical strip, then it is also within the horizontal strip. If you think about the meaning of this general result, you will see that we can make the following statement about the function f at $x = 3$:

> Given any positive number p, then we can determine a positive number d such that, for all values of x in the interval $|x - 3| < d$, it will be the case that $|f(x) - f(3)| < p$.

A careful examination of the graph should further convince you that we could make similar statements about any point on the graph, except for the point $(1, f(1))$.

Let us now consider the behavior of $f(x)$ for values of x near 1 in terms of the graph of f. First, we draw the horizontal line through $(1, f(1))$ and then construct a horizontal strip of width 1 that is symmetric with respect to this line. Next, we draw the vertical line $x = 1$ through $(1, f(1))$, as in Figure 2-44. We now observe that no matter how thin a

84 FUNCTIONS

vertical strip we construct that is symmetric with respect to $x = 1$, there will be points of the graph of f in this strip that are not in the horizontal strip.

Figure 2-44

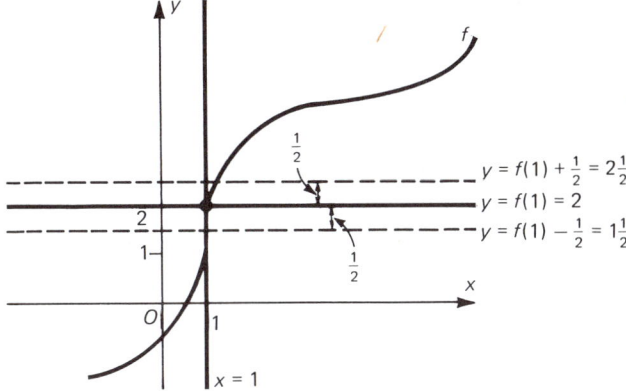

We can now summarize the difference in behavior of $f(x)$ for values of x near 3 and the behavior of $f(x)$ for values of x near 1 with the following statements.

1. Given any positive number p, $|f(x) - f(3)|$ will be less than p for all values of x sufficiently close to 3.
2. It is not true that, given any positive number p, $|f(x) - f(1)|$ will be less than p for all values of x sufficiently close to 1.

The sort of behavior described in this example is the basis upon which we refine our informal definition of continuity of a function f at a point a in its domain.

Definition 2.10a Let f be a function that is defined in an open interval containing the point a. Then f is said to be **continuous** at a provided that, given any positive number p, there is a positive number d such that,

$$\text{if } |x - a| < d, \text{ then } |f(x) - f(a)| < p.$$

As Definition 2.10a indicates, we speak of a function as being continuous at certain points of its domain and discontinuous at other points. This is because continuity is a property that a function has or fails to have at each point in its domain.* A function is said to be continuous on a subset of its domain if it is continuous at every point of this subset. A function that is continuous on its entire domain is called a continuous function.

*Hence, continuity is said to be a "pointwise" concept.

2.10 EXERCISES

For each of Exercises 1–4, state the values of x in R for which the given function is discontinuous.

1. $f:f(x) = \dfrac{6}{x+1}.$

2. $g:g(x) = \dfrac{x}{x^2-4}.$

3. $F:F(x) = \dfrac{x^2}{|x|}.$

4. $G:(x) = \dfrac{x-2}{2-x}.$

5. Consider the function $f:f(x) = \begin{cases} -1 \text{ when } x < 0 \\ 0 \text{ when } x = 0 \\ 1 \text{ when } x > 0. \end{cases}$
   a) Are there any real numbers for which f is undefined?
   b) Give values for $f(0.1)$, $f(0.01)$, $f(0.001)$, and $f(0.0001)$.
   c) Give values for $f(-0.1)$, $f(-0.01)$, $f(-0.001)$, and $f(-0.0001)$.
   d) Is f continuous at $x = 0$? Explain your answer.

6. Consider the function $f:f(x) = \begin{cases} \dfrac{x-1}{|x-1|} \text{ when } x \neq 1, \\ 0 \text{ when } x = 1. \end{cases}$
   a) Are there any real numbers for which f is undefined? What is $f(1)$?
   b) Give values for $f(0.9)$, $f(0.99)$, $f(0.999)$, and $f(0.9999)$.
   c) Give values for $f(1.1)$, $f(1.01)$, $f(1.001)$, and $f(1.0001)$.
   d) Is f continuous at $x = 1$? Explain your answer.

For each of the functions in Exercises 1–12, find the largest value of d that makes the given statement true.

7. $f:f(x) = 5x$ If $|x - 0| < d$, then $|f(x) - f(0)| < 0.5$.
8. $f:f(x) = -10x$ If $|x - 1| < d$, then $|f(x) - f(1)| < 0.1$.
9. $f:f(x) = 4x - 3$ If $|x - 2| < d$, then $|f(x) - f(2)| < 0.01$.
10. $f:f(x) = \tfrac{2}{3}x + b$ If $|x + 3| < d$, then $|f(x) - f(-3)| < 0.001$.
11. $f:f(x) = 2x + 3$ If $|x - a| < d$, then $|f(x) - f(a)| < 0.001$.
12. $f:f(x) = -5x + b$ If $|x - a| < d$, then $|f(x) - f(a)| < 0.001$.

For each of Exercises 13–16, express in terms of p the largest value of d that makes the given statement true for the given function.

13. $f:f(x) = 3x - 7$ If $|x - 8| < d$, then $|f(x) - f(8)| < p$.
14. $f:f(x) = 3x - 7$ If $|x - a| < d$, then $|f(x) - f(a)| < p$.
15. $f:f(x) = -2x + b$ If $|x - a| < d$, then $|f(x) - f(a)| < p$.
16. $f:f(x) = -\tfrac{2}{3}x + b$ If $|x - a| < d$, then $|f(x) - f(a)| < p$.
17. Consider the linear function $f:f(x) = 2x + 4$.

   a) Suppose $p > 0$. Show that if $|x - a| < \dfrac{p}{2}$, then $|f(x) - f(a)| < p$.

b) Explain how this shows that f is continuous at every point in its domain.

18. Consider the general linear function $f: f(x) = mx + b$, for all $x \in R$, with $m \neq 0$.

   a) Suppose $p > 0$. Show that if $|x - a| < \dfrac{p}{|m|}$, then $|f(x) - f(a)| < p$.

   b) Explain how this shows that any linear function is a continuous function.

19. Prove that any constant function is a continuous function.

2.11 The Limit of a Function

You have undoubtedly noticed that there is a definite similarity between the concept of a function being continuous at a point and that of a sequence having a limit. To bring this similarity into sharper focus we compare the definitions side by side.

$\lim\limits_{n \to \infty} x_n = X$	f is continuous at a
Given any positive number ϵ, there is an N such that if $n > N$, then $|x_n - X| < \epsilon$.	Given any positive number p, there is a d such that if $|x - a| < d$, then $|f(x) - f(a)| < p$.

After carefully comparing the two definitions and remembering that $\lim\limits_{n \to \infty} x_n = X$ is read as "the limit of x_n as n approaches infinity is X" it should seem reasonable, by analogy, that the statement "f is continuous at a" could also be phrased as "the limit of $f(x)$ as x approaches a is $f(a)$." We shall do this, and we shall use the symbol $\lim\limits_{x \to a} f(x) = f(a)$ to denote the statement "the limit of $f(x)$ as x approaches a is $f(a)$."

Definition 2.11a The symbol $\lim\limits_{x \to a} f(x) = f(a)$ means "For every $p > 0$, there is a $d > 0$ such that if $|x - a| < d$, then $|f(x) - f(a)| < p$." That is $\lim\limits_{x \to a} f(x) = f(a)$ is equivalent to the statement "f is continuous at a."

We take special note of the fact that a function must be defined at a point to be continuous there. In fact it must be defined in an open interval containing the point.

It can happen that although a function f is not defined at a point a, values of $f(x)$ get very close to some number l for all values of x close to a, *but unequal to* a. That is, if p is any positive number, then $|f(x) - l| < p$ for all x sufficiently close to a, *but unequat to* a. For example $f(x) = \dfrac{x^2 - a}{x - a}$ is not defined at $x = a$, but for all values of $x \neq a$, $f(x) = \dfrac{x^2 - a}{x - a} = \dfrac{(x - a)(x + a)}{x - a} = x + a$. It is easy to see that given any positive number p there is a positive number d, such that if $0 \neq |x - a| < d$, then $|f(x) - 2a| = |(x + a) - 2a| = |x - a| < p$. In fact it is clear that $d = p$ suffices. In this situation we will say that the $\lim_{x \to a} f(x) = 2a$. Note that f is not defined at a and therefore is *not* continuous there. We are saying by the statement $\lim_{x \to a} f(x) = 2a$ that the values of $f(x)$ approach $2a$ as x get close to a but is *never equal to* a. We define $\lim_{x \to a} f(x) = l$, as follows:

Definition 2.11b The symbol $\lim_{x \to a} f(x) = l$, means "For every $p > 0$, there is a $d > 0$, such that if $0 \neq |x - a| < d$, then $|f(x) - l| < p$."

Take note that $0 \neq |x - a|$, means $x \neq a$. In the definition above where we have written $0 \neq |x - a| < d$, it is more usual to write $0 < |x - a| < d$. It is clear that $0 < |x - a|$ also requires that $x \neq a$. We have written it as we have for emphasis.

We now emphasize the following: A function f can have a limit at a, that is $\lim_{x \to a} f(x) = l$, without being continuous there. To be continuous at a, f must be defined there, i.e., $f(a)$ is some number, and further $\lim_{x \to a} f(x) = f(a)$.

There is another point to be emphasized here. Namely, a careful reading of the definition of $\lim_{x \to a} f(x) = l$ makes it clear that it can happen that $\lim_{x \to a} f(x) = l$ and $f(a)$ is either not defined or is unequal to l. Note, however, that $f(x)$ must be defined at all points, other than a of some open interval containing a. Such situations will be considered in the exercises.

2.11 EXERCISES

1. The function $f : f(x) = \dfrac{1 - 2x}{x^2 + 1}$ is continuous for all real numbers. Find the value of each of the following.

   a) $\lim_{x \to -5} f(x)$ b) $\lim_{x \to 0} f(x)$ c) $\lim_{x \to \frac{1}{2}} f(x)$ d) $\lim_{x \to -\frac{2}{3}} f(x)$

For each of Exercises 2–5, state whether f is continuous at a. Then determine the value of $\lim_{x \to a} f(x)$, if it exists.

2. $f(x) = \dfrac{x}{x^2 - 1}$; $a = 0$.

3. $f(x) = \dfrac{x^2 + x}{x}$; $a = 0$.

4. $f(x) = \dfrac{x^2 - 25}{x^2 + 10x + 25}$; $a = -5$.

5. $f(x) = \dfrac{3x^2 - 8x - 3}{x^2 - 5x + 6}$; $a = 3$.

6. Given $f: f(x) = \dfrac{|x|}{x}$.

    a) For what value of x is f discontinuous?
    b) If $x > 0$, what is the value of $f(x)$?
    c) If $x < 0$, what is the value of $f(x)$?
    d) Does $\lim_{x \to 0} f(x)$ exist? If so, what is its value?

7. Given $g: g(x) = \begin{cases} |x| & \text{for } x \neq 0 \\ 1 & \text{for } x = 0. \end{cases}$

    a) Is g continuous at $x = 0$?
    b) Does $\lim_{x \to 0} g(x)$ exist? If so, what is its value?

8. Given $h: h(x) = \begin{cases} x + 1 & \text{for } x \geq 0 \\ x - 1 & \text{for } x < 0. \end{cases}$

    a) What is the value of $h(0)$? Is h continuous at $x = 0$?
    b) Does $\lim_{x \to 0} h(x)$ exist? If so, what is its value?

9. Given $F: F(x) = \begin{cases} -x & \text{for } x \leq 2 \\ x - 4 & \text{for } x > 2. \end{cases}$

    a) What is the value of $F(2)$? Is F continuous at $x = 2$?
    b) Does $\lim_{x \to 2} F(x)$ exist? If so, what is its value?

10. Given $G: G(x) = \dfrac{\sqrt{x + 9} - 3}{x}$.

    a) For what value of x is $G(x)$ not defined?
    b) Show that for $x \neq 0$,

    $$G(x) = \dfrac{1}{\sqrt{x + 9} + 3}$$

    [Hint: rationalize the numerator.]

    c) Does $\lim_{x \to 0} G(x)$ exist? If so, what is its value?

2.12 Algebra of Functions

In dealing with sequences we found that there were some very natural definitions for the sum, difference, product and quotient of sequences. Since the concept of a function is a generalization of the concept of a sequence, it would seem clear that we need only extend our definitions of algebraic operations on sequences to functions. We do this, but we need some added conditions, since, while sequences all have the same domain, functions in general do not. The manner in which we take care of this problem should be clear from the following definitions.

Definition 2.12a If f and g are functions, the **sum of f and g**, denoted by $f + g$, and the **product of f and g**, denoted by $f \cdot g$, are defined as follows:

$$f + g = \{(x, f(x) + g(x)) \mid f(x) + g(x) \text{ is defined}\}$$
$$f \cdot g = \{(x, f(x) \cdot g(x)) \mid f(x) \cdot g(x) \text{ is defined}\}.$$

When the domains of f, $D(f)$, and g, $D(g)$, are R, the words "is defined" in the definition are superfluous. When these domains are subsets of R, however, the expressions $f(x) + g(x)$ and $f(x) \cdot g(x)$ will be defined only when x is in both $D(f)$ and $D(g)$. From this fact, we see that the domains of $f + g$ and $f \cdot g$ are the intersection of the domains of f and g. That is, $D(f + g) = D(f) \cap D(g)$ and $D(f \cdot g) = D(f) \cap D(g)$. It can occur that $D(f) \cap D(g)$ is the empty set, $\emptyset$. In this case, the operations $f + g$ and $f \cdot g$ are not defined.

Example 1 If f is the linear function $f(x) = 2x + 3$, $x \in R$, and g is the linear function $g(x) = -2x + 5$, $x \in R$, then $(f + g)(x) = f(x) + g(x) = 2x + 3 + (-2x + 5) = 8$. That is, $f + g$ is the constant function h, $h(x) = 8$. We also see that

$$(f \cdot g)(x) = f(x) \cdot g(x) = (2x + 3)(-2x + 5) = -4x^2 + 4x + 15.$$

It is quite standard to denote the product of a constant function $h(x) = c$ and a function f, by cf instead of $h \cdot f$.

Example 2 If $h(x) = 2$ and $f(x) = x^2$ then $h \cdot f = 2f$ and $2f(x) = 2x^2$.

The definitions of the difference and product of functions are equally simple.

Definition 2.12b If f and g are functions, the **difference of f and g**, denoted by $f - g$, and the **quotient of f and g**, denoted by $\dfrac{f}{g}$, are defined as follows:

$$f - g = \{(x, f(x) - g(x)) \mid f(x) - g(x) \text{ is defined}\}$$
$$\frac{f}{g} = \left\{\left(x, \frac{f(x)}{g(x)}\right) \,\middle|\, \frac{f(x)}{g(x)} \text{ is defined}\right\}.$$

Note in particular that $\frac{f(x)}{g(x)}$ is defined only when $x \in D(f) \cap D(g)$ and $g(x) \neq 0$.

Example 3 If $f(x) = 2x + 3$ and $g(x) = -2x + 5$, $x \in R$, then $(f - g)(x) = f(x) - g(x) = 2x + 3 - (-2x + 5) = 4x - 2$. We also see that $\frac{f(x)}{g(x)} = \frac{2x + 3}{-2x + 5}$ provided $x \neq \frac{5}{2}$, that is provided $-2x + 5 \neq 0$.

When we consider the quotient $\frac{u}{g}$, where u is the unit function, $u(x) = 1$, it is fairly standard to denote $\frac{u}{g}$ by $\frac{1}{g}$ and call $\frac{1}{g}$ the **reciprocal function of g.**

Example 4 If $g(x) = 2x + 4$, $\frac{u}{g}(x) = \frac{1}{g(x)} = \frac{1}{2x + 4}$, $x \neq -2$. Also $3 \cdot \frac{1}{g}(x) = \frac{3}{g}(x) = \frac{3}{g(x)} = \frac{3}{2x + 4}$, $x \neq -2$, and in general if c is any constant $\frac{c}{g}(x) = \frac{c}{g(x)}$.

As we proceed in our study of functions we shall learn that if we know certain properties of functions f and g and of their graphs, we can deduce certain properties of the functions $f + g, f \cdot g, f - g$ and $\frac{f}{g}$ and of their graphs. Our purpose in introducing the notion of algebraic operations on functions at this time is to be able to point out that the sum, difference, product and quotient of continuous functions are continuous where they are defined. This we shall do in the next section.

2.12 EXERCISES

For each of Exercises 1–6, write in simplest form the equation that determines $(f + g)(x)$ and the equation that determines $(f - g)(x)$. Then describe the domains of f, g, and $f + g$.

1. $f(x) = \dfrac{3x - 5}{x^2}$; $g(x) = \dfrac{3 + x}{x^2}$.

2. $f(x) = \dfrac{x^2 - 1}{1 + x}$; $g(x) = \dfrac{7}{1 + x}$.

3. $f(x) = \dfrac{3}{7x}$; $g(x) = \dfrac{5}{2x}$.

4. $f(x) = \dfrac{x + 9}{x - 2}$; $g(x) = \dfrac{2x - 3}{2 - x}$.

5. $f(x) = \dfrac{1}{5x}$; $g(x) = \dfrac{4 - x}{3x^2}$.

6. $f(x) = \dfrac{3}{2(x - 1)}$; $g(x) = \dfrac{-2}{x(x - 1)}$.

For each of Exercises 7–10, express $(f \cdot g)(x)$ *as a polynomial.*

7. $f(x) = \tfrac{1}{2}x + 2$; $g(x) = 2x - 4$.
8. $f(x) = x^2 - 0.3$; $g(x) = x^2 + 0.3$.
9. $f(x) = x + 1$; $g(x) = x^2 - x + 1$.
10. $f(x) = x^2 + 2$; $g(x) = x - 3$.

For each of Exercises 11–16, determine a polynomial expression for $g(x)$.

11. $f(x) = -8x$; $(f \cdot g)(x) = 16x^3 - 24x^2 - 64x$.
12. $f(x) = x + 3$; $(f \cdot g)(x) = x^2 + 15x + 36$.
13. $f(x) = 3x - 5$; $(f \cdot g)(x) = 6x^2 + 11x - 35$.
14. $f(x) = 2x - 1$; $(f \cdot g)(x) = 8x^3 - 4x + 1$.
15. $f(x) = x^2 - 2x + 1$; $(f \cdot g)(x) = x^4 - 4x^2 + 4x - 1$.
16. $f(x) = x - 1$; $(f \cdot g)(x) = x^3 - 1$.
17. If $f(x) = 3x - 1$ and $g(x) = 5x - 4$, find the value of x for which
$$\dfrac{f}{g}(x) = \tfrac{2}{3}.$$
18. If $f(x) = 4x + 3$ and $g(x) = 3x + 11$, find the value of x for which
$$\dfrac{f}{g}(x) = \tfrac{3}{4}.$$
19. If $f(x) = x^2 - 3$ and $g(x) = x^2 + 5$, find the values of x for which
$$\dfrac{f}{g}(x) = \tfrac{3}{7}.$$

2.13 Some Properties of the Limits of Functions

In studying sequences we learned that the limits of the sum, difference, product and quotient of convergent sequences were equal respectively to the sum, difference, product and quotient of these sequences.

Again since the concepts of a function and its limit are generalizations of the concept of a sequence and its limit, it would seem reasonable that there should be properties for functions analogous to the above properties for sequences. This is the case as we see in the following theorem.

Theorem 2.13a *If* $\lim_{x \to a} f(x) = A$ *and* $\lim_{x \to a} g(x) = B$ *then*

1. $\lim_{x \to a} [f(x) + g(x)] = A + B$
2. $\lim_{x \to a} [f(x) - g(x)] = A - B$
3. $\lim_{x \to a} [f(x) \cdot g(x)] = A \cdot B$
4. $\lim_{x \to a} \dfrac{f(x)}{g(x)} = \dfrac{A}{B}$, if $B \neq 0$.

The proofs of these statements are completely analogous to the corresponding statements for sequences. They merely involve a mimicking of the earlier proofs with some obvious changes, which will be considered in the exercises.

The results of our theorem can probably be best remembered as follows:

1. The limit of the sum of two functions is equal to the sum of their limits.
2. The limit of the difference of two functions is equal to the difference of their limits.
3. The limit of the product of two functions is equal to the product of their limits.
4. The limit of the quotient of two functions is equal to the quotient of their limits provided the limit of the denominator is unequal to zero.

Some important results of the above theorem and the definition of continuity are:

1. The sum and the difference of two continuous functions are continuous functions.
2. The product of two continuous functions is a continuous function.
3. The quotient of two continuous functions is a continuous function if the divisor is not equal to zero.

Most of the functions that we will study in the remainder of this book are continuous whenever they are defined. (If they are not, we will so indicate.) The functions studied in elementary calculus are, for the most part, continuous where they are defined, although there are some exceptions. The reason for this is that most of the functions studied in elementary calculus are constructed, by the operations of addition, multiplication, and so on, from the basic functions which we are studying in this book. Most of these operations preserve the property of continuity, and so the resulting functions are also continuous. Another way of saying this is that, very often, functions constructed from continuous functions inherit the property of continuity.

2.13 EXERCISES

In each of the following exercises, assume that $\lim_{x \to a} f(x) = A$ and $\lim_{x \to a} g(x) = B$.

1. Prove that $\lim_{x \to a} (f(x) + g(x)) = A + B$ as follows:

   a) Given $p/2 > 0$. Show that there are numbers $d_1 > 0$ and $d_2 > 0$ such that
   $$|f(x) - A| < p/2 \quad \text{if } |x - a| < d_1$$
   $$|g(x) - B| < p/2 \quad \text{if } |x - a| < d_2.$$

   b) Show that there is a number $d > 0$ such that
   $$|(f(x) + g(x)) - (A + B)| < p \quad \text{if } |x - a| < d.$$

   [Hints: choose d to be the smaller of the numbers d_1 and d_2, and use the property that $|x + y| \leq |x| + |y|$.]

2. Prove that $\lim_{x \to a} (f(x) - g(x)) = A - B$.

3. To prove that $\lim_{x \to a} (f(x) \cdot g(x)) = AB$, one must show that given $p > 0$, there is a $d > 0$ such that $|f(x) \cdot g(x) - AB| < p$ if $|x - a| < d$.

   a) Use the fact that
   $$|f(x) \cdot g(x) - AB| = |f(x) \cdot g(x) - Ag(x) + Ag(x) - AB|$$
   to show that
   $$|f(x) \cdot g(x) - AB| \leq |g(x)| \cdot |f(x) - a| + |A| \cdot |g(x) - B|.$$

   b) Show that there is a number $d_1 > 0$ such that $|g(x)| < \alpha = |B| + 1$ if $|x - a| < d_1$.

   c) Show that there is a number $d_2 > 0$ such that $|f(x) - A| < \dfrac{p}{2\alpha}$ if $|x - a| < d_2$.

   d) Show that there is a number $d_3 > 0$ such that $|A| \, |g(x) - B| < \dfrac{p}{2}$ if $|x - a| < d_3$.

   e) Show that that is a number $d > 0$ such that
   $$|f(x) \cdot g(x) - AB| < p \quad \text{if } |x - a| < d.$$

   [Hint: let d be the smallest of the numbers d_1, d_2, d_3, and use the inequality in part a.]

4. To prove that $\lim_{x \to a} \dfrac{1}{g(x)} = \dfrac{1}{B}$ ($B \neq 0$), one must show that given $p > 0$, there is a $d > 0$ such that

$$\left| \frac{1}{g(x)} - \frac{1}{B} \right| < p \qquad \text{if } |x - a| < d.$$

a) Show that

$$\left| \frac{1}{g(x)} - \frac{1}{B} \right| = \frac{|g(x) - B|}{|B| \cdot |g(x)|}.$$

b) Show that there is a number $d_1 > 0$ such that $|g(x)| > \dfrac{|B|}{2}$ if $|x - a| < d_1$.

[Hint: because $\lim_{x \to a} g(x) = B$, there is a number d_1 such that

$$\bigl| |g(x)| - |B| \bigr| < \frac{|B|}{2} \qquad \text{if } |x - a| < d_1.]$$

c) Given $p_1 > 0$, show there is a $d_2 > 0$ such that $|g(x) - B| < p_1$ if $|x - a| < d_2$.

d) Let d be the smaller of the numbers d_1 and d_2, and let $p = \dfrac{2p_1}{|B|^2}$ and show that

$$\left| \frac{1}{g(x)} - \frac{1}{B} \right| < p \qquad \text{if } |x - a| < d.$$

e) Prove that $\lim_{x \to a} \dfrac{1}{g(x)} = \dfrac{1}{B}$.

5. Use the results of Exercises 3 and 4 to prove that

$$\lim_{x \to a} \frac{f(x)}{g(x)} = \frac{A}{B}, \qquad B \neq 0.$$

2.14 Symmetry

When we study the graphs of functions, symmetry—or the lack of it—will be of interest to us. You may have noticed that the graphs of some of the functions we have considered were symmetric with respect to the y-axis. For example, the graph of the function $f: f(x) = x^2$, $x \in R$, and the graph of the absolute value function are both symmetric with respect to the y-axis.

To review the concept of symmetry, we first need to discuss the idea of the *reflection* of a point. Suppose that we are given a point P and a line l in a plane. If P is not on the line l, then there is a point P' in the plane that is called the *reflection of P in l*. Point P' is the second endpoint of the segment $\overline{PP'}$ that is perpendicular to l and is bisected by l. If point P is on line l, then P is its own reflection in l. Thus, each point in l is its own reflection in l.

Clearly, if P' is a reflection in l of P, then P is a reflection in l of P'. The term "reflection" is an apt one; for, if a vertical mirror were erected along l in Figure 2-45 and if you were to stand on the same side of the mirror as P, then the reflection of P would appear to be in the same position as P'. This is an experiment that you can perform with a small pocket mirror.

Figure 2-45

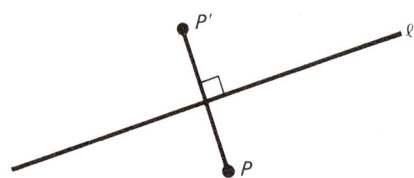

We can now use the idea of reflection in a line to define symmetry about a line as follows: Let l be a line and let S be a set of points in a plane. Then S is said to be *symmetric with respect to l* provided that, given any point P in S, there is a point P' in S such that P' is a reflection in l of P.

You should analyze for yourself *how* the definition just given applies to the following examples.

Example 1 A circle is symmetric with respect to the line of any diameter.

Example 2 A square is symmetric with respect to either of the lines of its diagonals and with respect to a line through the midpoints of a pair of opposite sides.

Example 3 The x-axis is symmetric with respect to the y-axis, and the y-axis is symmetric with respect to the x-axis.

Example 4 If you take a sheet of paper, splatter ink on it, and then fold the paper in half, you will find on unfolding it that you have a configuration which is symmetric with respect to the line of the fold.

If a set of points is symmetric with respect to a line, that line is said to be a "line of symmetry" for the set. For each of the figures shown in Figure 2-46, it is possible to draw at least one line of symmetry. In some cases, you should be able to draw several.

Figure 2-46

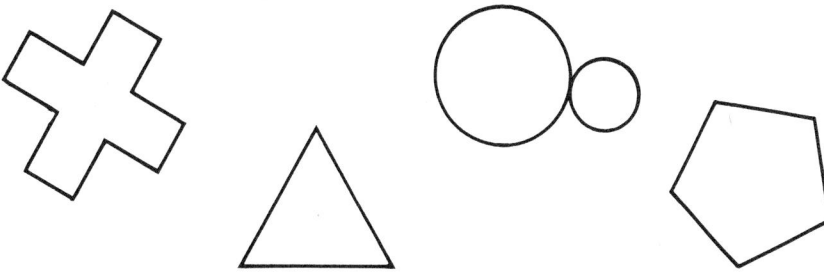

We will be particularly interested in sets of points that are symmetric with respect to the y-axis. It is not difficult to show that a set of points in the plane is symmetric with respect to the y-axis, if, and only if, for every point (a, b) in the set, the point $(-a, b)$ is also in the set. (We leave this as an exercise.)

Some important functions have graphs which are symmetric with respect to certain points in the plane, in particular, that are symmetric with respect to the origin.

Now let us consider symmetry about a point in the plane. Suppose that we are given a point Q in the plane. The reflection in Q of a point P in the plane is Q itself if $P = Q$; that is, Q is its own reflection in Q. If P is different from Q, then the reflection of P in Q is the point P' such that Q is the midpoint of segment $\overline{PP'}$. (See Figure 2–47.)

Figure 2-47

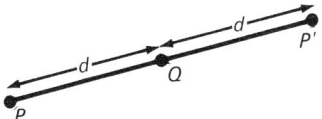

Clearly, if P' is a reflection in Q of P, then P is a reflection in Q of P'. We can use the idea of reflection in a point to define symmetry with respect to a point. Let Q be a point and let S be a set of points in a plane. Then S is said to be *symmetric with respect to Q* provided that, given any point P in S, the reflection of P in Q, denoted by P', is also in S.

Example 5 Some examples of symmetry with respect to a point are as follows: A circle is symmetric with respect to its center. A square is symmetric with respect to the intersection of its diagonals. A segment is symmetric with respect to its midpoint. Which of the figures in Figure 2–46 are symmetric with respect to some point?

In our study of graphs of functions, we will be particularly interested in sets of points that are symmetric with respect to the origin of the Cartesian plane, $(0, 0)$. As Figure 2–48 indicates, if a point P is in the first quadrant, then its reflection in the origin, point P', is in the third quadrant. Also, if the point is in the third quadrant, then its reflection in the origin is in the first quadrant. Similarly, the reflection in the origin of a point in the second quadrant is in the fourth quadrant, and the reflection of a point in the fourth quadrant is in the second quadrant.

Figure 2-48

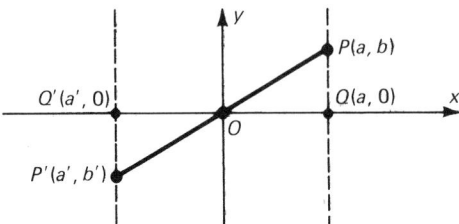

In Figure 2-48 points P and P' are symmetric with respect to the origin. Point Q is the intersection of the line $x = a$ with the x-axis, and point Q' is the intersection of the line $x = a'$ with the x-axis. We can now determine how the coordinates of P and P' are related.

The right triangles OPQ and $OP'Q'$ are clearly congruent, since $\overline{OP}$ and $\overline{OP'}$ are equal in length (Why?) and $\angle POQ \cong \angle P'OQ'$ (Why?). Thus, the lengths of $\overline{PQ}$ and $\overline{P'Q'}$ are equal, as are the lengths of $\overline{OQ}$ and $\overline{OQ'}$. It follows that $b' = -b$ and $a' = -a$. Thus, the point that is a reflection in the origin of (a, b) is the point whose coordinates are the additive inverses of those of (a, b), namely, $(-a, -b)$.

Example 6 The reflection in the origin of $(1, 2)$ is $(-1, -2)$, and the reflection of $(-3, -\pi)$ is $(3, \pi)$.

The discussion for points in the second or fourth quadrants is completely analogous, and, again, the reflection of (a, b) is $(-a, -b)$.

Example 7 The reflection in the origin of $(-2, 3)$ is $(2, -3)$, and the reflection of $(-\sqrt{2}, \frac{1}{2})$ is $(\sqrt{2}, -\frac{1}{2})$.

Similarly, if (a, b) is a point on one of the coordinate axes, its reflection in the origin can be shown to be $(-a, -b)$. The general result is that the reflection in the origin of an arbitrary point (a, b) is the point $(-a, -b)$. Thus, a set of points in the plane is symmetric with respect to the origin, if, and only if, for every point (a, b) in the set, the point $(-a, -b)$ is also in the set.

2.14 EXERCISES

1. Copy the figures shown in Figure 2–46 and draw all the possible lines of symmetry for each figure.

In each of Exercises 2–7, the coordinates of two points that are reflections of each other in a line are given. For each exercise, graph the points in the Cartesian plane and draw the line. Then give an equation for the line.

2. (2, 3), (8, 3)
3. $(-6, \frac{17}{4})$, $(-6, -\frac{7}{2})$
4. (2, −2), (−4, −2)
5. (1, −12.4), (1, −1.2)
6. (0, 5), (5, 0)
7. (0, 6), (−6, 0)

For each point listed below, give the coordinates of a) the reflection of the point in the x-axis and b) the reflection of the point in the y-axis.

8. (5, 5)
9. (−2, 3)
10. (−4, −1)
11. (8, −6)
12. $(0, 2 + \sqrt{3})$
13. $(2 - \sqrt{3}, 0)$

For each pair of points listed below, find the coordinates of the point Q so that the given points are reflections of each other in point Q.

14. (−4, −5), (3, −5)
15. (7, −2.5), (7, 1.5)
16. (−2, 2), (4, 2)
17. (−1, −12.4), (−1, 1.2)
18. (0, 4), (4, 0)
19. (0, 5), (−5, 0)

For each point given below, determine the coordinates of its reflection in the origin.

20. (5, 5)
21. (−2, 3)
22. (−4, −1)
23. (8, −5)
24. $(0, 2 + \sqrt{3})$
25. $(2 - \sqrt{3}, 0)$

Determine whether each of the following sets is symmetric with respect to the x-axis, the y-axis, or the origin.

26. $\{(x, 5) \mid x \in R\}$
27. $\{(-3, y) \mid y \in R\}$

28. $\{(x, -2x) \mid x \in R\}$

29. $\{(x, 2x + 1) \mid x \in R\}$

30. $\left\{\left(x, \dfrac{1}{x}\right) \mid x \in R \text{ and } x \neq 0\right\}$

31. $\{(x, |x|) \mid x \in R\}$

32. $\{(x, x^2) \mid x \in R\}$

33. $\{(x, \sqrt{x}) \mid x \in R \text{ and } x \geq 0\}$

34. Prove that a set of points is symmetric with respect to the y-axis if and only if for every point (a, b) in the set, the point $(-a, b)$ is in the set.

35. Prove that if a set of points is symmetric with respect to the x-axis and also with respect to the y-axis, then the set of points is symmetric with respect to the origin.

2.15 Symmetric Functions

We have already found one method of classifying functions—in terms of whether they are increasing or decreasing functions. The following definition classifies functions in terms of the symmetry of their graphs.

Definition 2.15a A function whose graph is symmetric with respect to either the y-axis or the origin is said to be a **symmetric function.** A function whose graph is symmetric with respect to the y-axis is said to be an **even function,** and one whose graph is symmetric with respect to the origin is said to be an **odd function.**

Figure 2–49 illustrates graphs of functions that are either even or odd. You will note that the graph of the domain of each of these functions is symmetric with respect to the y-axis (and with respect to the origin as well). The reason for this becomes apparent in the following argument: If a function f is symmetric with respect to the y-axis, then $(a, b) \in f$ implies that $(-a, b) \in f$. If f is symmetric with respect to the origin, then $(a, b) \in f$ implies that $(-a, -b) \in f$. In either case, if a belongs to the domain of f, then $-a$ belongs to the domain of f. Thus, in either case, if the graph of the domain, which is the projection of the function on the x-axis, contains the point $(a, 0)$, then it also contains the point $(-a, 0)$. This implies that the graph of the domain of a symmetric function is symmetric with respect to the y-axis.

Thus, an immediate test one can apply to a function to determine whether or not it *might* be a symmetric function is to determine whether or not the graph of its domain is symmetric with respect to the y-axis.

Figure 2-49

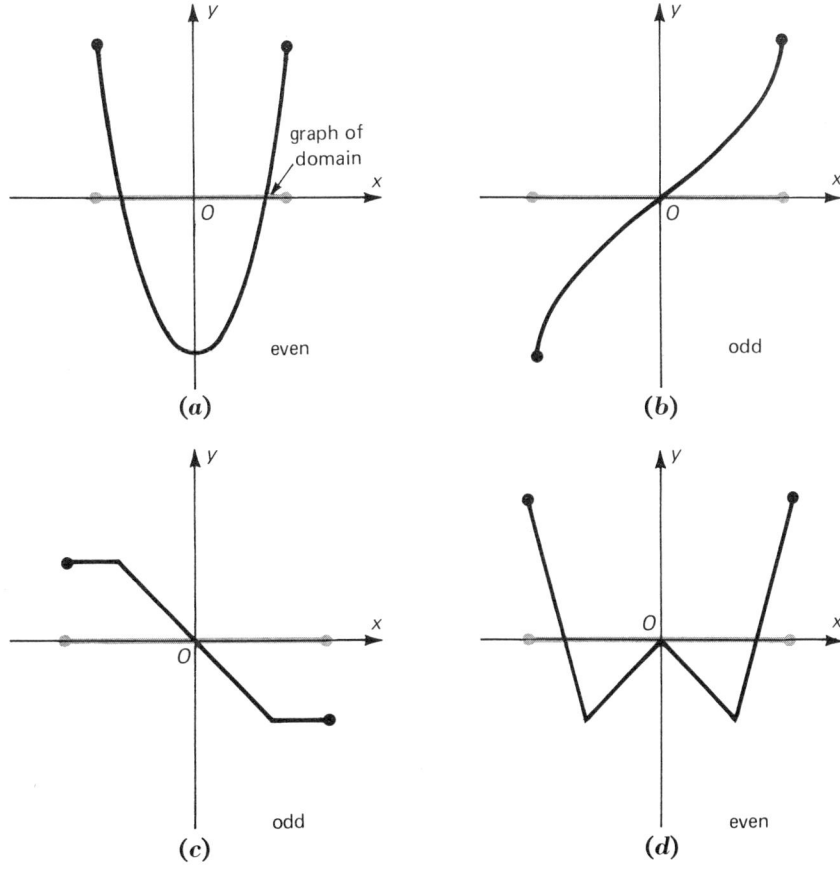

If the graph of the domain of a function is *not* symmetric with respect to the y-axis, then the function *cannot* be a symmetric function. In other words, if a function is symmetric, then it is *necessary* that the graph of its domain be symmetric with respect to the y-axis. Symmetry of the graph of the domain with respect to the y-axis is *not sufficient*, however, to insure that the function is symmetric, as is evidenced in the graph shown in Figure 2–50. As you see, the graph of the domain of this function is symmetric with respect to the y-axis, but the function is not symmetric with respect to either the origin or the y-axis.

Figure 2-50

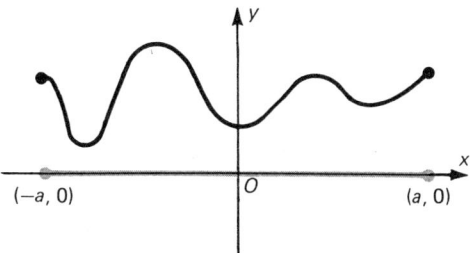

2.15 SYMMETRIC FUNCTIONS

For the graph of a function to be symmetric with respect to the y-axis, not only is it necessary for $-a$ to belong to the domain of the function whenever a belongs to the domain, but it is also necessary for a and $-a$ to have the same image in the range of the function. The following theorem establishes that these two conditions are both necessary and sufficient for the graph of a function to be symmetric with respect to the y-axis, that is, for a function to be an even function.

Theorem 2.15a *The following conditions are necessary and sufficient for a function f to be an even function:*

1. *If a is in the domain D of f, then $-a$ is also in D.*
2. *For every $a \in D$, $f(a) = f(-a)$.*

Proof We have already shown that, if f is symmetric with respect to the y-axis, then $a \in D$ implies that $-a \in D$. This is true because symmetry with respect to the y-axis requires that, if $(a, b) \in f$, then $(-a, b) \in f$. But this requirement also implies the second condition of the theorem, since $(a, b) \in f$ means that $f(a) = b$ and $(-a, b) \in f$ means that $f(-a) = b$, and, hence, $f(a) = f(-a)$. Thus we see that the two conditions of Theorem 2.15a are necessary.

Now let us assume that conditions 1 and 2 of the theorem are satisfied by a function f. We must show that these conditions are sufficient to imply that f is an even function. If (a, b) is a point in the graph of f, then the first condition imples that there must be a point $(-a, c)$ in the graph of f. Now $(a, b) \in f$ means that $b = f(a)$ and $(-a, c) \in f$ means that $c = f(-a)$. But condition 2 of the theorem implies that $c = b$; that is, $(-a, c) = (-a, b)$. Therefore, if $(a, b) \in f$, then $(-a, b) \in f$. But this means that the graph of f is symmetric with respect to the y-axis and, therefore, f is an even function. □

We now state an analogous theorem giving the necessary and sufficient conditions for a function to be symmetric with respect to the origin.

Theorem 2.15b *The following conditions are necessary and sufficient for a function f to be an odd function:*

1. *If a is in the domain D of f, then $-a$ is also in D.*
2. *For every $a \in D$, $f(a) = -f(-a)$.*

The proof of this theorem is quite similar to that of Theorem 2.15a and is left for you as an exercise.

Notice that, if n is an even positive integer, then the function f defined by
$$f : f(x) = x^n, \quad \text{for all } x \in R,$$

is an even function. This follows from the fact that, whenever n is an even positive integer, $(-x)^n = x^n$ for all $x \in R$. Similarly, if n is an odd positive integer, then the function f defined by

$$f : f(x) = x^n, \quad \text{for all } x \in R,$$

is an odd function. This is so because, whenever n is an odd positive integer, $(-x)^n = -(x^n)$ for all $x \in R$. These results may explain why the first type of function just described is called an even function and the second is called an odd function.

2.15 EXERCISES

In Exercises 1–8, determine $f(a)$, $f(-a)$, and $-f(-a)$. Then indicate whether the function is even, odd, or neither even nor odd. (Assume the largest possible domain for each function.)

1. $f(x) = -5$.
2. $f(x) = -10x^2$.
3. $f(x) = -\dfrac{1}{x^3}$.
4. $f(x) = x^2 - 6x$.
5. $f(x) = x^2 - 9$.
6. $f(x) = -x^2 + 4x - 5$.
7. $f(x) = x^2 - \dfrac{1}{x}$.
8. $f(x) = (x - 2)^2$.

Each of the following figures shows one half of the graph of an even function. Copy and complete the graph.

9.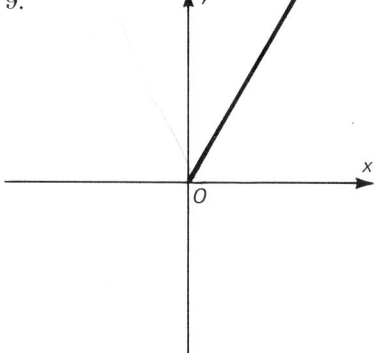

10.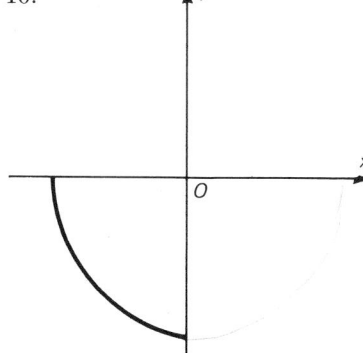

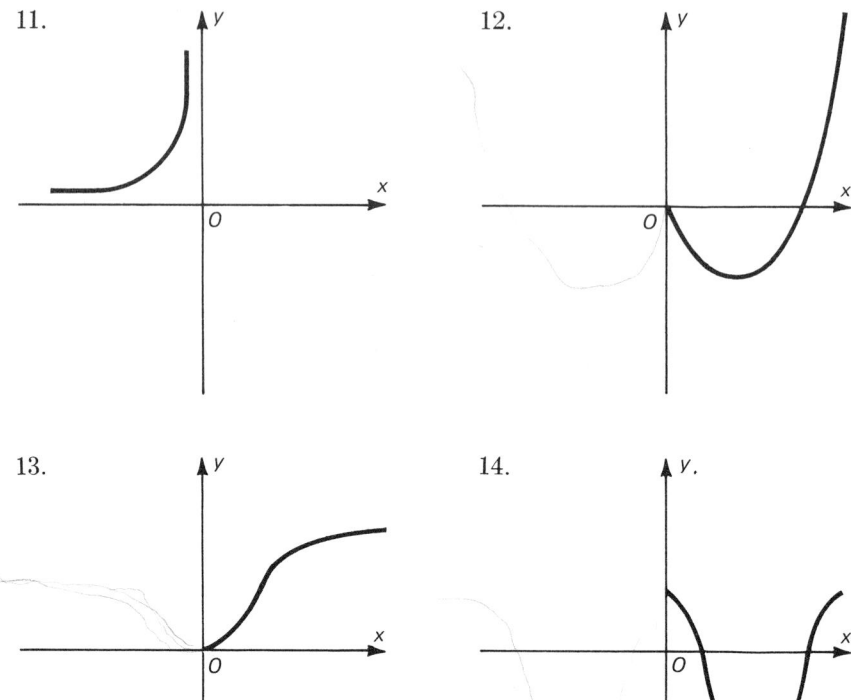

Each of the following figures shows one half of the graph of an odd function. Copy and complete the graph.

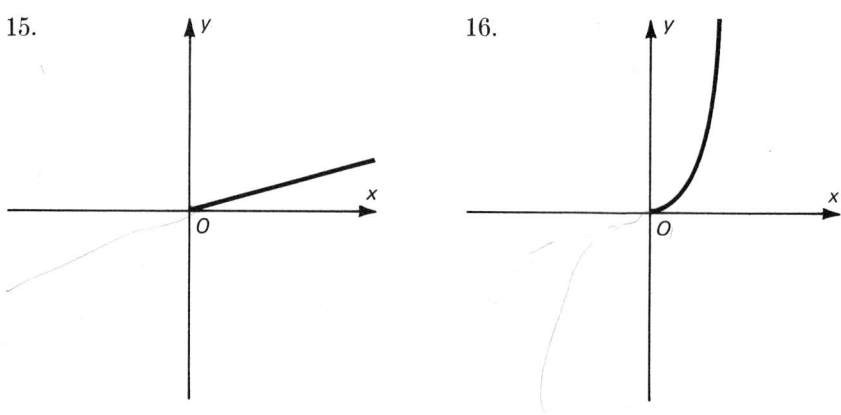

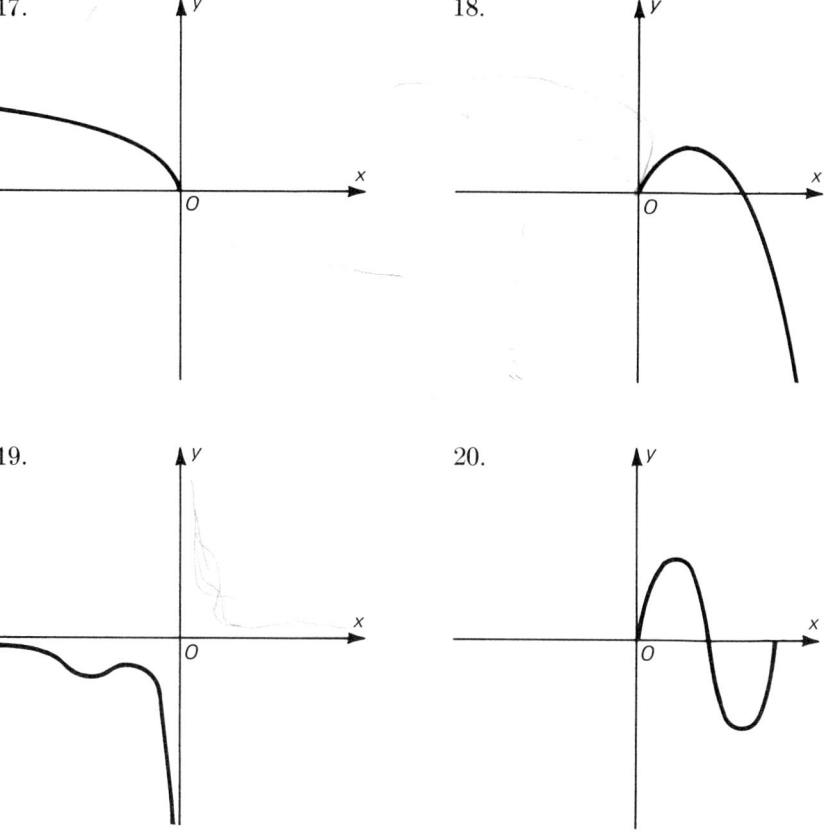

21. Prove that any constant function $f:f(x) = c$ for $x \in R$ is an even function.
22. Prove that any linear function $f:f(x) = mx$ for $x \in R$ is an odd function.
23. Prove that the absolute value function $f:f(x) = |x|$ for $x \in R$ is an even function.
24. Prove Theorem 2.15b.

2.16 The Graph of g^{-1} Is a Reflection of the Graph of g

We will now investigate the relationship between the graphs of a function g and its inverse function g^{-1}. To help determine this relationship, it will be helpful to consider first the identity function I and its inverse I^{-1}. The identity function is defined by

$$I:I(x) = x, \quad \text{for all } x \in R.$$

In ordered-pair notation,

$$I = \{(x, x) \mid x \in R\}.$$

We see immediately that $I = I^{-1}$; that is, I is its own inverse. The graph of I is the line passing through the origin that bisects the angles formed by the axes in the first and third quadrants of the Cartesian plane. (See Figure 2-51.)

Figure 2-51

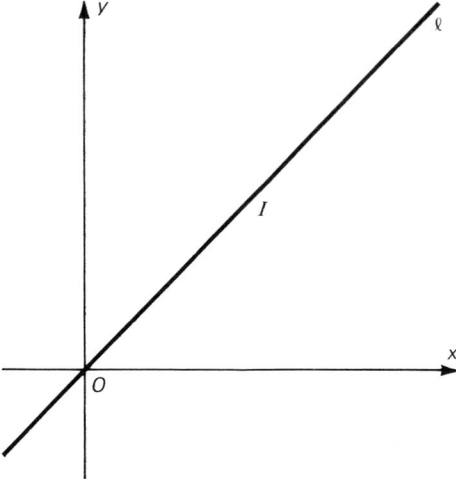

Now line l, which is the graph of I (also, of course, the graph of I^{-1}), has a very interesting property. This property is that the reflection of any point (a, b) in l is the point (b, a). This is shown by the following argument.

From our earlier work, we know that point (a, b) is the intersection of the vertical line defined by $x = a$ and the horizontal line $y = b$. Similarly, (b, a) is the intersection of the vertical line $x = b$ and the horizontal line $y = a$, In Figure 2-52, we have chosen an arbitrary point (a, b) in the first quadrant. Remember that line l is the graph of I.

Figure 2-52

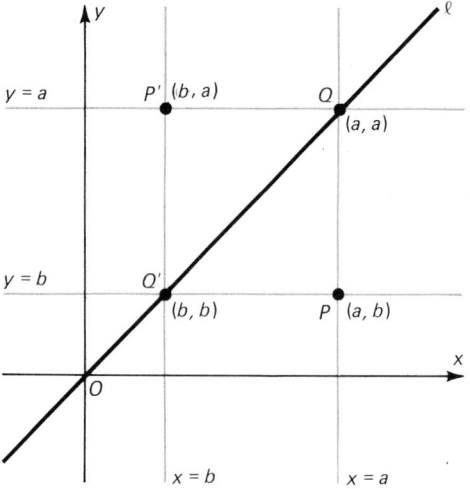

Notice that the horizontal line $y = a$ and the vertical line $x = a$ intersect at (a, a), which lies in l. Similarly, the lines $x = b$ and $y = b$ intersect at (b, b) in l. It is clear that $PQP'Q'$ is a rectangle. Why? Use the distance formula to show that the length of $\overline{PQ}$ is equal to the length of $\overline{P'Q}$. Therefore, $PQP'Q'$ is not only a rectangle, but it is a square. Why?

Figure 2-53

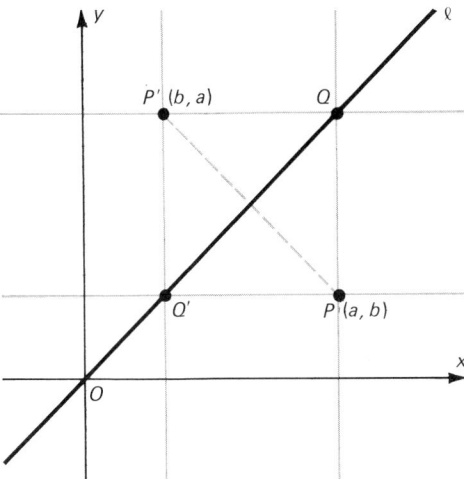

Next, we draw the diagonal $\overline{P'P}$, as shown in Figure 2–53. From a theorem in geometry concerning the diagonals of a square, we know that $\overline{P'P}$ is perpendicular to $\overline{Q'Q}$ and is bisected by $\overline{Q'Q}$. Can you state the theorem? This means that the line l is a perpendicular bisector of $\overline{P'P}$ and, therefore (see page 96), that $P'(b, a)$ is the reflection in l of $P(a, b)$. Hence, the reflection of any point (a, b) in line l, which is the graph of the identity function I, is the point (b, a).

This result leads to a very interesting observation. Since, if a function f has an inverse f^{-1}, then

$$f^{-1} = \{(b, a) \mid (a, b) \in f\},$$

we see that every point in the graph of f^{-1} is a reflection in the line l of a point in the graph of f, and conversely. Thus, the graph of f^{-1} is a reflection in the graph of I of the graph of f. Figures 2–54(a) and (b) on page 108 illustrate this property of graphs and their inverses.

There is another interesting way to see the connection between the graphs of f and f^{-1}. Suppose that we were to draw the graph of f on a pane of glass, as in Figure 2–55. If we then rotate the pane of glass through an angle of 180° about the line l, as in (a) the result will be as in (b). The horizontal axis of (a) has become the vertical axis of (b), the vertical axis of (a) has become the horizontal axis of (b), and the graph of f has become the graph of f^{-1}.

Figure 2-54

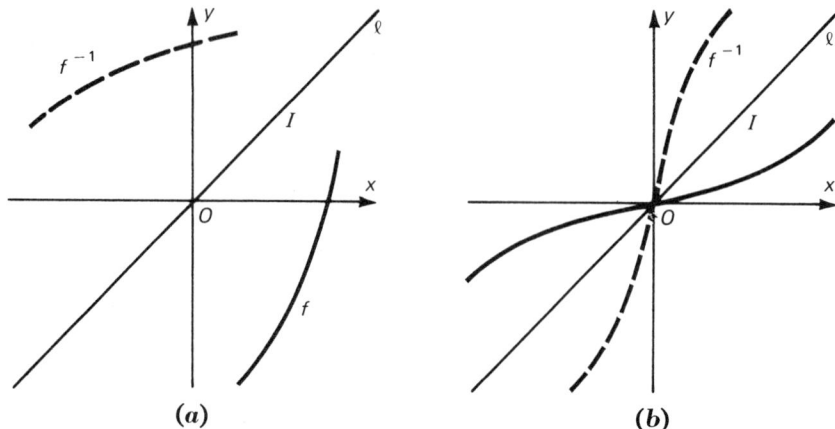

(a)   (b)

Figure 2-55

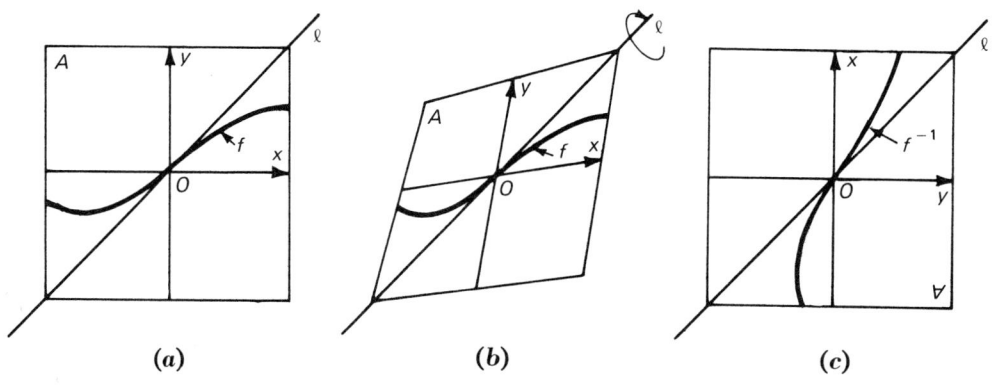

(a)   (b)   (c)

2.16 EXERCISES

Each of Exercises 1–8 shows the graph of a function f. If the function has an inverse, copy the graph of the function and sketch the graph of its inverse on the same set of axes. If the function does not have an inverse, indicate a way of forming a restriction f so that the restriction has an inverse.

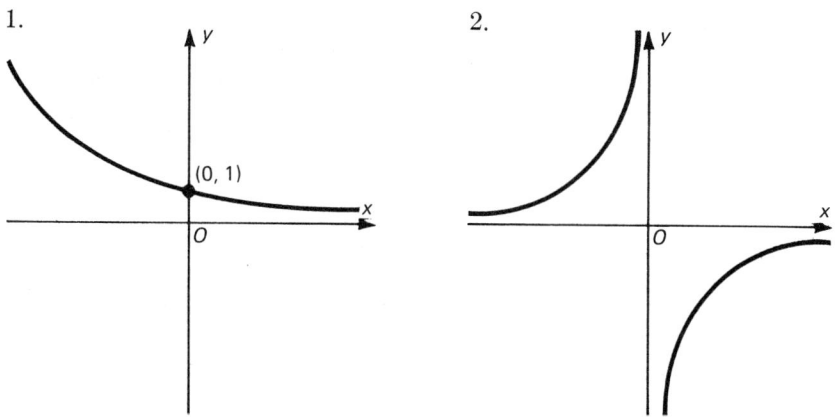

3.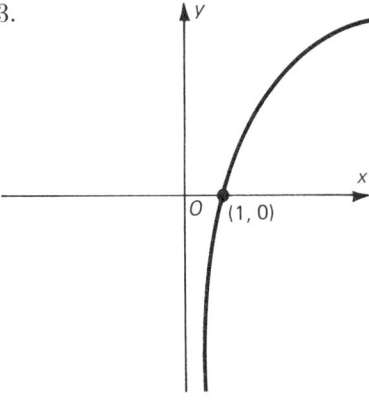
4.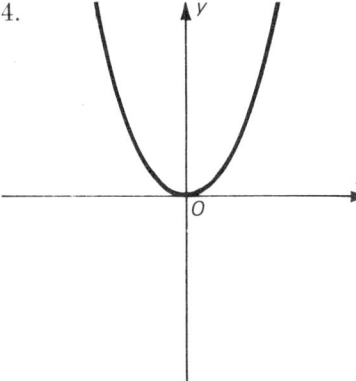
5.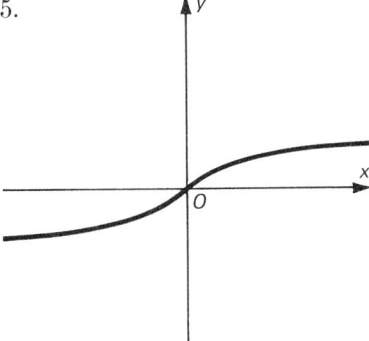
6.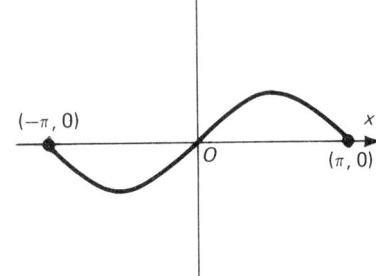
7.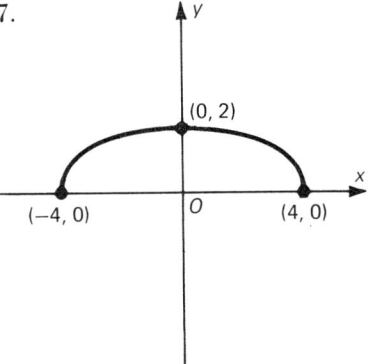
8.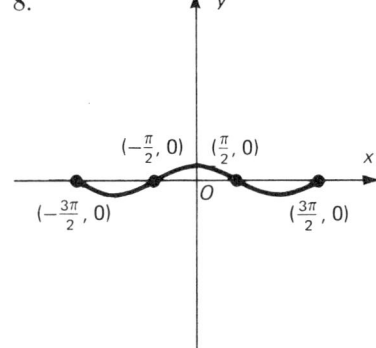

9. If the domain and range of f are each R and $f = f^{-1}$, what relationship exists between the graph of f and the line determined by $y = x$?

10. a) Prove that the inverse of a strictly increasing function is a strictly increasing function.
    b) Prove that the inverse of a strictly decreasing function is a strictly decreasing function.

2.16 REFLECTION OF THE GRAPH OF g 109

3

Quadratic Functions

Introduction

As you will see in the next chapter, constant and linear functions and the functions to be studied in this chapter, quadratic functions, are all included in a broad class of functions called polynomial functions. Like constant and linear functions, quadratic functions are particularly important because of their applicability to certain kinds of practical and scientific problems. In this chapter besides studying quadratic functions, we will also study certain operations on functions which are extremely useful in studying many classes of functions.

3.1 The Quadratic Function $q(x) = x^2$

In this section, we will begin our study of quadratic functions by discussing in detail the characteristics of a specific quadratic function and its graph. However, we will first give a general definition of a quadratic function.

Definition 3.1a If a, b, and c are real numbers, with $a \neq 0$, then the function Q defined by

$$Q(x) = ax^2 + bx + c, \quad x \in R,$$

is a **quadratic function.**

We will now study at length the special quadratic function q that is defined by
$$q(x) = x^2, \quad x \in R.$$
This is, of course, the quadratic function in which $a = 1$ and $b = c = 0$. The graph of the function q has the following properties:

1. No part of the graph of q lies below the x-axis, because $q(x)$ is never negative.
2. The graph of q is symmetric with respect to the y-axis, since q is an even function, that is, since $q(-x) = (-x)^2 = x^2 = q(x)$.
3. The graph of q has the origin as its common x- and y-intercept, and there are no other intercepts.

We will now discuss some of the less obvious properties of the graph of q. On pages 54-55, we considered the functions f and g defined by $f(x) = x^2$, $x \geq 0$, and $g(x) = x^2$, $x \leq 0$. We now see that f and g are restrictions of q and that the graph of q is the union of the graphs of f and g. In our earlier work, we showed that f is an increasing function and that g is a decreasing function. As a result, we can say that q is increasing for $x \geq 0$ and that q is decreasing for $x \leq 0$.

Since the graph of q is symmetric with respect to the y-axis, we need only determine what the graph of either f or g looks like to know what the graph of q will be like. The graph of q is the graph of either f or g, together with its reflection in the y-axis.

On page 54, we pointed out that the graph of f does *not* look like the one in Figure 3-1. The reason is simple—the graph shown is obviously not the graph of an increasing function. But why could not this portion of the graph of q look like one of the graphs of the increasing functions in Figure 3-2?

Figure 3-1

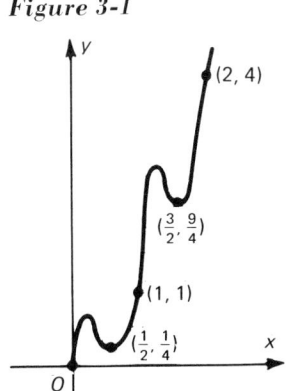

Figure 3-2

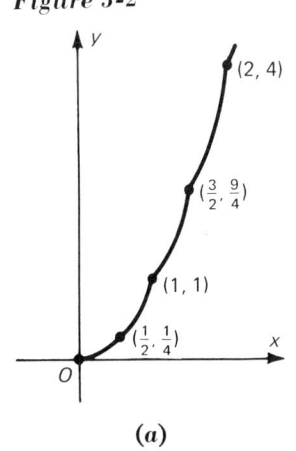

(a)

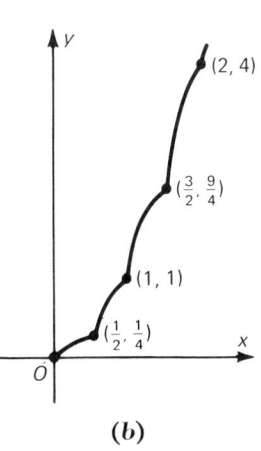

(b)

We will show that the graph cannot have either of these appearances, since it has the property of being *concave upward*. A graph is said to be *concave upward* if, given arbitrary points P and Q on the graph, each point of the graph between the vertical lines through P and Q is below the point on line segment $\overline{PQ}$ with the same abscissa. This idea is illustrated in Figure 3-3.

Figure 3-3

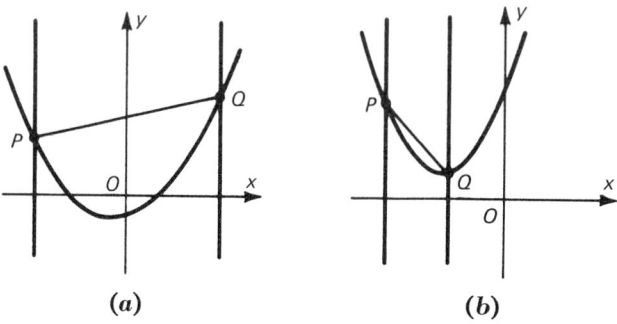

(a)   (b)

The illustrations in Figure 3-4 are examples of graphs that are *not* concave upward because it is possible to locate points P and Q in each graph so that some points of the graph between the vertical lines through P and Q are above the line segment $\overline{PQ}$.

Figure 3-4

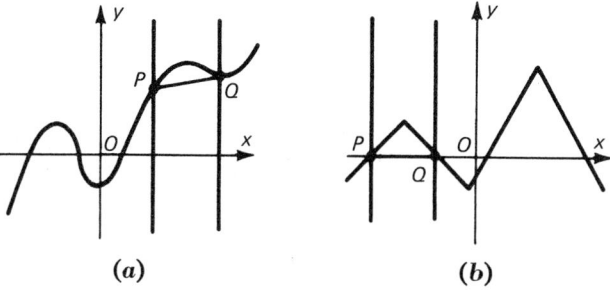

(a)   (b)

Similarly, we say that the graph of a function is *concave downward* if, given any two points P and Q on the graph, each point of the graph between the vertical lines through P and Q is above the point on line segment $\overline{PQ}$ with the same abscissa. The illustrations in Figure 3-5 are examples of graphs that are concave downward.

Figure 3-5

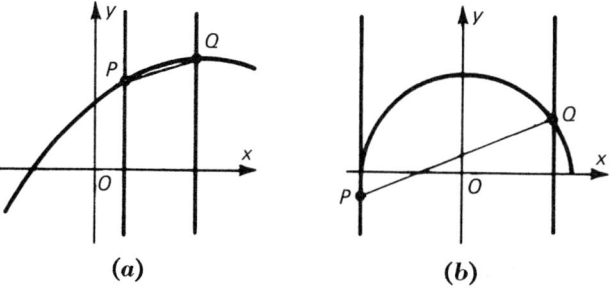

(a)   (b)

The illustrations in Figure 3–4 of graphs that are not concave upward are also illustrations of graphs that are not concave downward. You should verify this by locating points P and Q so that not every point in the portion of the graph between the vertical lines through P and Q is above segment $\overline{PQ}$.

We will sometimes speak of a *portion* of a graph as being concave upward or concave downward. For example, the graph in Figure 3–6 is concave upward for $x \leq a$ and concave downward for $x \geq a$.

Figure 3-6

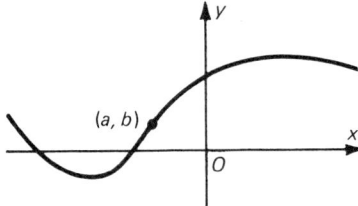

We will now prove that the graph of the function q defined by $q(x) = x^2$ is concave upward. Let (c, c^2) and (d, d^2), with $c < d$, be any two points on the graph of q. We know that the linear equation defining the line through these two points is of the form

$$k(x) = mx + b.$$

Since (c, c^2) and (d, d^2) must satisfy this equation, we have $c^2 = cm + b$ and $d^2 = dm + b$. Solving these equations for m and b, we obtain $m = c + d$, and $b = -cd$. Hence, the equation of the line through these two points is $k(x) = (c + d)x - cd$.

Now $q(x) - k(x)$ is difference of the ordinates of the points on the graphs of q and k that have x for their abscissa. (See Figure 3–7.) If $q(x) - k(x)$ is positive, the point in the graph of q is above the point in the graph of k. If $q(x) - k(x)$ is negative, the point in the graph of q is

Figure 3-7

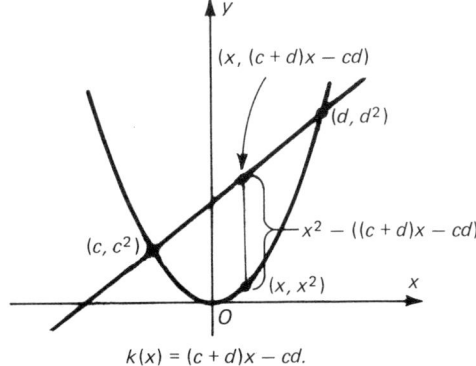

$k(x) = (c + d)x - cd.$

below the point in the graph of k. Of course, if $q(x) - k(x) = 0$, the two graphs intersect. Since $q(x) = x^2$ and $k(x) = (c + d)x - cd$, we obtain

$$q(x) - k(x) = x^2 - ((c + d)x - cd) = x^2 - (c + d)x + cd$$
$$= (x - c)(x - d).$$

Suppose that x is the abscissa of any point in the graph of q that is between (c, c^2) and (d, d^2). Then $c < x < d$. That is, $x - c > 0$ and $x - d < 0$, and $(x - c)(x - d)$ is negative. Hence, $q(x) - k(x)$ is negative. But this means that, if $c < x < d$, the points in the graph of q are below the corresponding points in the graph of k. That is, when $c < x < d$, the graph of q is below the line segment joining (c, c^2) and (d, d^2). But we chose the points (c, c^2) and (d, d^2) in the graph of q quite arbitrarily. We, therefore, conclude that, for *any* two points in the graph of q, the portion of the graph between these two points is below the line segment joining the two points. Hence, the graph of q is concave upward.

It should now be an easy matter for you to see that the graph of q could not look like either of the graphs in Figure 3–2. For the first graph, if you choose the points $(\frac{5}{4}, \frac{25}{16})$ and $(\frac{7}{4}, \frac{49}{16})$, for example, you can see that the point $(\frac{3}{2}, \frac{9}{4})$ in this graph is above the segment joining these two points. Give a specific example to explain why the second graph is not concave upward.

By computing a few values of $q(x)$ and using the knowledge that we have gained about the graph of q, we can draw a very reasonable approximation of a portion of this graph, as shown in Figure 3–8.

Figure 3-8

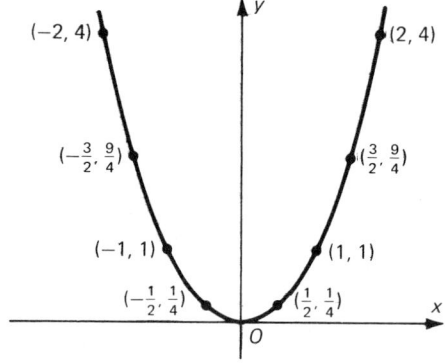

In the following discussion, we will observe several properties of the graph of q. These properties will lead us to see that the graph of the function q belongs to a special class of curves called *parabolas*.

As you know, every point on the graph of q has coordinates of the form (x, x^2). If $(0, a)$ is a fixed point on the y-axis, as shown in Figure 3–9,

Figure 3-9

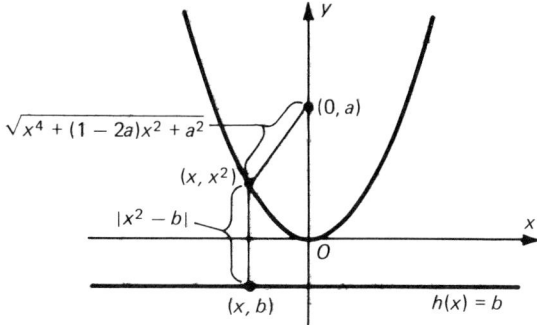

then, by the distance formula, the distance d_1 from (x, x^2) to $(0, a)$ is given by

$$d_1 = \sqrt{(x-0)^2 + (x^2 - a)^2}$$
$$= \sqrt{x^4 + (1-2a)x^2 + a^2}.$$

The distance d_2 from a point (x, x^2) on the graph of q to the horizontal line determined by $h(x) = b$, also shown in Figure 3-9, is given by $d_2 = |x^2 - b|$. But let us rewrite this so that we can decide when $d_1 = d_2$. From our knowledge of absolute values, we know that d_2 can be written as follows:

$$d_2 = \sqrt{(x^2 - b)^2}$$
$$= \sqrt{x^4 - 2bx^2 + b^2}.$$

Now we can determine when these two distances, d_1 and d_2, will be equal. Since $d_1 = \sqrt{x^4 + (1-2a)x^2 + a^2}$ and $d_2 = \sqrt{x^4 - 2bx^2 + b^2}$, d_1 and d_2 will be equal if we choose a and b such that

$$1 - 2a = -2b$$
and
$$a^2 = b^2.$$

The requirement that $a^2 = b^2$ means, of course, that either $a = b$ or $a = -b$. If $a = b$, then the requirement that $1 - 2a = -2b$ is clearly not satisfied. If $a = -b$, then the requirement that $1 - 2a = -2b$ becomes $1 - 2(-b) = -2b$, or $1 + 2b = -2b$. This is equivalent to $4b = -1$, or $b = -\frac{1}{4}$. Thus, if we choose $b = -\frac{1}{4}$ and $a = -b = \frac{1}{4}$, then

$$d_1 = d_2 = \sqrt{x^4 + \tfrac{1}{2}x^2 + (\tfrac{1}{4})^2}.$$

Thus, when $d_1 = d_2$, d_1 is the distance from an arbitrary point (x, x^2) on the graph of q to the point $(0, \tfrac{1}{4})$, and d_2 is the distance from (x, x^2) to the horizontal line $h(x) = -\tfrac{1}{4}$. This means that *any* point (x, x^2) in the graph of q is *equidistant* from the point $(0, \tfrac{1}{4})$ and the horizontal line

$h(x) = -\frac{1}{4}$. Notice that this does *not* mean that *all* points in the graph of q are the same distance from $(0, \frac{1}{4})$ and the line $h(x) = -\frac{1}{4}$; rather, it means that *each* point is the same distance from the point $(0, \frac{1}{4})$ as it is from the line $h(x) = -\frac{1}{4}$. Conversely, every point (x_1, y_1) that is equidistant from the point $(0, \frac{1}{4})$ and the line $h(x) = -\frac{1}{4}$ is in the graph of q. You can verify this by showing that, if $\sqrt{x_1^2 + (y_1 - \frac{1}{4})^2} = |y_1 - (-\frac{1}{4})|$, then $y_1 = x_1^2$. (See Exercise 16, page 117.)

Figure 3-10

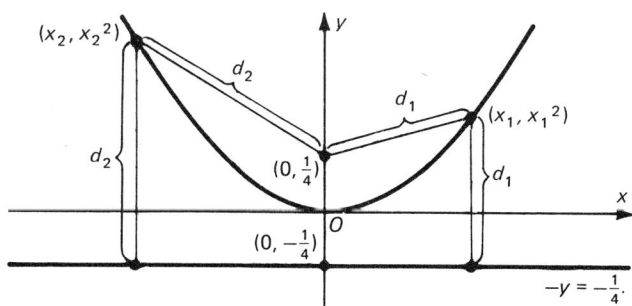

Because the graph of q has these properties, it is a parabola. The parabola, which is a classical curve in geometry with a rich and interesting history, is historically defined as the locus (path) of a point moving in the plane in such a manner that its distance from a fixed point P in the plane is always equal to its distance from a fixed line l in the plane, where point P is not contained in l. The point P is called the *focus* of the parabola and the line l is called the *directrix* of the parabola. We see from this definition that the graph of q is a parabola having the point $(0, \frac{1}{4})$ for its focus and the line defined by $h(x) = -\frac{1}{4}$ for its directrix.

Viewing the graph of q in the kinetic terms of a moving point tracing out a parabola tends to make us think of it as a smooth, unbroken, continuous curve. It can be verified that q is a continuous function by appealing to our earlier discussion of continuity. The identity function, defined by $I(x) = x$, $x \in R$, is continuous. This fact, together with the fact that the product of two continuous functions is a continuous function, tells us that $I \cdot I = I^2 = q$ is a continuous function. This justifies our intuitive feeling that q is a continuous function, having a graph with no breaks or jumps.

3.1 EXERCISES

1. a) Express the distance between the point (x_0, y_0) and the point $(0, 4)$.
   b) Express the distance between the point (x_0, y_0) and the line $y = -4$.

c) Use the definition given on page 116 to write an equation for the parabola whose focus is (0, 4) and whose directrix is $y = -4$.
d) Simplify the equation you wrote for part c by squaring both sides and combining like terms.
e) Sketch the parabola, its focus, and its directrix.

In each of the exercises below, follow the pattern of Exercise 1 to write an equation for the parabola with the given focus and directrix. Then sketch the curve.

2. Focus: (0, 2); directrix: $y = -2$.
3. Focus: (3, 0); directrix: $x = -3$.
4. Focus: (0, −1); directrix: $y = 1$.
5. Focus: (−2, 0); directrix: $x = 2$.
6. Focus: (2, $\frac{1}{2}$); directrix: $y = -\frac{1}{2}$.
7. a) Prove that the equation for the parabola with focus (0, p) and directrix $y = -p$ is $x^2 = 4py$.
   b) Is this parabola the graph of a function?
   c) What are its domain and its range if $p > 0$? If $p < 0$?
8. a) Prove that the equation for the parabola with focus (p, 0) and directrix $x = -p$ is $y^2 = 4px$.
   b) Is this parabola the graph of a function? Is it the graph of the union of two functions, $g \cup -g$?
   c) What are the domains and ranges of g and $-g$ if $p > 0$? If $p < 0$?

Use the results of Exercises 7 and 8 to help you find the coordinates of the focus and the equation for the directrix for each of the following parabolas. Then sketch the parabola.

9. $\{(x, y) \mid x^2 = 20y\}$
10. $\{(x, y) \mid x^2 = -20y\}$
11. $\{(x, y) \mid y^2 = 2x\}$
12. $\{(x, y) \mid y^2 = -16x\}$
13. $\{(x, y) \mid x^2 = \frac{1}{2}y\}$
14. $\{(x, y) \mid y^2 = -\frac{4}{3}x\}$
15. An isosceles right triangle is inscribed in the parabola which is the graph of $q:q(x) = x^2$, with the vertex of the right angle at (0, 0).
    a) What are the coordinates of the other two vertices?
    b) Determine the lengths of the sides of the triangle.
    c) What is the area of the triangle?
16. Prove that every point (x_1, y_1) that is equidistant from the point (0, $\frac{1}{4}$) and the line $h(x) = -\frac{1}{4}$ is on the graph of $q:q(x) = x^2$. That is, prove that if $\sqrt{x_1^2 + (y_1 - \frac{1}{4})^2} = |y_1 - (-\frac{1}{4})|$, then $y_1 = x_1^2$.
17. Prove that the distance of a point (x, x^2) on the graph of $q:q(x) = x^2$ from either the focus or directrix of the graph is $x^2 + \frac{1}{4}$.

3.2 Multiplication of a Function by a Real Number

In the preceding section we studied, in considerable detail, the quadratic function q, defined by $q(x) = x^2$. We would like to use the information that we have gathered about q to give us similar information about the general quadratic function Q, defined by $Q(x) = ax^2 + bx + c$. Before we can do this we need to investigate the relationship between the graphs of sums and products of functions and the graphs of the functions themselves. For example, we would like to know the relationship of the graph of q to that of aq, where a is a constant. Also, we might like to consider the relationship of the graph of q to the graph of $aq + b$, where a and b are constants. The relationship of the graph of $Q(x) = ax^2 + bx + c$ to the graphs of $aq(x) = ax^2$ and the linear function $l(x) = bx + c$ could also be revealing. For this reason we shall temporarily divert our attention from the general quadratic to consider some relationships between the graphs of functions and the graph of a function obtained from them by an algebraic operation. These relationships will also be useful in our study of other functions.

The simplest example of multiplication of functions is when one of the functions is a constant function. In fact, we consider this as a special case of multiplication of functions and often refer to it as the multiplication of a function by a real number. If the constant function is $g(x) = a$, where $a \in R$, and f is the other function, we denote their product by af. For example, if $a = 2$, the product af is denoted by $2f$, or if $a = -\frac{1}{2}$, af is written $-\frac{1}{2}f$. If the function f has a domain D, and $a \in R$, then

$$af = \{(x, af(x)) \mid x \in D\}$$

From this characterization it follows that af is a function and that it has the same domain as f. If we consider f and af as mappings that map the subset D of the real-number line into the real-number line, we get the following interpretation. If f maps x onto y, then af maps x onto ay. For example, if f maps x onto 1, then $3f$ maps x onto 3 and $-3f$ maps x onto -3 as shown in Figure 3-11.

Figure 3-11

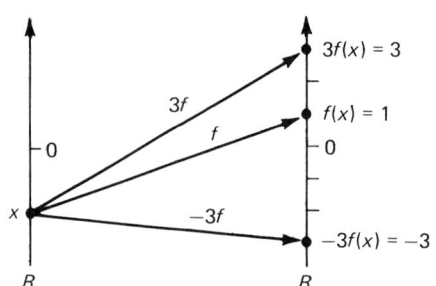

If the range of a function f is the closed interval from 1 to 2, then the range of $2f$ is the closed interval from 2 to 4. If g is a function whose range is the open interval from -1 to π, then $\sqrt{3}g$ has for its range the open interval from $-\sqrt{3}$ to $\sqrt{3}\pi$, and $-\sqrt{3}g$ has for its range the open interval from $-\sqrt{3}\pi$ to $\sqrt{3}$. In general, if a is a positive real number different from 1, the range of af is the range of f "stretched away from," or "compressed towards," the origin so that the distance of every point in the range from the origin is multiplied by a. The range is "stretched" if $a > 1$; the range is "compressed" if $a < 1$. If a is negative and different from -1, the range of af is the range of f reflected in the origin and then "stretched away from," or "compressed towards," the origin so that the distance of every point in the range from the origin is multiplied by $|a|$. In this case, the range is "stretched" if $|a| > 1$, and the range is "compressed" if $|a| < 1$. Figure 3–12 illustrates these results for a specific example. Of course, if a is 1, the range is unchanged; if a is 0, the range consists of the single point 0; and if a is -1, the range is simply reflected in the origin.

Figure 3-12

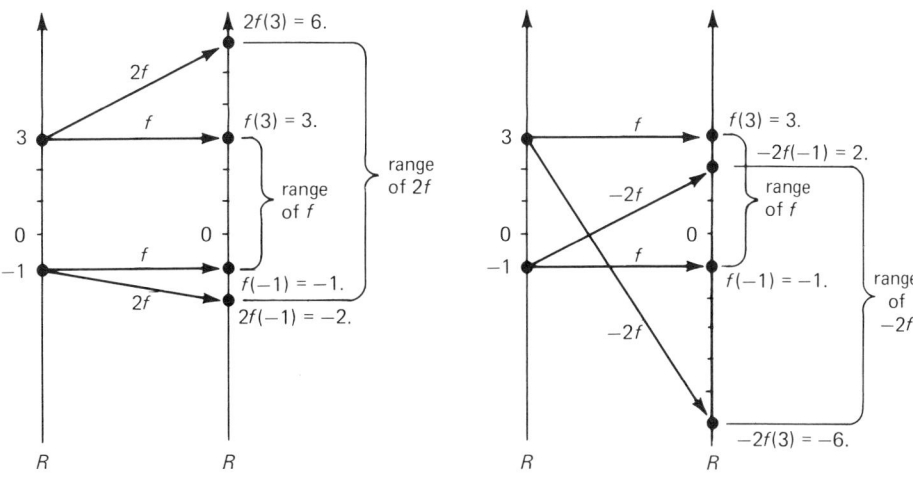

$f: f(x) = x, -1 \leqslant x \leqslant 3.$

What does the graph in the Cartesian plane of this new function af look like? As you might expect, changes in the graph of the function correspond to the changes just described for the range of the function. To illustrate graphical changes, we will consider some special values for a.

Example 1 If $a = 2$, then, to every point $(x, f(x))$ belonging to f, there corresponds the point $(x, 2f(x))$ of $2f$. The graph of $2f$ can be thought of in the following way: If you take the graph of f and stretch it away from the x-axis, keeping the x-axis fixed, so that every ordinate is doubled, then you have the graph of $2f$, as shown in Figure 3–13,

Figure 3-13

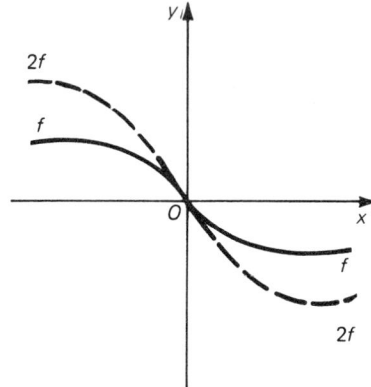

Example 2 If $a = \frac{1}{2}$, the graph of $\frac{1}{2}f$ is the graph of f compressed towards the x-axis so that every ordinate is $\frac{1}{2}$ as large as the ordinate of the corresponding point in the graph of f. (See Figure 3–14.)

Figure 3-14

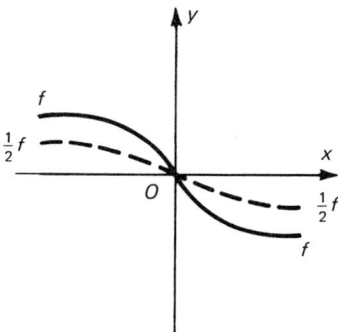

In summary, these examples show that, if a is positive, and different from 1, the graph of af is the graph of f stretched or compressed about the x-axis, depending upon whether a is greater than or less than 1. In the following examples, we will see what happens to the graph of f if a is negative.

Example 3 If $a = -1$, then $(-1)f = \{(x, -y) \mid (x, y) \in f\}$. From our studies of symmetry, we see that the graph of $(-1)f$ is just the reflection of the graph of f in the x-axis, as illustrated in Figure 3-15. We will denote the function $(-1)f$ by the symbol $-f$.

Figure 3-15

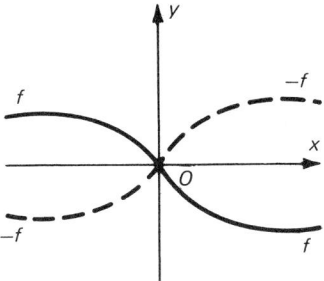

Now suppose that $a < 0$ and $a \neq -1$. In this case, $a = -|a|$; therefore, the function $af = -|a|f$. Thus, if a is negative and $a \neq -1$, we can think of the graph of af as being obtained from that of f by two consecutive operations. The first operation is the stretching or compressing of the graph of f about the x-axis, according to whether $|a|$ is greater than 1 or less than 1. In this stretching or compressing, every point $(x, f(x))$ in the graph of f is shifted to the point $(x, |a|f(x))$, and the graph of f is converted into that of $|a|f$. The second operation is the reflection in the x-axis of the graph of $|a|f$ to obtain the graph of $-|a|f = af$.

Example 4 Figure 3-16 shows what happens to the graph of a function f when it is multiplied by -2. Notice that the graph of f is first stretched when f is multiplied by $|-2|$. This graph, indicated as a broken curve, is then reflected in the x-axis to obtain the graph of $-2f$.

Figure 3-16

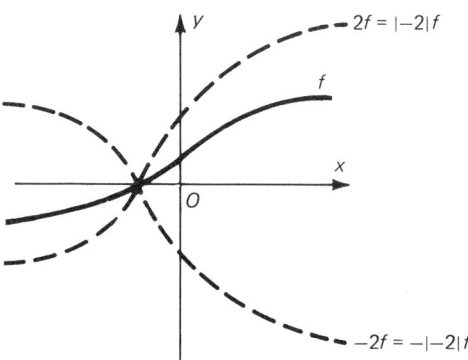

3.2 EXERCISES

In each of Exercises 1–4, copy the following graph of the given function f. Then, on the same set of axes, sketch the graph of the indicated multiple of the function.

1.

2.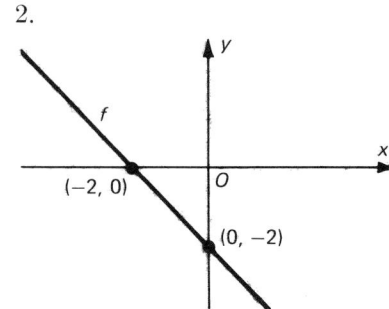

3.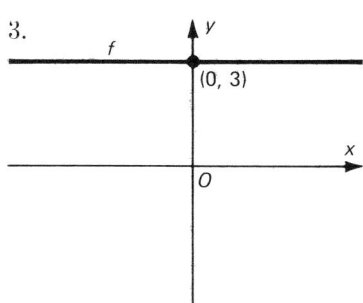

4.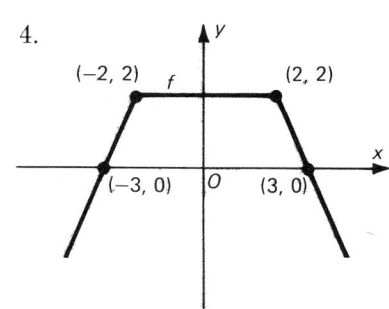

1. Sketch $-f$.
2. Sketch $-3f$.
3. Sketch $\tfrac{2}{3}f$.
4. Sketch $\tfrac{1}{3}f$.

In each of Exercises 5–10, the graphs of two functions, f and af are shown. For each exercise, use the graphs to determine which of the following conditions is satisfied by the real number a.

(a) $a > 1$. (b) $0 < a < 1$. (c) $-1 < a < 0$. (d) $a < -1$.

5.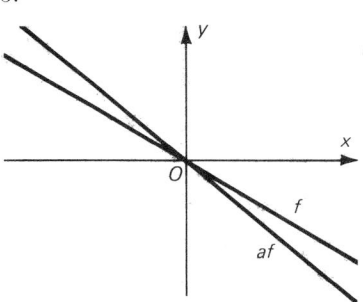

6.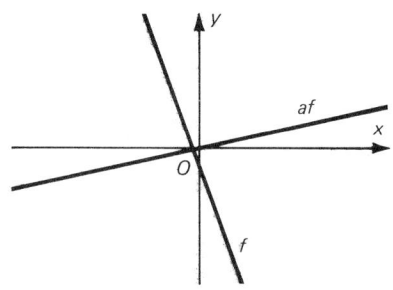

122 QUADRATIC FUNCTIONS

7.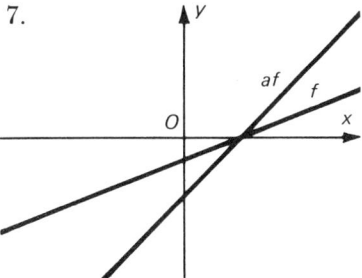
8.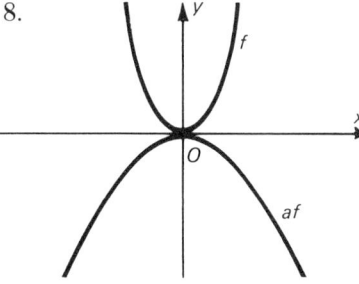
9.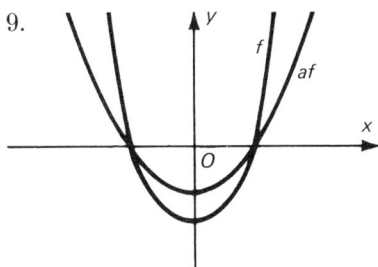
10.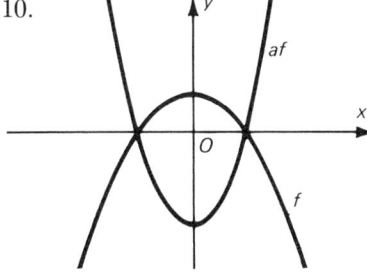

11. Prove that, if $a \neq 1$ and the graphs of f and af intersect at a point on the y-axis, then $f(0) = 0$.
12. Prove the following: If $a \neq 1$ and the graphs of f and af intersect, then the point of intersection is on the x-axis.

Draw the graph of each of the following functions.

13. $f : f(x) = 2|x|$.
14. $g : g(x) = -|x|$.
15. $h : h(x) = \frac{1}{3}|x|$.
16. $F : F(x) = -3|x|$.
17. $G : G(x) = |-\frac{1}{2}x|$.
18. $H : H(x) = -2|-3x|$.

19. What is the relationship between the graphs of the functions f and af if $a = 1$? If $a = 0$?

3.3 More about Quadratic Functions

We now turn to the function aq where $a \in R$ and q is the quadratic function defined by $q(x) = x^2$, $x \in R$. We let aq be defined by $q_a(x) = ax^2$.

From our work on multiplication of functions by real numbers (pages 118–121), we know that the graph of $q_{-1} = (-1)q$ is just the reflection in the x-axis of the graph of $q = q_1$. The graph of q_{-1} is shown in Figure 3–17.

Figure 3-17

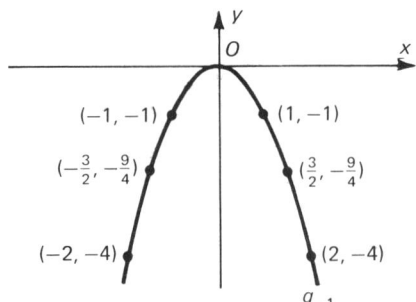

In general, the graph of q_{-a} is the reflection in the x-axis of the graph of q_a.

We know that the graph of $q = q_1$ consists of the set of all points having coordinates (x, x^2), where $x \in R$. The graph of $q_a = a \cdot q$ consists of the set of all points having coordinates (x, ax^2), where $x \in R$. Thus, if a is positive, the graph of q_a is the graph of q stretched vertically or compressed vertically by a factor of a about the fixed x-axis, depending upon whether a is greater than or less than 1. (See Figure 3–18.)

From our earlier comments about the graph of q_{-a}, we see that the graphs of q_{-2} and $q_{-\frac{1}{2}}$ have the appearance of those in Figure 3–19.

Figure 3-18

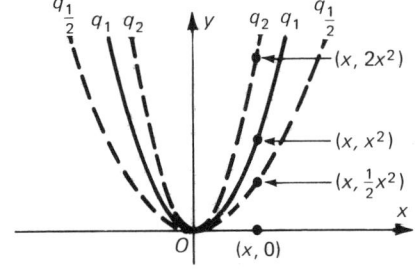

Figure 3-19

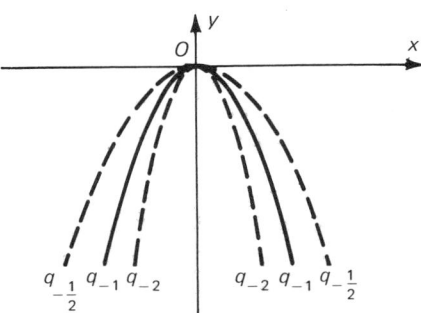

The graph of q_0, $q_0(x) = 0 \cdot x^2 = 0$, is the x-axis. (Explain why q_0 is not a quadratic function. What kind of function is it?) Thus, we see that, if we regard a as a parameter taking on real number values, then the equation

$$q_a(x) = ax^2, \quad x \in R,$$

defines a family of functions. The graphs of these functions form a one-parameter family of curves which pass through the origin and are symmetric with respect to the y-axis. If a is positive, the graph of q_a is concave upward; and if a is negative, the graph of q_a is concave downward. We can use some terminology, which although not very precise, is very descriptive. If the absolute value of a is less than 1, the graph of q_a is "fat"; and if the absolute value of a is greater than 1, the graph of q_a is "thin."

You will recall that on page 115 we established that the graph of $q = q_1$ is a parabola with the point $(0, \frac{1}{4})$ for its focus and the line defined by $h(x) = -\frac{1}{4}$ for its directrix. Now the multiplication of the function q by the nonzero real number a, where $|a| \neq 1$, produces a function whose graph has a different shape from that of q, since such a multiplication involves a vertical stretching or compression. Hence, one could legitimately ask, "Is the graph of q_a also a parabola?"

If the graph of q_a is a parabola, from what we already know about the symmetry of the graph of q_a, it would seem reasonable that the focus of the parabola would be on the y-axis and that the directrix would be perpendicular to the y-axis. If the graph of q_a is indeed a parabola, then the distance from the focus to the origin is equal to the distance from the origin to the directrix. Let us therefore conjecture that the graph of q_a is a parabola with focus at the point $(0, k)$ and directrix defined by $h(x) = -k$.

If our conjecture is correct, we must be able to find a real number k such that an arbitrary point (x_0, ax_0^2) on the graph of q_a is equidistant from $(0, k)$ and $h(x) = -k$. The distance from (x_0, ax_0^2) to $(0, k)$ is

$$\sqrt{x_0^2 + (ax_0^2 - k)^2},$$

and the distance from (x_0, ax_0^2) to the line $h(x) = -k$ is $|ax_0^2 - (-k)|$. If these distances are to be equal, we must find a real number k such that

$$\sqrt{x_0^2 + (ax_0^2 - k)^2} = |ax_0^2 - (-k)|.$$

See Figure 3-20.

Figure 3-20

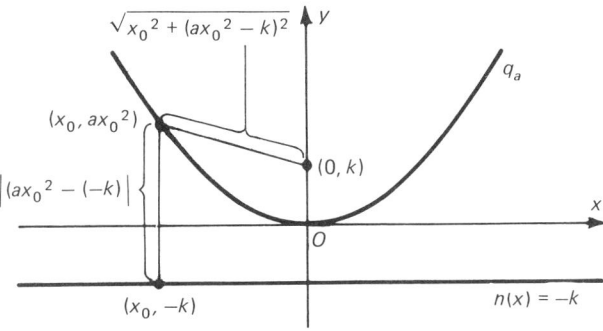

You should be able to show algebraically that this equation simplifies to

$$x_0^2 = 4akx_0^2.$$

Therefore, if the two distances are to be equal, $4ak$ must be equal to 1, and the desired real number is $k = \dfrac{1}{4a}.$

This tells us that all the points in the graph of q_a are in the parabola which has the point $\left(0, \dfrac{1}{4a}\right)$ for its focus and the line defined by $h(x) = -\dfrac{1}{4a}$ for its directrix. Now we must show that all the points in this parabola are in the graph of q_a.

Let us consider an arbitrary point (x_0, y_0) of the parabola. By definition, (x_0, y_0) is equidistant from $\left(0, \dfrac{1}{4a}\right)$ and the line $h(x) = -\dfrac{1}{4a}$, and thus it follows that

$$\sqrt{x_0^2 + \left(y_0 - \dfrac{1}{4a}\right)^2} = \left| x_0 - \left(-\dfrac{1}{4a}\right) \right|.$$

Again, you should be able to show that this equation simplifies to

$$y_0 = ax_0^2,$$

which clearly shows that $(x_0, y_0) = (x_0, ax_0^2)$ is a point in the graph of q_a. Of course, since (x_0, y_0) was selected arbitrarily, this means that every point in the parabola is a point in the graph of q_a.

3.3 EXERCISES

Each of Exercises 1–4 illustrates a graph of $q:q(x) = x^2$ and a graph of $q_a:q_a(x) = ax^2$. For each exercise, indicate which of the following statements is true.

(a) $a > 1$.   (b) $0 < a < 1$.   (c) $-1 < a < 0$.
(d) $a = -1$.   (e) $a < -1$.

1.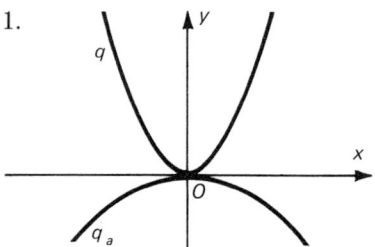

2.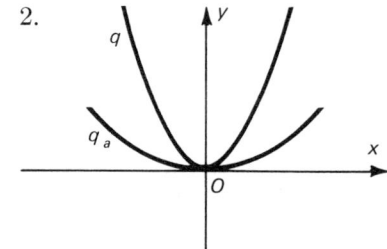

3.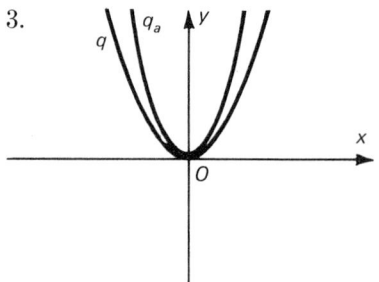

4.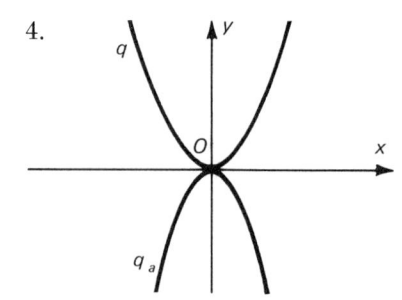

In each of Exercises 5–10, sketch the graph of $q:q(x) = x^2$. Then on the same set of axes sketch the graph of q_a.

5. $q_a:q_a(x) = 2x^2$.
6. $q_a:x \to \frac{1}{3}x^2$.
7. $q_a:y = -5x^2$.
8. $q_a:q_a(x) = -\frac{1}{5}x^2$.
9. $q_a:x \to -x^2$.
10. $q_a:y = 4x^2$.

The graphs of functions are useful tools in solving inequalities. For example, the graphs of $q_a:q_a(x) = 2x^2$ and $f:f(x) = 6 - x$ below clearly indicate that the values of x which satisfy the inequality $2x^2 < 6 - x$ consist of the following set: $\{x \mid -2 < x < \frac{3}{2}\}$.

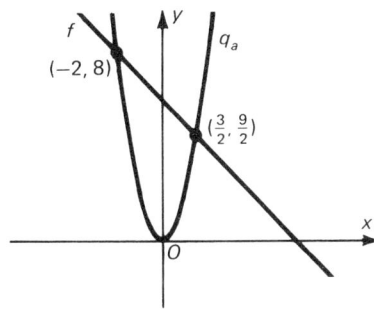

In each of Exercises 11–16, use the graphical technique indicated above to find the set of solutions for the inequality.

11. $\frac{1}{4}x^2 < 9$.
12. $27 - 9x^2 > 0$.
13. $x^2 < 3 - 2x$.
14. $x + 6 < x^2$.
15. $2x^2 \geq 3 - 5x$.
16. $x^2 \leq 6x - 8$.

17. An equilateral triangle is inscribed in the graph of $q_a:q_a(x) = \frac{1}{4}x^2$ with one vertex of the triangle at $(0, 0)$.
    a) Determine the coordinates of the other two vertices of the triangle.
    b) What is the length of each side of the triangle?
    c) Find the area of the triangle.
18. Repeat Exercise 7 for the equilateral triangle inscribed in the graph of $q_a:q_a(x) = ax^2$.

3.4 Addition of a Constant to a Function

One of the simplest types of addition of functions is when one of the functions is a constant function. If f is an arbitrary function, with domain D, and $C(x) = C$, $x \in R$, is a constant function, then from our definition of the sum of functions, we see that

$$f + C = \{(x, f(x) + C) \mid x \in D\}$$

The relationship of the graphs of f and $f + C$ should be quite clear. Since the coordinates of the graph of $f + C$ are the same as those of f except that the ordinate of $f + C$ is the ordinate of f plus the number C, we see that the graph of $f + C$ is just the graph of f raised or lowered $|C|$ units, according as C is positive or negative. Figure 3–21 illustrates this fact.

Figure 3-21

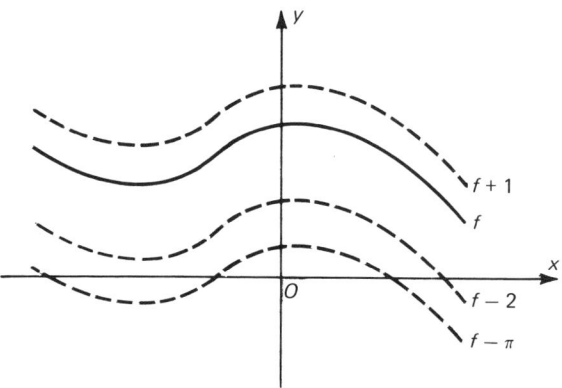

Thus, the graph of $Q(x) = ax^2 + C$, $C \in R$, is just the graph of $q_a(x) = ax^2$, which is raised or lowered vertically, according as C is positive or negative.

3.4 EXERCISES

Draw the graph of each of the following functions.

1. $f : f(x) = x^2 + 4$.
2. $g : g(x) = 2x^2 - 8$.
3. $h : h(x) = \frac{1}{2}x^2 + 1$.
4. $F : F(x) = 16 - x^2$.
5. $G : G(x) = -3x^2 + 6$.
6. $H : H(x) = -\frac{1}{4}x^2 - 4$.
7. $f : f(x) = |x| - 5$.
8. $g : g(x) = 2|x| + 3$.
9. $F : F(x) = 7 - |x|$.
10. $G : G(x) = -3|x| - 2$.

11. Show that if $ac > 0$, then the graph of $Q(x) = ax^2 + c$ does not intersect the x-axis.
12. Show that if $ac < 0$, then the graph of $Q(x) = ax^2 + c$ has two x-intercepts. Give the coordinates of these intercepts.

3.5 Composition of Functions

Before we discuss the most general type of quadratic function, we need to discuss the operation of composition of functions. It is different from the operations of addition and multiplication of functions in that it cannot be related to one of the familiar arithmetical operations.

Suppose that we are given two functions f and g that have for their domains the set of all real numbers, R. If we consider a particular number a, then $f(a)$ is a real number, because f is a real-valued function. Now, since $g(x)$ is also a real-valued function with domain R, $g(f(a))$ is a real number. If we think of f and g as mappings of the real-number line into itself, we could illustrate this process for a specific real number a, as shown in Figure 3–22.

Figure 3-22

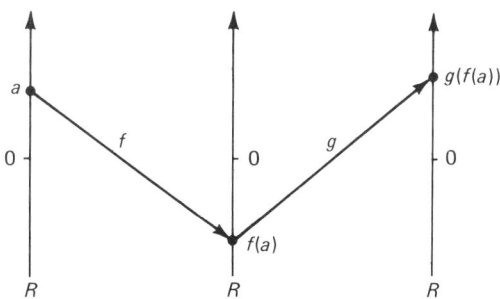

We have mapped a from the first real-number line onto $g(f(a))$ in the third number line, but in two steps. First, with the mapping f, we mapped a from the first line onto $f(a)$ in the second line. Then, with the mapping g, we mapped $f(a)$ from the second line onto $g(f(a))$ in the third line. Every point x in the first line is mapped onto the point $g(f(x))$ in the third line by this two-step procedure. That is,

$$x \to f(x) \to g(f(x)).$$

If we now think of x as going directly from the first line to $g(f(x))$ on the third line, then we have a mapping, say, h, of the first line into the third line defined by

$$h(x) = g(f(x)), \quad \text{for } x \in R.$$

This mapping h defines a function which makes a unique real number $g(f(x))$ correspond to each real number x.

Example 1 Consider the functions f and g, defined as follows:

$$f: f(x) = x + 1, \quad \text{for } x \in R,$$
$$g: g(x) = x^2, \quad \text{for } x \in R.$$

The number 1 in the first line is mapped by f onto $f(1) = 1 + 1 = 2$ in the second line. Then $f(1) = 2$ is mapped by g onto $g(f(1)) = g(2) = 2^2 = 4$ in the third line. This mapping and the mappings of a few other points under functions f and g are shown in Figure 3-23.

Figure 3-23

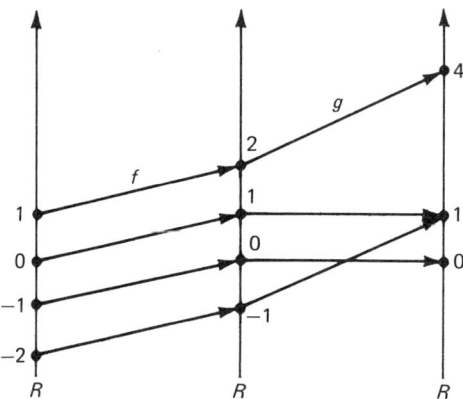

If we showed this two-step mapping as a single mapping from the first to the third line, the illustration would look like the one in Figure 3-24. This single mapping defines the function

$$h: h(x) = g(f(x)) = (x + 1)^2, \quad \text{for } x \in R.$$

We see that, given two functions f and g that are defined on R, we can combine them to produce a new function h, defined by $h(x) = g(f(x))$, for all $x \in R$. This method of producing a new function from two given functions is called the *composition of functions*, and the new function is called a *composite* of the given functions.

Figure 3-24

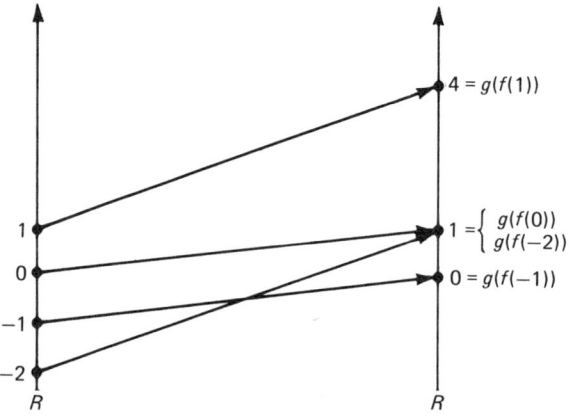

In introducing this concept, we considered two functions that have R for their domains. We did this to simplify our preliminary discussions; and we will want to define the composition of two functions whose domains may not be all of R. The essential point is that $g(f(x))$ be defined. For $g(f(x))$ to be defined, not only must x be in the domain of f, but $f(x)$ must be in the domain of g.

Definition 3.5a If f and g are functions, then the function defined by:
$$\{(x, g(f(x))) \mid g(f(x)) \text{ is defined}\},$$
is called a **composite** of f and g and is denoted by the symbol $g \circ f$.

By Definition 3.5a, $g \circ f$ is the function $g \circ f : (g \circ f)(x) = g(f(x))$, for x in the domain of f such that $f(x)$ is in the domain of g. We note that $f \circ g$ would be a composite of f and g defined by: $f \circ g : (f \circ g)(x) = f(g(x))$, for x in the domain of g such that $g(x)$ is in the domain of f.

To see that $g \circ f$ and $f \circ g$ are not, in general, the same function, we need only return to our first example and observe that:
$$(f \circ g)(1) = f(g(1)) = f(1^2) = 1 + 1 = 2,$$
$$(g \circ f)(1) = g(f(1)) = g(1 + 1) = g(2) = 2^2 = 4.$$

This means that the operation of composition of functions is not in general a commutative operation. There are some specific cases, however, where composition is commutative. For example if a function g has an inverse g^{-1} then from Theorem 2.5b we see that:
$$g^{-1}(g(x)) = I(x) \quad \text{for all } x \text{ in the domain of } g$$
and
$$g(g^{-1}(x)) = I(x) \quad \text{for all } x \text{ in the range of } g.$$

Thus if the domain of g is the same as the range of g we have
$$g \circ g^{-1} = g^{-1} \circ g.$$

There is a definite relationship between the graph of $g \circ f$ and the graphs of f and g. In general this relationship is somewhat complicated and difficult to describe. There is a special case, in which we are particularly interested, where the relationship is quite simple to describe. This is the case $g \circ f_\alpha$, where f_α is a linear function of the form $f_\alpha(x) = x + \alpha$, $x \in R$, and g is an arbitrary function with domain D. In this case $g \circ f = \{(x, g(x + \alpha)) \mid x \in D\}$ and we have
$$(g \circ f_\alpha)(x) = g(f_\alpha(x)) = g(x + \alpha).$$

[In the ensuing discussion, we will refer to the graph of $g(x + \alpha)$ where it would be more precise to refer to the graph of $g \circ f_\alpha$, where f_α is defined by $f_\alpha(x) = x + \alpha$. This lapse in preciseness will make the discussion briefer and, thus, easier to follow.]

First, let us consider the case where $\alpha = 1$ and see what we can determine about the graph of $g(x + 1)$. It is clear that

if $x = 0$, then $g(x + 1) = g(0 + 1) = g(1)$;
if $x = 1$, then $g(x + 1) = g(1 + 1) = g(2)$;
if $x = 2$, then $g(x + 1) = g(2 + 1) = g(3)$;

and so on. (We are assuming that g is defined for the values considered.) Thus we see that, whatever the function g may be, $g(x + 1)$ takes on the same values as $g(x)$, but one unit "earlier." For example, if $x = 0$, then $g(x + 1) = g(1)$ but $g(x) = g(0)$. In terms of their graphs, this means that the graph of $g(x + 1)$ is the same curve as the graph of $g(x)$, but it is shifted one unit to the left of the graph of $g(x)$. Figure 3-25 shows the relationship between the two graphs for a specific function g. We say that the graph of $g(x + 1)$ is a *left translate* of the graph of $g(x)$. It is the graph of g shifted, or translated, one unit to the left.

Figure 3-25

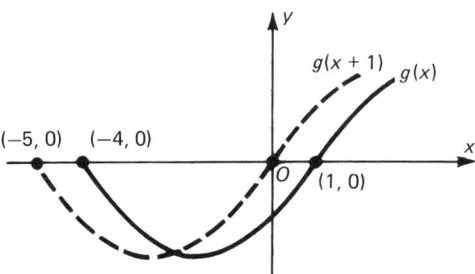

If we now let $\alpha = -1$, then $(g \circ f_\alpha)(x) = g(x - 1)$. It is easy to see that $g(x - 1)$ takes on the same values as $g(x)$, but one unit "later." The result is that the graph of $g(x - 1)$ is the graph of $g(x)$ translated one unit to the right, as shown in Figure 3-26. Hence, the graph of $g(x - 1)$ is a *right translate* of the graph of $g(x)$.

Figure 3-26

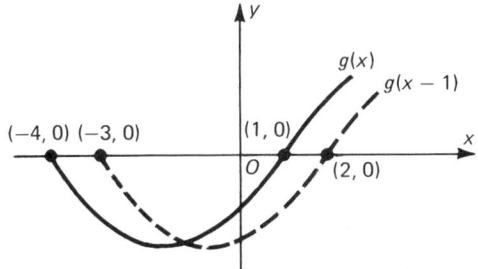

These two special examples set the pattern for the generalization that the graph of $g(x + \alpha)$ is the graph of $g(x)$ translated α units to the left if α is positive and translated $|\alpha|$ units to the right if α is negative. This concept will be of considerable importance to us in some of our later work.

3.5 EXERCISES

In Exercises 1 and 2, $f(x) = (x - 1)^2$ and $g(x) = 2x + 1$.

1. Find the value of each of the following.
   a) $(f \circ g)(1)$ b) $(g \circ f)(-2)$ c) $f(f(1))$ d) $g(g(-3.5))$
2. Find the solutions for each of the following.
   a) $(f \circ g)(x) = 16$.   b) $(g \circ f)(x) = 3$.   c) $g(g(x)) = -5$.
   d) $f(f(x)) = 9$.

For each of Exercises 3–7, express $(f \circ g)(x)$ and $(g \circ f)(x)$ as polynomials.

3. $f{:}y = 2x - 3;\ g{:}y = (x - 1)^2$.
4. $f{:}x \to x^2 + 1;\ g{:}x \to 3x - 1$.
5. $f{:}f(x) = -3;\ g{:}g(x) = 5x + 4$.
6. $f{:}y = 3x - 1;\ g{:}y = \dfrac{x + 1}{3}$.
7. $f{:}x \to 2x;\ g{:}x \to (x + 1)^3$.
8. If $f(x) = x + 2$, find a function $g{:}x \to g(x)$ such that $(f \circ g)(x) = x$.

Use the graphs of the functions f, g, and h given below in connection with Exercises 9–14.

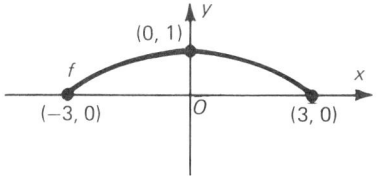
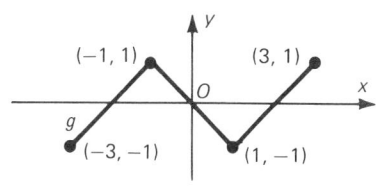

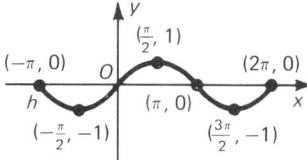

Sketch the graph of each of the functions defined in Exercises 9–14.

9. $y = f(x + 3)$.
10. $y = f(x) + 3$.

11. $y = g(x - 2)$.
12. $y = g(x) - 2$.
13. $y = h\left(x + \dfrac{\pi}{2}\right)$.
14. $y = h(x - \pi)$.
15. If the domain of g is $a \leq x \leq b$, what is the domain of $g(x + \alpha)$, where $\alpha \neq 0$?
16. If $f:f(x) = x + \alpha$, how is the graph of $f \circ g$ related to the graph of g? Explain.
17. a) Let $g:g(x) = |x|$ and $f:f(x) = x + 3$. Write an expression for $(g \circ f)(x)$.
    b) Draw the graph of $g \circ f$.
    c) On the same set of axes draw the graph of $h:h(x) = 4$.
    d) Using the graphs drawn in answer to parts b and c, find the set of solutions of $|x + 3| < 4$.

Use the graphical technique suggested in Exercise 17 to solve each of the following conditions.

18. $|x - 5| \geq 1$.
19. $|3.05 - x| > 7.02$.
20. $|2x + 4| \leq 1$. [Hint: $|2x + 4| = |2| \cdot |x + 2|$.]
21. $|5x - 2| > 3$.
22. Prove that if $c \geq 0$, $|x - a| < c$ is equivalent to $a - c < x < a + c$.
23. Prove that if $c \geq 0$, $|x - a| > c$ is equivalent to $x < a - c$ or $x > a + c$.

3.6 The General Quadratic Function

We now know a great deal about the graphs of quadratic functions defined by equations of the form $q_a(x) + C = ax^2 + C$, which are special cases of the general quadratic function Q defined by $Q(x) = ax^2 + bx + c$, where a, b, and c are real numbers and $a \neq 0$. It might appear that we need to go through a complete analysis to determine the graph of the more general function Q. Fortunately, this is not the case, as we shall presently show.

In what follows, we will want to consider specific examples of the general quadratic function Q. These special cases will be obtained by replacing a, b, and c by particular real numbers, with $a \neq 0$. To avoid confusion, we will adopt a convenient notation to denote specific examples of Q. The notation $Q_{3,2,1}$ will designate the function defined by replacing a by 3, b by 2, and c by 1 in $Q(x) = ax^2 + bx + c$. Thus,

$$Q_{3,2,1}(x) = 3x^2 + 2x + 1.$$

In general, if a, b, and c are any real numbers, with $a \neq 0$, then $Q_{a,b,c}$ designates the quadratic function defined by

$$Q_{a,b,c}(x) = ax^2 + bx + c.$$

In terms of this notation, $q = q_1 = Q_{1,0,0}$. Let us now discuss some examples in detail.

Example 1 Consider the quadratic function $Q_{1,2,1}$, defined by

$$Q_{1,2,1}(x) = x^2 + 2x + 1, \quad x \in R.$$

Note that $Q_{1,2,1}(x)$ can also be written as

$$Q_{1,2,1}(x) = (x+1)^2.$$

But, since $q(x) = x^2$, $q(x+1) = (x+1)^2$, and, hence,

$$Q_{1,2,1}(x) = q(x+1).$$

From our discussion of the translates of graphs, we know that the graph of $q(x+1)$ is the graph of q translated one unit to the left. Therefore, since $Q_{1,2,1}(x) = q(x+1)$, this means that the graph of $Q_{1,2,1}$ is the graph of q translated one unit to the left. (See Figure 3-27.)

Figure 3-27

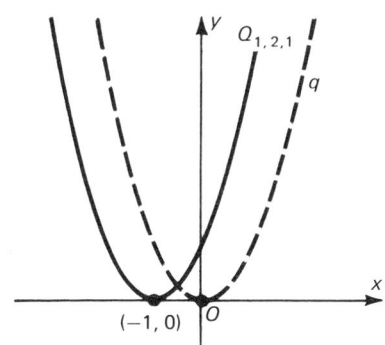

Example 2 Now consider a slightly different quadratic function, namely $Q_{1,2,3}$, defined by

$$Q_{1,2,3}(x) = x^2 + 2x + 3.$$

We note that $Q_{1,2,3}(x)$ can be written as

$$Q_{1,2,3}(x) = (x^2 + 2x + 1) + 2 = (x+1)^2 + 2.$$

Therefore,

$$Q_{1,2,3}(x) = q(x+1) + 2.$$

Now, again using what we know about translates of the graphs of functions, we see that the graph of the function $Q_{1,2,3}(x) = q(x+1) + 2$ is the graph of $q(x+1)$ translated upward 2 units. Furthermore, since the graph of $q(x + 1)$ is the graph of q translated one unit to the left, this means that the graph of $Q_{1,2,3}$ is the graph of q translated 1 unit to the left and then translated upward 2 units. (See Figure 3–28.)

Figure 3-28

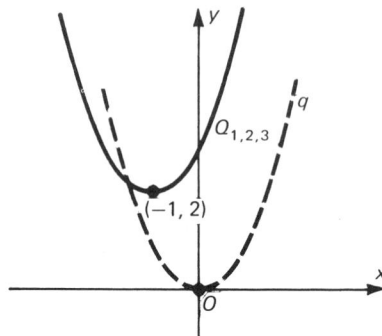

These examples suggest that, in general, the graph of $Q_{a,b,c}$ has the same shape as the graph of q_a, but that the two graphs are translates of each other. That is, by translating either one of the graphs vertically, horizontally, or perhaps both vertically and horizontally, you could superimpose it on the other graph.

To simplify our discussion of translations, we will review ideas discussed earlier and adopt the following terminology:

If α is a real number, then by an α *horizontal translation* we will mean a horizontal translation of $|\alpha|$ units; the translation is to the right if $\alpha > 0$ and to the left if $\alpha < 0$.

If β is a real number, then by a β *vertical translation*, we will mean a vertical translation of $|\beta|$ units. The translation is upward if $\beta > 0$ and downward if $\beta < 0$. For example, a -3 horizontal translation is a translation of 3 units to the left, and a 4 vertical translation is a translation of 4 units upward. Using this terminology, we can state some of the results given on pages 135–136 in the following way:

1. The graph of $f(x + \alpha)$ is a $-\alpha$ horizontal translate of the graph of f. As a specific example, the graph of $f(x + (-2))$ is a $-(-2) = 2$ horizontal translate of the graph of $f(x)$; hence, it is a translation of 2 units to the right.
2. The graph of $g(x) + \beta$ is a β vertical translate of the graph of g.
3. The graph of $h(x + \alpha) + \beta$ is a $-\alpha$ horizontal and a β vertical translate of the graph of h.

To prove that the graph of $Q_{a,b,c}$ is a translate of the graph of q_a, we need to show that, for appropriate values of α and β, the graph of $Q_{a,b,c}$ is an α horizontal and a β vertical translate of the graph of q_a. That is, we must show that

$$q_a(x + \alpha) + \beta = Q_{a,b,c}(x)$$

for all real numbers x. Now the above equation is equivalent to

$$a(x + \alpha)^2 + \beta = ax^2 + bx + c,$$

or

$$ax^2 + 2a \cdot \alpha \cdot x + a \cdot \alpha^2 + \beta = ax^2 + bx + c.$$

Since two polynomials are equal if and only if the coefficients of terms of the same degree are equal, we see that this equation will hold for all real values of x provided that

$$2a \cdot \alpha = b \quad \text{and} \quad a \cdot \alpha^2 + \beta = c.$$

Thus, $q_a(x + \alpha) + \beta = Q_{a,b,c}(x)$ if $\alpha = \dfrac{b}{2a}$ and $\beta = c - a \cdot \alpha^2 = c - \dfrac{ab^2}{4a^2} = \dfrac{4ac - b^2}{4a}$. (Remember that $a \neq 0$.)

We have just established the following important fact:

The graph of the general quadratic function $Q_{a,b,c}$ is a translate of the graph of q_a.

Specifically, the graph of $Q_{a,b,c}$ is the $-\dfrac{b}{2a}$ horizontal translate and the $\dfrac{4ac - b^2}{4a}$ vertical translate of the graph of q_a. In the graph in Figure 3-29, $a = 2$, $b = -12$, and $c = 16$, and, hence, $\alpha = -3$ and $\beta = -2$.

Rather than memorizing the fact that $\alpha = \dfrac{b}{2a}$ and $\beta = \dfrac{4ac - b^2}{4a}$ are the desired values for changing $Q_{a,b,c}(x)$ from the form $ax^2 + bx + c$ to the form $q_a(x + \alpha) + \beta$, one may use the process of *completing the square* to effect this change.

Figure 3-29

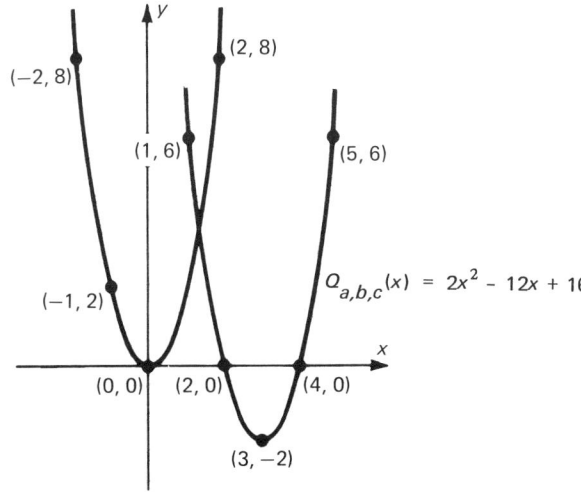

3.6 GENERAL QUADRATIC FUNCTION

Example 3 Let $Q_{a,b,c}$ be defined by

$$Q_{2,8,5}(x) = 2x^2 + 8x + 5.$$

First, we rewrite the right side of this equation, using the coefficient of x^2 as a common factor for the first two terms.

$$Q_{2,8,5}(x) = 2(x^2 + 4x) + 5.$$

To complete the square for the expression in the parentheses on the right side, we must add 4. However, because this expression is multiplied by 2, this is equivalent to adding 8 to the right side. To keep the equation correct, therefore, we must also subtract 8 from the right side. Hence, we have

$$Q_{2,8,5}(x) = 2(x^2 + 4x + 4) + 5 - 8,$$

which becomes

$$Q_{2,8,5}(x) = 2(x + 2)^2 - 3.$$

The graph of $Q_{2,8,5}$ may now be seen to be a -2 horizontal and a -3 vertical translate of the graph of $q_2 : q_2(x) = 2x^2$.

3.6 EXERCISES

Draw the graph $f : y = -3x + 3$. Then, on the same set of axes, draw the graphs of the functions listed below. For each function, write a defining equation in the form $y = mx + b$.

1. $g : y = f(x - 2)$.
2. $h : y = f(x) + 3$.
3. $r : y = f(x - 2) + 3$.
4. $s : y = f(x + 2) - 3$.

In each of Exercises 5–8, do the following: (a) Sketch the graph of q_a. (b) On the same set of axes, sketch the graph of F; explain how the graph of F is related to the graph of q_a; and write a defining equation for F in the form $F(x) = ax^2 + bx + c$. (c) On the same set of axes, sketch the graph of G; explain how the graph of G is related to the graph of F and how the graph of G is related to the graph of q_a; and write a defining equation for G in the form $G(x) = ax^2 + bx + c$.

5. $q_a(x) = x^2$; $F(x) = q_a(x + 5)$; $G(x) = q_a(x + 5) + 2$.
6. $q_a(x) = x^2$; $F(x) = q_a(x - 3)$; $G(x) = q_a(x - 3) - 4$.
7. $q_a(x) = 2x^2$; $F(x) = q(x + 6)$; $G(x) = q_a(x + 6) - 3$.
8. $q_a(x) = -x^2$; $F(x) = q_a(x - 4)$; $G(x) = q_a(x - 4) + 1$.

For each of the following functions, use the process of completing the square to express $Q_{a,b,c}(x)$ in the form $q_a(x + \alpha) + \beta$. Then sketch the graphs of q_a and $Q_{a,b,c}$ on the same set of axes.

9. $Q(x) = x^2 - 2x + 1$.
10. $Q(x) = x^2 + 4x + 6$.
11. $Q(x) = x^2 - 4x - 1$.
12. $Q(x) = x^2 + 6x + 6$.
13. $Q(x) = -x^2 - 2x - 1$.
14. $Q(x) = 3x^2 + 6x$.
15. $Q(x) = -x^2 - 6x - 8$.
16. $Q(x) = 2x^2 + 4x + 4$.

3.7 Analysis of the General Quadratic

We will now consider some of the consequences of the conclusions we have drawn in Section 3.6. In informal terms, a translate of a graph is just a vertical and horizontal "movement" of the graph that does not twist it or distort it in any way. Thus, since we know a great deal about the graph of any quadratic function q_a, where a is a nonzero real number, we also know a great deal about the graph of any quadratic function $Q_{a,b,c}$, since it is just a translate of q_a.

In particular, we know the following facts. Given any triple of real numbers, a, b, c, with $a \neq 0$, the quadratic function Q defined by

$$Q(x) = ax^2 + bx + c, \qquad x \in R,$$

has the following properties:

1. If $a > 0$, the graph of Q is concave upward; and if $a < 0$, the graph of Q is concave downward.
2. The graph of Q is continuous; that is, it does not have any "breaks" or "jumps."
3. The graph of Q is symmetric with respect to the vertical line
$$x = \frac{-b}{2a}.$$

To see this third property, we observe first that the graph of q_a is symmetric with respect to the line $x = 0$, and second that the line $x = \frac{-b}{2a}$ is the $\frac{-b}{2a}$ horizontal translate of the line $x = 0$. In other words, the line of symmetry of the graph of q_a is translated to the line of symmetry of the graph of Q.

4. The graph of Q is a parabola.

This property follows from the fact that, as we have already shown, the graph of q_a, with $a \neq 0$, is a parabola with focus $\left(0, \dfrac{1}{4a}\right)$ and directrix $h(x) = \dfrac{-1}{4a}$. Since the graph of Q is just a translate of the graph of q_a, it follows that the graph of Q is a parabola whose focus and directrix are translates of the focus and directrix of the graph of q_a.

There is a particular point on a parabola that has special significance; this point is the intersection of the parabola and its line of symmetry. This point is called the *vertex* of the parabola and is the midpoint of the segment of the line of symmetry that lies between the focus and the directrix of the parabola. The vertex of a parabola is the closest of all points on the parabola to both the focus and the directrix. We can roughly describe the vertex as the point on the parabola at which the parabola is most "pointed."

The vertex of the parabola that is the graph of a function q_a is the origin, $(0, 0)$. (Why?) It is the lowest point of the parabola if the parabola is concave upward, and the highest point of the parabola if the parabola is concave downward. (See Figure 3–30.)

Figure 3-30

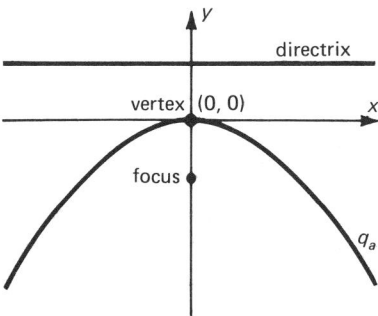

Since the graph of $Q_{a,b,c}$ is the $\dfrac{-b}{2a}$ horizontal translate and the $\dfrac{4ac - b^2}{4a}$ vertical translate of the graph of q_a, the point $\left(\dfrac{-b}{2a}, \dfrac{4ac - b^2}{4a}\right)$ is the vertex of the parabola which is the graph of $Q_{a,b,c}$. This follows because the point $\left(\dfrac{-b}{2a}, \dfrac{4ac - b^2}{4a}\right)$ is the $\dfrac{-b}{2a}$ horizontal translate and the $\dfrac{4ac - b^2}{4a}$ vertical translate of the vertex of q_a, $(0, 0)$. From the remarks made above about the graph of q_a, we can conclude that this point is on the line of symmetry of the graph of $Q_{a,b,c}$, which is the line $x = \dfrac{-b}{2a}$. This vertex is the low point of the graph if $a > 0$, and the high point of the graph if $a < 0$. (See Figure 3-31.)

Figure 3-31

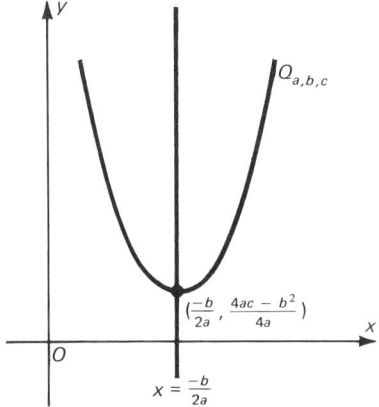

Often, high and low points of graphs are of special interest to us in solving problems. Of course, the graphs of functions other than quadratic functions may have high and low points. The following definition applies to any function f.

Definition 3.7a If there is a point $(k, f(k))$ on the graph of a function f that is at least as high as (that is, has an ordinate greater than or equal to that of) any other point on the graph of f, then the function f is said to have a **maximum** (or a maximum value) of $f(k)$ at k.

Similarly, if $(h, f(h))$ is a point on the graph of f that is at least as low as (that is, has an ordinate less than or equal to that of) any other point on the graph of f, then the function f is said to have a **minimum** (or a minimum value) of $f(h)$ at h.

Example 1 The function whose graph is shown in Figure 3-32 has a maximum of 5 at -3 and a minimum of -4 at 2.

Figure 3-32

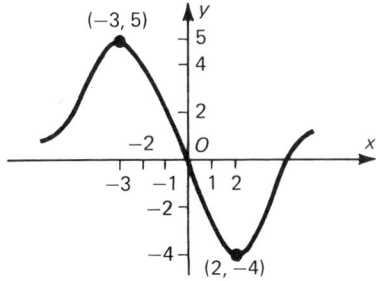

Using this terminology, we can say that if $a > 0$, then $Q_{a,b,c}$ has a minimum value of $\dfrac{4ac - b^2}{4a}$ at $\dfrac{-b}{2a}$, because $\left(\dfrac{-b}{2a}, \dfrac{4ac - b^2}{4a}\right)$ is the vertex of the graph of $Q_{a,b,c}$. Similarly, if $a < 0$, then $Q_{a,b,c}$ has a maximum value of $\dfrac{4ac - b^2}{4a}$ at $\dfrac{-b}{2a}$.

Now that we know how to determine the maximum or minimum value taken on by a quadratic function, we can apply this knowledge to solving problems like the following one.

Example 2 Suppose that you have 100 feet of fencing and that you want to fence a rectangular section with it in such a way as to enclose the largest possible area. If one dimension of the rectangle is x feet, then, because the total perimeter is to be 100 feet, the other dimension of the rectangle must be $50 - x$ feet. (See Figure 3-33.)

Figure 3-33

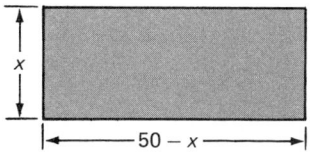

Thus, the area enclosed by the rectangle is
$$A = Q(x) = -x^2 + 50x, \qquad 0 < x < 50.$$

We require that x be greater than 0 and less than 50 because x represents the measure of a side of the rectangle, which obviously will not be 0. We want to find the value of x that will make A a maximum. To do this, we first complete the square to obtain
$$-x^2 + 50x = -(x^2 - 50x + 625) + 625$$
$$= -(x - 25)^2 + 625.$$

From this last expression, we can see that A will have its greatest value, 625, when $x = 25$. Thus, of all the rectangles with perimeter 100 feet, the rectangle that has the greatest area is the square whose side is 25 feet long. In the following exercises, you are asked to determine for all the rectangles having perimeter p, where p is a fixed number, what the dimensions of the rectangle having the largest area must be.

3.7 EXERCISES

The graph of each of the following quadratic functions in Exercises 1–6 is a parabola. For each parabola, carry out the following instructions: (a) Write a defining equation for its axis of symmetry. (b) Give the coordinates of its vertex. (c) Indicate whether the ordinate of the vertex is a maximum value or a minimum value of the function and justify your answer. (d) *Give the coordinates of the point where the parabola intersects the y-axis and give the coordinates of the point which is the image of this y-intercept in the axis of symmetry.* (e) *Plot the vertex and the y-intercept and its image, draw the axis of symmetry, and sketch the parabola.*

1. $Q:Q(x) = x^2 - 4x$.
2. $Q:Q(x) = 2x - x^2$.
3. $Q:x \to -x^2 + 6x + 7$.
4. $Q:x \to x^2 + 2x - 3$.
5. $Q:y = 2x^2 + 8x - 5$.
6. $Q:y = -2x^2 - 5x - 1$.

Let (x_1, y_1), (x_2, y_2), and (x_3, y_3) be the coordinates of three points, no two of which have the same abscissa. If the points are collinear, there is exactly one linear or constant function whose graph contains the points. If the points are not collinear, it is possible to determine a unique quadratic function whose graph contains the points. To illustrate, consider the points $(-1, -2)$, $(1, 4)$, and $(2, 13)$. Substituting these coordinates successively in the equation $Q_{a,b,c}(x) = ax^2 + bx + c$, we can produce the following system of three equations in a, b, and c:

$$-2 = a - b + c,$$
$$4 = a + b + c,$$
$$13 = 4a + 2b + c.$$

Solving this system for a, b, and c, we find that $a = 2$, $b = 3$, and $c = -1$. Hence, the quadratic function containing the three given points is defined by

$$Q(x) = 2x^2 + 3x - 1.$$

In each of Exercises 7–10 use the method just described to determine the quadratic function whose graph contains the three given points. Then find the equation of the axis of symmetry and the coordinates of the vertex of this graph. Finally, plot the three given points and the vertex, draw the axis of symmetry, and sketch the parabola.

7. $(0, 0)$, $(6, 0)$, $(3, 9)$
8. $(0, -12)$, $(4, 4)$, $(-1, -11)$
9. $(-4, 8)$, $(-2, 2)$, $(4, 8)$
10. $(-2, -3)$, $(-1, -3)$, $(1, 3)$

11. Determine the two numbers whose sum is 50 and whose product is a maximum.
12. A rectangular corral is to be constructed, using the straight bank of a stream as one side and fencing along the other three sides. If 400 feet of fencing are available, what are the dimensions of the

corral that can be constructed with the largest area? (Hint: $A = xy$ and $2x + y = 400$.)

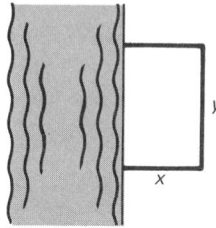

13. A piece of wire 20 cm. long is to be cut into two pieces, and each piece is to be bent into the shape of a square. Into what lengths should the wire be cut if the sum of the areas of the squares is to be a minimum?

14. A rectangle may be inscribed in an isosceles triangle by placing one side of the rectangle along the base of the triangle and selecting the other two vertices on the sides of the triangle, as shown in the illustration. Find the dimensions of the rectangle of greatest area that can be so inscribed in an isosceles triangle with base 20 inches and altitude 12 inches. (Hint: Use similar triangles to express y in terms of x.)

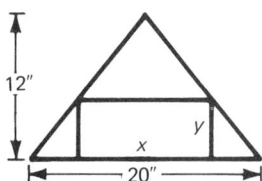

15. Determine the point on the line defined by $y = -2x + 10$ for which the square of its distance from the origin is a minimum. [Hint: In the illustration, $(d(O, P))^2 = x^2 + y^2$.]

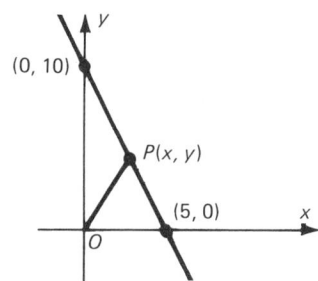

144   QUADRATIC FUNCTIONS

16. Prove that, of all the rectangles having perimeter p, where p is a fixed number, the rectangle with greatest area is a square.

17. A piece of wire 11 inches long is to be cut into two pieces. One piece is to be bent into the shape of an equilateral triangle, and the other bent into the shape of a square.

   a) If the length of the piece to be used for the triangle is represented by x, express the sum of the areas of the triangle and the square as a function of x. (Reminder: In an equilateral triangle, $A = \dfrac{s^2}{4}\sqrt{3}$.)

   b) If it is permissible to use all the wire for the triangle or all the wire for the square, what is the domain of this function?

   c) Is the graph of this function concave upward or concave downward?

   d) For what value of x would the sum of the areas be a minimum?

   e) For what value of x would the sum of the areas be a maximum? (Hint: Sketch the graph of the function.)

3.8 The Quadratic Equation—A Graphical Interpretation

Perhaps the process of completing the square and the appearance of the terms $\dfrac{-b}{2a}$ and $\dfrac{4ac - b^2}{4a}$ in the preceding discussion reminded you of the solutions of the *quadratic equation* $ax^2 + bx + c = 0$. These solutions are usually given as

$$x = \frac{-b \pm \sqrt{b^2 - 4ac}}{2a}.$$

As you may recall, this equation is called the **quadratic formula.** The expression $b^2 - 4ac$, which appears under the radical in the numerator on the right side of the formula, is called the **discriminant** of the quadratic equation $ax^2 + bx + c = 0$. You may remember that, if the discriminant is equal to zero, then the quadratic equation has just one solution, $x = \dfrac{-b}{2a}$. If the discriminant is positive, the quadratic equation has two solutions,

$$x = \frac{-b - \sqrt{b^2 - 4ac}}{2a} \quad \text{and} \quad x = \frac{-b + \sqrt{b^2 - 4ac}}{2a}.$$

If the discriminant is negative, the quadratic equation has no real solutions, since the square root of a negative number is not a real number.

The relationship of the quadratic equation and its solutions to the quadratic function is easily explained. Those real values of x for which $Q_{a,b,c}(x) = ax^2 + bx + c = 0$ are the abscissas of the points on the graph of $Q_{a,b,c}$ whose ordinates are 0. In other words, the solutions of the quadratic equation are the abscissas of the points where the graph crosses the x-axis. See Figure 3–34.

Figure 3-34

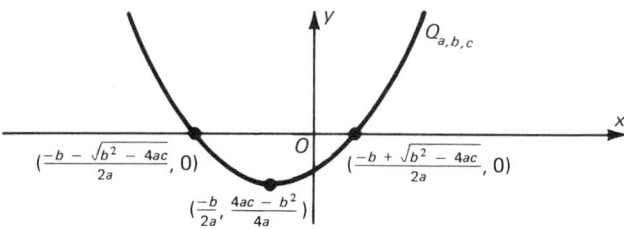

As we have remarked, if the discriminant $b^2 - 4ac > 0$, the quadratic equation has two solutions, and, hence, the graph of the function crosses the x-axis at two points. This can be related to what we already know about maximum and minimum values of the function. If the graph of the function is concave upward and if the minimum value is less than 0, that is, if $\dfrac{4ac - b^2}{4a} < 0$, then the vertex of the parabola is below the x-axis. In this case, the parabola intersects the x-axis in two points. (Explain why $\dfrac{4ac - b^2}{4a} < 0$ and the fact that the parabola is concave upward imply that $b^2 - 4ac > 0$.) Similarly, if the graph is concave downward and the vertex of the parabola is above the x-axis, the graph intersects the x-axis in two points. (Explain why this is true only if $b^2 - 4ac > 0$.)

Figure 3-35

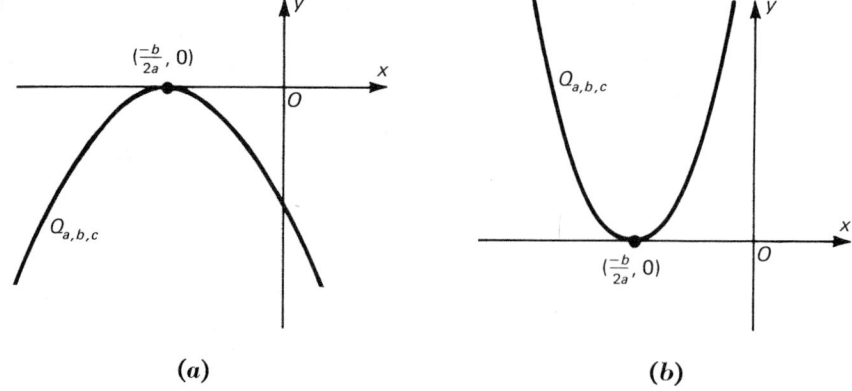

(a) (b)

Now, of course, the parabola that is the graph of $Q_{a,b,c}$ may not intersect the x-axis in two points. It may, as in Figure 3–35, intersect the x-axis in a single point. This will be the case when the discriminant, $b^2 - 4ac$, of the quadratic equation is equal to zero and, therefore, the quadratic equation has the single solution $x = \dfrac{-b}{2a}$. The single point of contact of the graph of $Q_{a,b,c}$ with the x-axis will, in this case, be $\left(\dfrac{-b}{2a}, 0\right)$. Again, this ties in with what we know about maximum and minimum values. If the vertex of a parabola is on the x-axis, then the maximum or minimum value, $\dfrac{4ac - b^2}{4a}$, corresponding to $x = \dfrac{-b}{2a}$ must be equal to 0. But if $\dfrac{4ac - b^2}{4a} = 0$, then $4ac - b^2 = 0$, which implies that $b^2 - 4ac = 0$.

Finally, the parabola that is the graph of $Q_{a,b,c}$ may fail to intersect the x-axis, as in Figure 3–36. When this is the case, the quadratic expression $Q_{a,b,c}(x) = ax^2 + bx + c$ has a negative discriminant and, therefore, no real solutions.

Figure 3-36

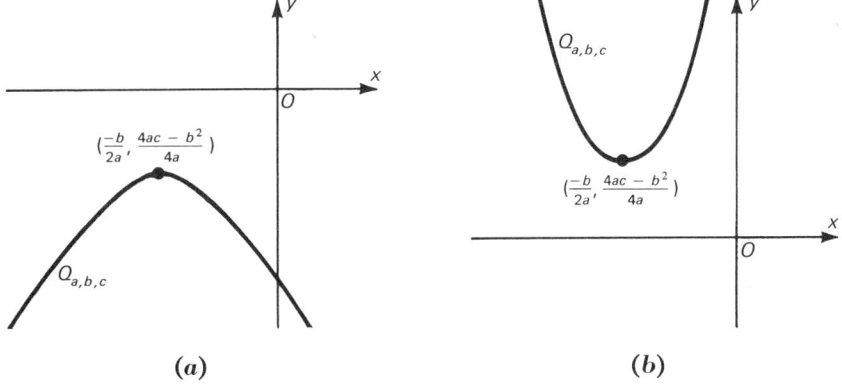

(a)   (b)

You should be able to show that if $b^2 - 4ac < 0$, then a parabola that is concave upward has its vertex above the x-axis and a parabola that is concave downward has its vertex below the x-axis.

We can summarize this discussion concerning the relation between the quadratic equation and the graph of the quadratic function as follows:

1. If $b^2 - 4ac > 0$, the graph of $Q_{a,b,c}$ intersects the x-axis in two distinct points, whose abscissas are the solutions, or roots, of the quadratic equation $Q_{a,b,c}(x) = 0$.
2. If $b^2 - 4ac = 0$, the graph of $Q_{a,b,c}$ intersects the x-axis in precisely one point, whose abscissa is the solution, or root, of the quadratic equation $Q_{a,b,c}(x) = 0$.
3. If $b^2 - 4ac < 0$, the graph of $Q_{a,b,c}$ does not intersect the x-axis, and the equation $Q_{a,b,c}(x) = 0$ has no real roots.

In the case where the parabola, which is the graph of $Q_{a,b,c}$, intersects the x-axis in two points, we have enough information available to give a rough approximation of the graph. We first locate the two points of intersection with the x-axis,

$$\left(\frac{-b - \sqrt{b^2 - 4ac}}{2a}, 0\right) \quad \text{and} \quad \left(\frac{-b + \sqrt{b^2 - 4ac}}{2a}, 0\right).$$

Then we construct the perpendicular bisector of the line segment joining these points. (You should be able to show that this line is the line of symmetry for the parabola.) Once the vertex of the parabola is located on this line, the essential shape of the parabola is easily seen. (See Figure 3-37.)

Figure 3-37

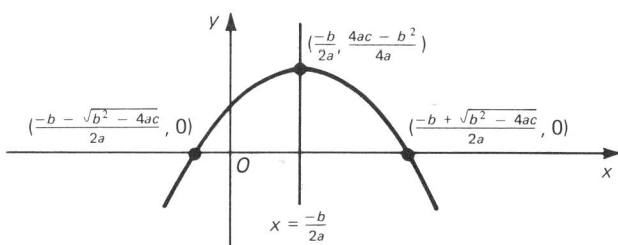

In the cases where the parabola does not cross the x-axis, it is still a simple matter to construct the line of symmetry $x = \dfrac{-b}{2a}$ and to locate the vertex of the parabola on this line. To locate two additional points on the parabola which are symmetric with respect to the line of symmetry, choose abscissas $-\dfrac{b}{2a} + k$ and $-\dfrac{b}{2a} - k$, where k is a constant, and then compute the ordinates. These three points together with the line of symmetry afford enough information to give a rough approximation of the shape of the parabola, as illustrated in Figure 3-38.

Figure 3-38

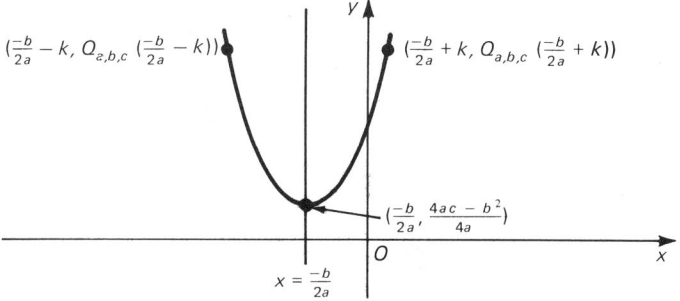

3.8 EXERCISES

Find the solutions, to the nearest tenth, of the following quadratic equations.

1. $x^2 + 3x - 3 = 0$.
2. $x^2 + 2x = 5$.
3. $x^2 = 4x + 2$.
4. $2x^2 + 2x - 3 = 0$.
5. $2x^2 = 5x - 1$.
6. $3x^2 = 1 - 5x$.

By examining the discriminant, indicate whether each of the following equations has two real solutions, one real solution, or no real solutions. If the equation has two real solutions, indicate whether these solutions are rational or irrational.

7. $x^2 + 7x + 10 = 0$.
8. $3x^2 - 7x - 4 = 0$.
9. $x^2 - 3x + 4 = 0$.
10. $4x^2 + 3x + 5 = 0$.
11. $2x^2 + 5x - 4 = 0$.
12. $4x^2 - 24x + 36 = 0$.
13. $x^2 + 2x - 10 = 0$.
14. $5x^2 - 11x + 2 = 0$.

The graph of a quadratic function $Q_{a,b,c}(x) = ax^2 + bx + c$ may be used to find the set of solutions for a quadratic inequality. For example, consider the inequality $3x^2 - 6x + 5 > 2x^2$, which can be simplified to

$$x^2 - 6x + 5 > 0.$$

We know that the graph of $Q_{a,b,c}(x) = x^2 - 6x + 5$ intersects the x-axis at $(1, 0)$ and $(5, 0)$, since $x = 1$ and $x = 5$ are the solutions of $x^2 - 6x + 5 = 0$. From the fact that $a > 0$, we know that the parabola is concave upward and, hence, resembles the one in the illustration. Obviously, $Q_{a,b,c}(x) > 0$ when $x < 1$ or $x > 5$. Thus, the set of solutions for the inequality

$$3x^2 - 6x + 5 > 2x^2$$

is $\{x \mid x < 1 \text{ or } x > 5\}$.

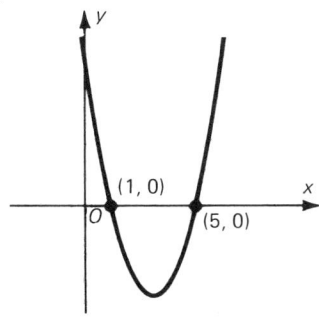

Use the method just described to find the set of real solutions for each of the following inequalities.

15. $x^2 - 6x + 8 < 0$.
16. $5 + 5x - x^2 > 1 - 2x^2$.
17. $3x^2 - 4x + 2 \leq 2x^2 - 1$.
18. $4x^2 + 1 \leq 4x$.
19. $5x + 2 > 3x^2$.
20. $6(x^2 + 1) > 13x$.
21. $x^2 + 4x + 5 \geq 0$.
22. $x^2 + 2x \leq x + 1$.

For each of the following exercises, determine the values of k for which the statement will be true.

23. The equation $x^2 + kx + 36 = 0$ has only one real root.
24. The equation $x^2 + kx + (k + 8) = 0$ has only one real root.
25. The equation $3kx^2 + 4kx + 4 = 0$ has no real roots.
26. The equation $kx^2 + 12x + k + 5 = 0$ has two real roots.

3.9 Tangent Lines

In geometry, you learned that, through each point P on a circle, there is a unique line in the plane of the circle that intersects the circle at P and is perpendicular to the radius of the circle with endpoint P. This unique line is defined to be the tangent line to the circle at the point P. Here, we will consider the problem of generalizing the concept of a tangent line so that we can speak of tangent lines to curves other than circles. We begin by considering what we might mean by a tangent line to a parabola, which, as we know, is the graph of a quadratic function.

At any point P on the graph of a quadratic function, it is possible to place a straightedge and adjust its direction so that the graph of the function near P is "close" to the straightedge.

As an example, we will now consider a point on the graph of q, defined by $q(x) = x^2$. To simplify our discussion, we choose the point at the origin. If you plot a few points of the graph of q, choosing points with small abscissas, it becomes apparent that the graph of q near the origin is very close to the x-axis. In Figure 3-39, a magnified portion of the graph of q near the origin is shown.

As you see, for small values of x, it is difficult to distinguish between the graph and the x-axis. We say that the x-axis is **tangent** to the graph of q at $(0, 0)$ or that the x-axis is a **tangent line** to the graph of q at $(0, 0)$. For future discussions, we need a more precise definition of what we mean by a tangent to a curve.

150 QUADRATIC FUNCTIONS

Figure 3-39

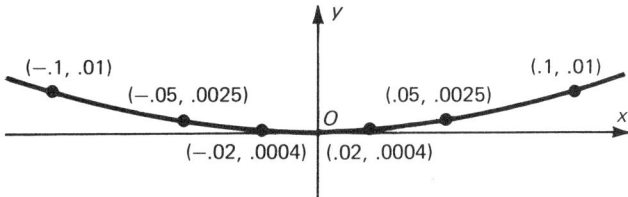

Roughly, what we mean by saying "the x-axis is tangent to the graph of q at $(0, 0)$" is that, if l is any line through the origin other than the x-axis, then, sufficiently near the origin, all points of the graph of q, except the point $(0, 0)$, will be closer to the x-axis than they are to l. It will turn out that, through each point of the graph of q, there is a unique line that is tangent to the graph at that point.

By considering various points on other continuous curves, you will probably come to the conclusion that continuous functions have tangent lines at most points of their graphs. As examples, in each of the drawings in Figure 3–40, the curve is very close to the line l near the point P, It would appear in each of these cases that, given any line l' through P, where l' is different from l, the points of the curve sufficiently near P, except P itself, will be closer to l than to l'. That is, in each of the cases, l appears to be the tangent to the curve at P.

Figure 3-40

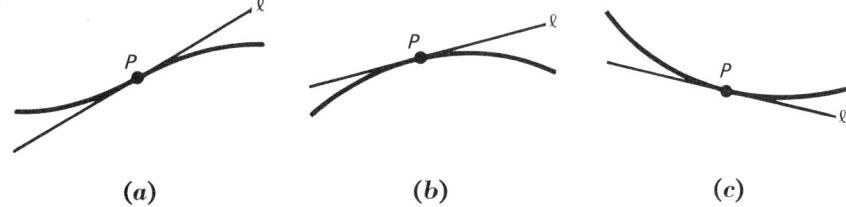

(a)   (b)   (c)

In the preceding discussion, we have implicitly been using the following informal definition of a tangent.

> A curve g has a *tangent line l* through the point P on g if there is a line l, such that all points on g sufficiently near P, except P itself, are closer to l than to any other line through P.

When we say a point P is closer to the line l than to the line l', we mean that its perpendicular distance from l is less than its perpendicular distance from l'. That is, $d(P, l) < d(P, l')$. (See page 77.)

Obviously, if a curve g and two lines l and l' all intersect at a point P, it does not make sense to say that g is closer to l than it is to l' at point P. Therefore, let us agree that, under these conditions, when we say that g is closer to a line l than to a line l' *near P*, we will mean that every point of g near P, other than P itself, is closer to l than it is to l'.

Now what do we mean by "near P"? If we draw a circle in the plane with center P, then the points inside the circle are nearer to P than are the other points in the plane. When a mathematician asserts that a curve has a certain property near P, he means that there is a circle of some specific radius with center P within which the curve has the stated property. Thus, to say that the curve g is closer to l than l' near P means that there is a circle with center P such that every point of g within that circle, except the point P, is closer to the line l than it is to l'.

Consider Figure 3–41. Inside the circle C_1, some of the points of g are obviously closer to l' than they are to l, while other points of g are closer to l than they are to l'. But, inside circle C_2, every point of g, other than P, is closer to l than it is to l'. We say, therefore, that near P, the curve g is closer to l than it is to l'.

Figure 3-41

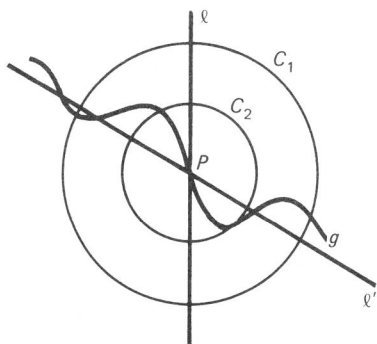

It should be clear from our definition that at every point on a line l, l is its own tangent line. We now return to the question—What is the equation of a tangent line to the graph of a quadratic equation?

We start by considering a point $P(p, ap^2)$ on the parabola whose equation is $q_a(x) = ax^2$. A line through P with slope m has the equation

$$l_m(x) = m(x - p) + ap^2$$

or
$$l_m(x) = mx + (-mp + ap^2).$$

What is the distance from an arbitrary point $Q(x_0, ax_0^2)$ on the graph of q_a to the line l_m? (See Figure 3–42.) From earlier work, (see page 77), we know that $d(Q(x_0, ax_0^2), l_m)$ is given by

$$\frac{|mx_0 + (-mp + ap^2) - ax_0^2|}{\sqrt{1 + m^2}}.$$

Simplified, this gives

$$d(Q, l_m) = \frac{|m - a(p + x_0)|}{\sqrt{1 + m^2}} |p - x_0|.$$

Figure 3-42

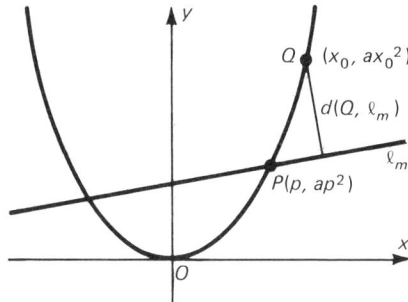

We see that the distance from Q to l_m is a multiple of $|p - x_0|$, and that multiple is

$$M = \frac{|m - a(p + x_0)|}{\sqrt{1 + m^2}}.$$

As Q gets close to P, x_0 gets close to p, and, therefore, $a(p + x_0)$ gets close to $2ap$.

Now carefully consider the following two statements:

1. If m is *any* number other than $2ap$, (i.e., $m - 2ap \neq 0$), then as x_0 gets close to p, M approaches the *constant* $\dfrac{|m - 2ap|}{\sqrt{1 + m^2}}$ which is greater than zero.

2. If $m = 2ap$, then as x_0 gets close to p, M approaches 0.

From these two statements, we deduce that if Q is sufficiently close to P, and, therefore, x_0 is close to p, then Q will be closer to l_{2ap} than it is to l_m if $m \neq 2ap$.

This means that l_{2ap} is the *tangent line* to the graph of q_a at the point (p, ap^2).

$$l_{2ap}(x) = 2ap(x - p) + ap^2$$

or

$$l_{2ap}(x) = 2apx - ap^2.$$

Since l_{2ap} is the tangent line to the graph of q_a at the point P, we know that if l is any other line through P, then all points on the graph of q_a sufficiently near P are closer to l_{2ap} than they are to l. That is, if we try to approximate the graph of q_a at P by a line through P, then the line l_{2ap} gives the best approximation. For this reason, the tangent line to a graph at a point P is sometimes called the "best linear approximation" to the graph at P.

3.9 EXERCISES

In the following exercises find the slope and an equation of the tangent line to the graph of the given function at the given point. Then sketch the graph of the function and draw the tangent.

1. $f:f(x) = x^2; (-2, 4)$.
2. $g:g(x) = -x^2; (3, -9)$.
3. $h:h(x) = -2x^2; (-1, -2)$.
4. $F:F(x) = 3x^2; (0, 0)$.
5. $G:G(x) = -\frac{1}{2}x^2; (4, -8)$.
6. $H:H(x) = \frac{1}{4}x^2; (2, 1)$.

In each of the following exercises, an equation for a function and the slope of a tangent to the graph of the function are given. For each exercise find the coordinates of the point of tangency. Then sketch the function and draw the tangent.

7. $f:f(x) = x^2; m = 8$.
8. $g:g(x) = 3x^2; m = 6$.
9. $F:F(x) = -x^2; m = 6$.
10. $G:G(x) = -\frac{1}{2}x^2; m = -3$.

11. Show that the slope of the tangent line drawn to the graph of $Q:Q(x) = ax^2 + c$ at the point $(x_0, Q(x_0))$ is $2ax_0$. [Hint: How is the graph of Q related to the graph of $q_a:q_a(x) = ax^2$?]

12. Show that the slope of the tangent line drawn to the graph of $Q:Q(x) = q_a(x + \alpha) + \beta$ at the point $(x_0, Q(x_0))$ is $2a(x + \alpha)$.

Use the results of Exercises 11 and 12 to find the slope and an equation of the tangent line drawn to the graph of each of the following functions at the given point. Then sketch the graph of the function and draw the tangent.

13. $f:f(x) = 2x^2 - 3; (2, 5)$.
14. $g:g(x) = \frac{1}{3}(x - 2)^2; (-1, 3)$.
15. $F:F(x) = -(x + 3)^2 + 8; (1, -8)$.
16. $G:G(x) = x^2 - 2x + 4; (0, 4)$.

3.10 The Slope Function

Some graphs, like that of q_a, have a tangent line at every point $(x, f(x))$ of the graph. At each such point, the tangent line to the graph has a slope. It is common practice to refer to the slope of the tangent line as the "*slope of the graph.*" Now corresponding to every x in the domain of f for which the graph of f has a tangent line at $(x, f(x))$, there is the real number which is the slope of the graph of f. We shall designate this number by $f'(x)$. From our definition of function, it should be clear that $\{(x, f'(x)) \mid \text{there is a tangent line to the graph of } f \text{ at } (x, f(x))\}$ is a function. We call this the **derivative function** of f or the **slope function** of f.

Example 1 The slope function of q_a is q_a', $q_a'(x) = 2ax$, since at every point (x, ax^2) on the graph of q_a there is a tangent line to the graph with slope $2ax$.

Most of the functions we shall be interested in have slope functions. These slope functions can often give us valuable information about the function from which they were derived.

Let us consider the general problem of determining the slope function or derivative f' of the function f. If $(x_0, f(x_0))$ is a point on the graph of f, then the equation of a line through this point with slope m is given by:

$$l_m(x) = m(x - x_0) + f(x_0) = mx + (f(x_0) - mx_0).$$

By the formula for the distance from a point to a line, we have the distance from an arbitrary point $P(x, f(x))$ on the graph of f to l_m is given by:

$$d(P, l_m) = \frac{|mx + (f(x_0) - mx_0) - f(x)|}{\sqrt{1 + m^2}}$$

or

$$d(P, l_m) = \left|\frac{f(x) - f(x_0)}{x - x_0} - m\right| \frac{|x - x_0|}{\sqrt{1 + m^2}}.$$

Thus the distance from P to l_m is a multiple, that varies with x, of $\frac{|x - x_0|}{\sqrt{1 + m^2}}$, namely:

$$d(P, l_m) = K(x) \frac{|x - x_0|}{\sqrt{1 + m^2}}$$

where

$$K(x) = \left|\frac{f(x) - f(x_0)}{x - x_0} - m\right|.$$

We now assume that $\lim_{x \to x_0} \frac{f(x) - f(x_0)}{x - x_0} = \alpha$, where α is some real number. This happened for example when $f(x) = x^2$; $\lim_{x \to x_0} \frac{x^2 - x_0^2}{x - x_0} = \lim_{x \to x_0} (x + x_0) = 2x_0$. This means that $K(x) \to |\alpha - m|$ as $x \to x_0$, and therefore:

1. If $m = \alpha$, $K(x) \to 0$ as $x \to x_0$
2. If $m \neq \alpha$, $K(x) \to k = |\alpha - m| > 0$ as $x \to x_0$.

From this we can conclude that $d(P, l_\alpha) < d(P, l_m)$ if $m \neq \alpha$ for all x sufficiently close to x_0. But this means l_α is the tangent line to the graph of f at $(x_0, f(x_0))$.

We summarize this in a theorem:

Theorem 3.10a If $\lim\limits_{x \to x_0} \dfrac{f(x) - f(x_0)}{x - x_0}$ exists then the graph of f has a tangent line at $(x_0, f(x_0))$ with slope equal to this limit. In this event we denote $\lim\limits_{x \to x_0} \dfrac{f(x) - f(x_0)}{x - x_0}$ by $f'(x_0)$.

Example 2 If $f(x) = x^3$,

$$\lim_{x \to x_0} \frac{x^3 - x_0^3}{x - x_0} = \lim_{x \to x_0} (x^2 + xx_0 + x_0^2) = 3x_0^2.$$

Thus $f'(x_0) = 3x_0^2$. Since x_0 was arbitrary this means $f'(x) = 3x^2$ for any $x \in R$, and f' is the slope function or derivative of the function f.

3.10 EXERCISES

Use Theorem 3.10a to prove each of the following statements.

1. If $f(x) = c$, then $f'(x) = 0$.
2. If $f(x) = mx + b$, then $f'(x) = m$.
3. If $f(x) = ax^2 + c$, then $f'(x) = 2ax$.
4. If $f(x) = ax^2 + bx + c$, then $f'(x) = 2ax + b$.

For each of the following exercises use the results of Exercise 4 to determine the equation of the tangent to the graph of the function at the given point. Then sketch the graph of the function and draw the tangent.

5. $Q(x) = 2x^2 - 3$; $(\tfrac{1}{2}, -\tfrac{5}{2})$.
6. $Q(x) = 12x - 4x^2$; $(2, 8)$.
7. $Q(x) = -x^2 + 5x + 6$; $(-2, -8)$.
8. $Q(x) = \tfrac{1}{2}x^2 - 3x + 8$; $(4, 4)$.

In each of the following exercises, an equation for a function and the slope of a tangent to the graph of the function are given. For each exercise, find the coordinates of the point of tangency. Then sketch the graph of the function and draw the tangent.

9. $Q(x) = 9 - 2x^2$; $m = 8$.
10. $Q(x) = \tfrac{1}{2}x^2 + 3x$; $m = -1$.
11. $Q(x) = x^2 - 2x + 3$; $m = 4$.
12. $Q(x) = -2x^2 + 6x - 1$; $m = -2$.

Suppose that we are given a quadratic function $Q:Q(x) = ax^2 + bx + c$ and the coordinates of a point R not on the graph of Q. We may use the fact that $2ax_0 + b$ is the slope of the tangent at any point $(x_0, ax_0^2 + bx_0 + c)$ on the graph of Q to determine if there are any points on the graph of Q at which a tangent may be drawn that will contain R. For example, consider the function $Q: Q(x) = x^2 - 4x + 3$ and the point $R = (0, -6)$. If there is a point $(x_0, Q(x_0))$ on the graph of Q at which the tangent drawn to Q also contains $(0, -6)$, we can use the slope formula, $m = \dfrac{y_2 - y_1}{x_2 - x_1}$, to obtain the slope of this line. Thus, the slope is $\dfrac{(x_0^2 - 4x_0 + 3) - (-6)}{x_0 - 0}$. However, from the slope function Q', we also know that the slope of the tangent is $2x_0 - 4$. Hence, we have the equation

$$\frac{x_0^2 - 4x_0 + 9}{x_0} = 2x_0 - 4.$$

This equation simplifies to the quadratic equation $x_0^2 - 9 = 0$, whose set of solutions is $\{-3, 3\}$. Thus, -3 and 3 are the abscissas of the points in the graph at Q at which tangents through $(0, -6)$ may be drawn. Substituting these values in $Q(x) = x^2 - 4x + 3$, we see that the two points are $(-3, 24)$ and $(3, 0)$. Since we know the coordinates of two points in each of these tangents, we can easily determine that the equations of these lines are $y = -10x - 6$ and $y = 2x - 6$, respectively.

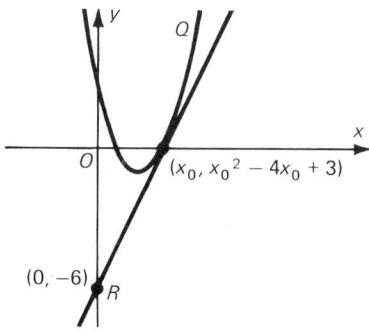

In each of the following exercises, an equation for a function and the coordinates of a point not on the graph of the function are given. For each exercise, determine the coordinates of any point on the graph of the function at which a tangent may be drawn which passes through the given point. Then write the equations of any such tangent lines. Finally, sketch the graph of the function and draw the tangents.

13. $Q(x) = x^2 - 4$; $(0, -8)$.
14. $Q(x) = x^2 - 6x$; $(1, -9)$.
15. $Q(x) = -x^2 + 4x - 4$; $(4, -3)$.
16. $Q(x) = 4x^2 - 4x - 3$; $(0, -4)$.

4

Polynomial Functions

Introduction

The constant, linear, and quadratic functions that you studied in Chapters 2 and 3 are special cases of the general category of functions that are called *polynomial functions*. In this chapter we will analyze polynomial functions in much the same way that we analyzed linear and quadratic functions. This means that we will discuss solutions and methods of determining solutions of polynomial equations, and we will discuss the characteristics of polynomial functions and their graphs, such as slope at a given point, maximum and minimum values, x- and y-intercepts, regions of increase and decrease, and regions of concavity.

4.1 Polynomial Functions

We begin our study with the definitions of a polynomial and a polynomial function.

Definition 4.1a   A **polynomial in x** is an expression of the form

$$a_n x^n + a_{n-1} x^{n-1} + \ldots + a_1 x + a_0,$$

where n is a nonnegative integer, $a_0, a_1, \ldots, a_n$ are numbers, with $a_n \neq 0$.

If $a_0, a_1, \ldots, a_n$ are real numbers, then the polynomial expression is called a **polynomial over the field of real numbers.** The numbers $a_0, a_1, \ldots, a_n$ are called the *coefficients* of the terms of the polynomial. The integer n is called the *degree* of the polynomial. The following are examples of polynomials of degree 3, 8, and 17, respectively.

Example 1 $3x^3 + 4x^2 + 7$

Example 2 $-\sqrt{3}\, x^8$

Example 3 $x^{17} + 3$

Definition 4.1b   A **polynomial function of a single real variable** x is a function defined by an equation of the form

$$P(x) = a_n x^n + a_{n-1} x^{n-1} + \ldots + a_1 x + a_0,$$

where n is a nonnegative integer and $a_0, a_1, \ldots, a_n$ are numbers, with $a_n \neq 0$.

The positive integer n is called the degree of the polynomial function P. If the domain of P is R and $a_0, a_1, \ldots, a_n$ are real numbers, then P is called a **real polynomial function.** Unless we specify otherwise, the polynomial functions that we will consider in this book will be real polynomial functions.

Quadratic functions, which are defined by equations of the form $Q(x) = ax^2 + bx + c$, are polynomial functions of degree 2; and linear functions, which are defined by equations of the form $l(x) = mx + b$, are polynomial functions of degree 1. The constant functions, except for the function f defined by $f(x) = 0$, $x \in R$, are polynomial functions of degree zero. There are reasons why it is convenient to call the function f, defined by $f(x) = 0$, the **zero polynomial function,** but we do not assign it any degree. (As you will see, for a polynomial function P of degree n, there are at most n distinct values of x for which $P(x)$ is equal to 0, but the zero polynomial function is equal to 0 for all values of x.)

In dealing with a function f, it is often useful to be able to determine all of the values of x in the domain of f for which $f(x) = 0$. In practical applications, these values may be the solutions to the problem being discussed. If c is in the domain of f and $f(c) = 0$, then $(c, 0)$ is a point on the graph of f where the graph intersects the x-axis. When we determine the values of x for which $f(x) = 0$, we are determining the *roots* or *solutions* of the equation $f(x) = 0$. If we think of the function f as a mapping of the real-number line into the real-number line and if a number c in the domain of f is a solution of $f(x) = 0$, then c is mapped by f onto the number 0. This fact probably inspired the following definition for the domain elements that are mapped by f onto zero.

Definition 4.1c If a is an element in the domain of a function f such that $f(a) = 0$, then a is said to be a **zero** of the function f.

The zeros of a polynomial function are also called the zeros of the polynomial used in defining the function. Notice that polynomial functions of degree zero have no zeros. (Remember that the zero polynomial function is not a polynomial function of degree zero; it has *no* degree.) We also know that a polynomial function P of degree one (a linear function), which is defined by an equation of the form $l(x) = mx + b$, with $m \neq 0$, has precisely one zero, namely $x = \dfrac{-b}{m}$.

A polynomial function Q of degree two (a quadratic function), which is defined by an equation of the form $Q(x) = ax^2 + bx + c$, with $a \neq 0$, has two, one, or no real zeros. If its discriminant $b^2 - 4ac$ is positive, it has two real zeros; if its discriminant is 0, it has one real zero; if its discriminant is negative, it has no real zeros. If $b^2 - 4ac > 0$, the two zeros are $\dfrac{-b \pm \sqrt{b^2 - 4ac}}{2a}$; and if $b^2 - 4ac = 0$, the single zero of Q is $\dfrac{-b}{2a}$.

We see that the zeros of a quadratic function Q can always be expressed in terms of the coefficients a, b, and c by means of formulas involving only the rational operations (addition, subtraction, multiplication, and division) and the extraction of roots. Similarly, the zeros of a polynomial function of degree three, defined by $p(x) = a_3 x^3 + a_2 x^2 + a_1 x + a_0$, can be expressed in terms of the coefficients a_3, a_2, a_1, and a_0, by formulas involving only the rational operations and the extraction of roots. Analogous formulas exist which give the zeros of polynomial functions of degree four in terms of their coefficients. Such formulas were known in the early part of the sixteenth century. However, all attempts to obtain similar formulas for polynomial functions of degree greater than four proved fruitless.

In the early part of the nineteenth century, a Norwegian mathematician, Niels Henrik Abel, proved that no such formulas exist for polynomial functions of degree greater than four. This does not mean that methods for finding the zeros of polynomial functions of degree greater than four do not exist. There are such methods. What Abel proved was that if the degree of a polynomial

$$a_n x^n + a_{n-1} x^{n-1} + \ldots + a_1 x + a_0$$

is greater than four, then there are no formulas, expressed solely in terms of the coefficients a_n, a_{n-1}, ..., a_1, a_0 and involving only a finite number of rational operations and the extraction of roots, which give the zeros of *every* polynomial of that degree.

As a matter of fact, the formulas for finding the zeros of polynomial functions of degrees three and four are complicated, and we will not consider them. We will, however, develop other methods for locating the zeros of a polynomial of any degree. Before we are able to do this, we will need to learn more about the characteristics of the general polynomial function.

We first state some basic properties of polynomial functions:

1. If P is a polynomial function and a is a real number, then aP is a polynomial function. Further, if $a \neq 0$, then the degree of aP is the same as that of P, and P and aP have the same zeros.
2. If P_1 and P_2 are polynomial functions, then $P_1 \cdot P_2$ is a polynomial function. Further, if neither P_1 nor P_2 is the zero polynomial function, then the degree of $P_1 \cdot P_2$ is the sum of the degrees of P_1 and P_2, and the set of zeros of $P_1 \cdot P_2$ is the union of the set of zeros of P_1 and the set of zeros of P_2.
3. Two polynomial functions are equal if and only if they are of the same degree and their corresponding coefficients are equal.
4. If P_1 and P_2 are polynomial functions, then $P_1 + P_2$ is a polynomial function. Further, if P_1 and P_2 do not have the same degree, then the degree of $P_1 + P_2$ is the same as the degree of P_1 or of P_2, whichever is greater. If P_1 and P_2 have the same degree and if $P_1 \neq -P_2$ (that is, if the coefficients of P_1 are *not* the additive inverses of the corresponding coefficients of P_2), then the degree of $P_1 + P_2$ is the exponent of the highest power of x for which the sum of the corresponding coefficients of P_1 and P_2 is different from zero. If $P_1 = -P_2$, then $P_1 + P_2$ is the zero polynomial.

Although statement 3 would appear to be obvious, we are not in a position to prove it at this time, and we assume its validity.

Using properties of the real-number system, you should be able to carry out proofs of statements 1, 2, and 4.

4.1 EXERCISES

In each of the Exercises 1–4, state whether the given equation defines a polynomial function and, if so, state the degree of the function.

1. $f(x) = 10 - 3x^5 + x^{11}$.
2. $g(x) = \sqrt{2}$.
3. $h(x) = x^2 - 2\sqrt{x} + 5$.
4. $F(x) = 2x^2 - x + \dfrac{1}{x}$.

For each of the Exercises 5–8, determine the value of the function as indicated.

5. $g(-2)$, where $g(x) = 2x^3 - x^2 + 4x - 5$.
6. $F(\frac{1}{2})$, where $F(x) = 3x^3 + 2x^2 + 6x - 3$.
7. $G(2.5)$, where $G(x) = 4x^3 - 3x^2 - 2x - 13$.
8. $H(-\sqrt{2})$, where $H(x) = x^3 + x^2 - 2x - 2$.

In each of the Exercises 9 and 10, write equations that define $f + g$ and $f \cdot g$. Then state the degrees of the functions f, g, $f + g$, and $f \cdot g$.

9. $f : f(x) = 3x^3 + 2x + 1$; $g : g(x) = x^2 - 3x + 5$.
10. $f : f(x) = \frac{1}{2}x^4 + \frac{1}{3}x^3 + \frac{2}{3}x$; $g : g(x) = -\frac{1}{2}x^4 - \frac{1}{3}x^3$.

For each of the Exercises 11–13, determine the values of a and b so that the given polynomial functions will be equal.

11. $f : f(x) = ax^3 + 9x$; $g : g(x) = 6x^3 + bx^2 + 9x$.
12. $f : f(x) = (a + b)x^2 + 7x + 16$; $g : g(x) = 11x^2 + (a - b)x + 16$.
13. $f : f(x) = 6x^4 + (b - a)x^2 - 11x + 3$; $g : g(x) = abx^4 - x^2 - 11x + 3$.

We have used the following theorem concerning real numbers to determine the zeros of a quadratic function: If $xy = 0$, then $x = 0$ or $y = 0$. This theorem may be used to derive the following, more general, theorem: If the product of two or more real numbers is zero, then at least one of the numbers is zero. Use this general theorem to find the real zeros of the polynomial function P defined by the given equation in each of Exercises 14–17.

14. $P(x) = (x - 3)(x^2 + 2)$.
15. $P(x) = (9x^2 - 4)(x + 1)$.
16. $P(x) = (2x + 1)(x^2 - 2x - 15)$.
17. $P(x) = (2x^2 + x - 3)(6x^2 + 5x - 6)$.

For each of the Exercises 18–25, write $P(x)$ in factored form. Then find the real zeros of function P.

18. $P(x) = x^3 - 4x$.
19. $P(x) = x^3 - 3x^2 + 2x$.
20. $P(x) = 2x^3 + x^2 - 6x$.
21. $P(x) = x^4 - 81$.
22. $P(x) = x^4 - 10x^2 + 9$.
23. $P(x) = x^3 - 8$.
24. $P(x) = x^6 - 1$.
25. $P(x) = x^6 - 9x^3 + 8$.
26. Given that P is a polynomial function and that a is a real number, prove each of the following.
    a) aP is a polynomial function.
    b) If $a \neq 0$, the degree of aP is the same as that of P.
    c) If $a \neq 0$, P and aP have the same zeros.

4.2 Graphing Polynomial Functions

We studied the polynomial functions of degree zero, one, and two in considerable detail in Chapters 2 and 3. It is obvious that we cannot study polynomial functions of each degree in the same fashion. The fact that we have considered the constant, linear, and quadratic functions as special cases can be justified on the grounds that they occur more often in application than do polynomial functions of greater degree. We will now concentrate on techniques for gaining information about polynomial functions of any degree.

When a mathematician or scientist is working on a problem related to a specific function f, he is interested in the behavior of $f(x)$ as x takes on various values in the domain of f. He is particularly interested in values of x at which there are basic changes in the behavior of $f(x)$, or in sets of values of x for which $f(x)$ has a certain type of behavior. In order to solve his problem he will usually need to be able to answer questions like the following.

1. For what values of x is $f(x)$ defined; that is, what is the domain of f? Is f discontinuous anywhere?
2. If $f(0)$ is defined, what is its value?
3. What values does $f(x)$ take on as x takes on values in the domain of f; that is, what is the range of f? For what values of x is $f(x) = 0$?
4. For what values of x is $f(x)$ positive, and for what values of x is $f(x)$ negative?
5. For what regions in the domain of f is $f(x)$ increasing as x increases? For what regions is $f(x)$ decreasing as x increases?
6. Are there values x_0 such that, for all values of x sufficiently near x_0, $f(x_0) \geq f(x)$? Are there values x_0 such that, for all values of x sufficiently near x_0, $f(x_0) \leq f(x)$? In other words, does the graph of f have local maximum and minimum points and, if so, where are these points?
7. What is the behavior of $f(x)$ as x becomes large, and what is the behavior of $f(x)$ as $|x|$ becomes large, but $x < 0$?
8. Is there a greatest value of $f(x)$? Is there a least value of $f(x)$?

It is not always possible to answer exactly all of those questions requiring numerical answers. However, there are straightforward techniques, applicable to all functions considered in this book, which will produce answers to these questions to any desired degree of accuracy. For some functions, including many polynomial functions, we will be able to give exact answers to these questions.

One of the greatest advantages to be gained from an ability to graph functions is that a graph of a function expresses, in a compact and unified manner, the answers to the preceding questions. Hence, given a graph of a function, we can usually read off the answers to these questions very quickly. For example, consider the graph of a function f shown in Figure 4-1. (The arrowheads at the ends of the graph indicate that the graph continues to rise, both to the left and to the right.)

Figure 4-1

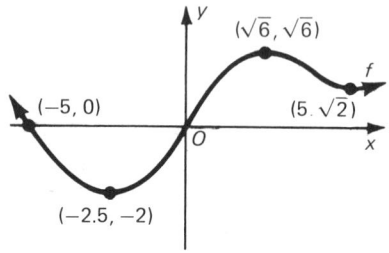

By examining this graph, you should be able to answer for this function all the questions on page 163. Now consider the graph in Figure 4-2 and give the answers to the questions for the function g that it represents.

The answers to the eight questions should be the same for both graphs. Thus, the information contained in these eight answers is *not* sufficient to characterize a function.

Figure 4-2

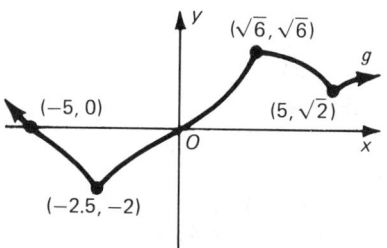

If, however, we can answer questions 1 through 8 not only for f, but also for its slope function f', we *can* draw a good approximation of the graph of f. [Recall that $f'(x)$ is equal to the slope of the tangent line to the graph of f at $(x, f(x))$, if this tangent line exists.] Thus, to be able to draw a good approximation to the graph of a polynomial function P, we will need to be able to determine its slope function and to answer the questions on page 163 for both P and P'. One of our first goals will be, given a polynomial function P, to determine its slope function P'.

In earlier courses, a common procedure that you may have used in graphing a function f is first to compute a set of coordinate pairs for points in the graph of the function (usually with the abscissas forming a sequence of integers). These points are then plotted and connected with a smooth curve to represent the graph of the function. This method is suitable for first- and second-degree polynomial functions, since we know in advance that the graph will be a line or a parabola.

However, a mechanical procedure such as this can often give a very false impression of the nature of a polynomial function of degree three or greater. For example, this method applied to the graphing of the polynomial function p defined by

$$p(x) = 32x^3 - 6x - 1$$

might produce the graph in Figure 4–3, with the points $(-2, -245)$ and $(2, 243)$ not plotted because there is not enough space. A second attempt might result in Figure 4–4.

Figure 4-3

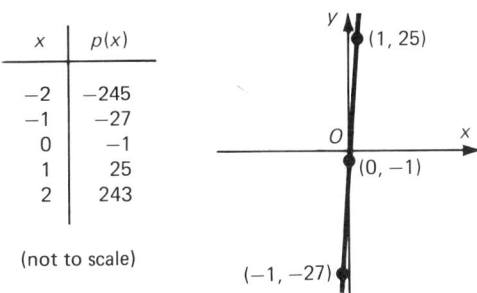

Figure 4-4

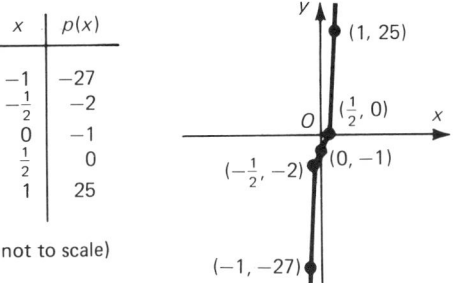

A better approach to the graphing of p involves some preliminary analysis of the functions p and p'. The results of this analysis enable us to plot four points, the abscissas of which are values at which major changes occur in the behavior of the function. Other facts learned in the analysis enable us to connect these points with a curve which, although not intended to give an accurate description of the function at every point, does give a correct characterization of the behavior of p over its entire domain. (See Figure 4–5 on page 166.)

Some of the properties of p and p' which were used to produce the graph in Figure 4-5 are listed below. In this chapter, you will see how properties such as these can be determined and what their significance is.

Figure 4-5

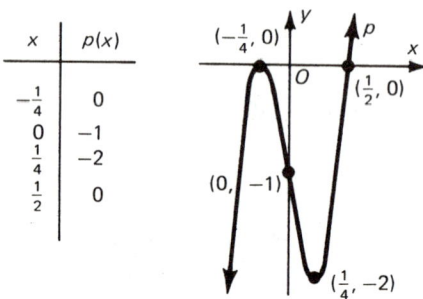

1. The function p is continuous with domain and range R, and its slope function is defined by $p'(x) = 96x^2 - 6$, with domain R.
2. The function p is an increasing function for $x < -\frac{1}{4}$, a decreasing function for $-\frac{1}{4} < x < \frac{1}{4}$, and an increasing function for $x > \frac{1}{4}$. The function p' is a decreasing function for $x < 0$ and is an increasing function for $x > 0$.
3. The zeros of p are $-\frac{1}{4}$ and $\frac{1}{2}$. The zeros of p' are $-\frac{1}{4}$ and $\frac{1}{4}$.

The basic lesson to be learned from this example is that, instead of graphing a function to learn about its behavior and characteristics, we learn about the behavior and characteristics of the function by analysis and then record our findings in a compact manner by graphing the function. This, like many generalizations, is *not* entirely true, but nearly so. There are occasions when a painstaking, detailed plotting of a function is the best method for learning about the nature of the function, although such a procedure is quite uncommon unless done with a computer. There are other occasions when, after some preliminary analysis, a rough graph of the function is made in order to get approximations of functional values to be used in further analysis.

Our immediate goal will be to describe various analytical techniques for determining values of x at which basic changes in the behavior of a function f occur and for determining characteristics of the behavior of f over certain regions of its domain.

4.2 EXERCISES

1. Use the graph of the function f that is shown to answer each of the following questions. The arrowheads indicate that the graph continues to rise to the left and continues to fall to the right.

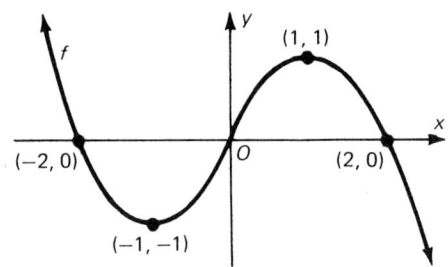

a) What is the domain of f? Is f discontinuous anywhere?
b) What is the range of f?
c) For what values of x is $f(x) = 0$? For what values of x is $f(x) < 0$? For what values of x is $f(x) > 0$?
d) What is the behavior of $f(x)$ as x becomes large? As $|x|$ becomes large when $x < 0$?
e) For what values of x is $f'(x) = 0$? For what values of x is $f'(x) < 0$? For what values of x is $f'(x) > 0$?

Follow the pattern of Exercise 1 for the functions whose graphs are shown in Exercises 2–5.

2.

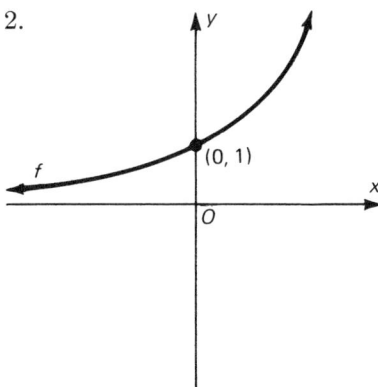

3.

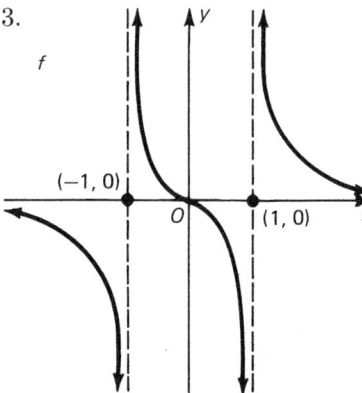

4.

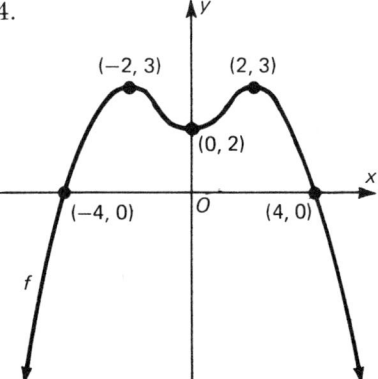

5.

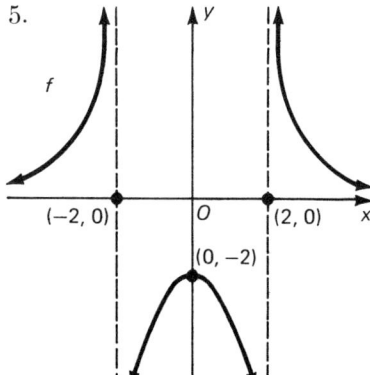

4.2 GRAPHING POLYNOMIAL FUNCTIONS

For each of Exercises 6, 7, and 8, sketch the graph of a function with all of the stated properties.

6. a) The function f is continuous and its slope function f' is defined for all x, $x \in R$. The range of f is R^+.
   b) $f(0) = 1$.
   c) f' has no zeros. $f'(x) > 0$ for all $x \in R$.
   d) As x becomes large, $f(x)$ becomes large. As $|x|$ becomes large, but $x < 0$, $f(x)$ becomes small, approaching closer and closer to zero.

7. a) The function g is continuous and its slope function g' is defined for all $x \in R$. The range of g is the set of all real numbers greater than or equal to -5.
   b) $g(0) = -2$. The zeros of $g(x)$ are -1 and 5.
   c) $g(x) > 0$ for $\{x \mid x < -1 \text{ or } x > 5\}$.
   $g(x) < 0$ for $\{x \mid -1 < x < 5\}$.
   d) The zero of $g'(x)$ is 2. $g'(x) > 0$ for $\{x \mid x > 2\}$. $g'(x) < 0$ for $\{x \mid x < 2\}$.
   e) As $|x|$ becomes large, $g(x)$ becomes large.

8. a) The function h is continuous and its slope function h' is defined for all $x \in R$. The range of h is R.
   b) $h(0) = 3$. $h(4) = -3$. The zeros of $h(x)$ are -2, 2, and 6.
   c) $h(x) > 0$ for $\{x \mid -2 < x < 2 \text{ or } x > 6\}$.
   $h(x) < 0$ for $\{x \mid x < -2 \text{ or } 2 < x < 6\}$.
   d) The zeros of $h'(x)$ are 0 and 4. $h'(x) > 0$ for $\{x \mid x < 0 \text{ or } x > 4\}$. $h'(x) < 0$ for $\{x \mid 0 < x < 4\}$.
   e) As x increases and becomes large, $h(x)$ becomes large. As $|x|$ becomes large, but $x < 0$, $|h(x)|$ becomes large, but $h(x) < 0$.

Describe the behavior of $P(x)$ for large values of x and for large values of $|x|$, when $x < 0$, in each of the following exercises.

9. $P(x) = 3x^3 + 7x - 11$.
10. $P(x) = -3x^6 + x^3 - 1$.
11. $P(x) = -5x^5 - 4x^2 - x$.
12. $P(x) = 4x^4 - 3x^3 + 2x - 1$.

As Exercises 9–12 indicate, for large values of x, $P(x) = a_n x^n + \ldots + a_1 x + a_0$ "behaves" like $f(x) = a_n x^n$. For this reason, the behavior of $P(x)$, as $|x|$ becomes large, can be analyzed simply by investigating the behavior of the function f defined by $f(x) = a_n x^n$. For each of the following cases, describe the behavior of $f(x)$ when x becomes large and also when $|x|$ becomes large, but $x < 0$.

13. $a_n > 0$ and n is even.
14. $a_n > 0$ and n is odd.
15. $a_n < 0$ and n is even.
16. $a_n < 0$ and n is odd.

4.3 Continuity and the Intermediate Value Theorem

We have seen that the constant, linear, and quadratic functions are continuous functions. It is not difficult to show that every polynomial function is a continuous function, as our informal proof of the following theorem demonstrates.

Theorem 4.3a *All polynomial functions are continuous functions.*

In proving this theorem, we need the following properties of functions:

1. The identity function I, defined by
$$I(x) = x, \quad \text{for all } x \in R,$$
is a continuous function.
2. The product of a real number and a continuous function is a continuous function.
3. The product of two continuous functions is a continuous function.
4. The sum of two continuous functions is a continuous function.
5. The constant function, defined by
$$f(x) = a_0, \quad \text{for all } x \in R,$$
is a continuous function.

We now use these properties to draw some conclusions. The product of I with itself, $I \cdot I = I^2$, defined by
$$I^2(x) = I(x) \cdot I(x) = x \cdot x = x^2,$$
is a continuous function, since it is the product of two continuous functions.

Since I^2 is a continuous function, $I \cdot I^2 = I^3$, defined by
$$I^3(x) = I(x) \cdot I^2(x) = x \cdot x^2 = x^3,$$
is a continuous function.

If we continue this process, we see that, if n is a positive integer, the nth power of the function I, I^n, defined by
$$I^n(x) = \underbrace{I(x) \cdot I(x) \cdot \ldots \cdot I(x)}_{n \text{ terms}} = \underbrace{x \cdot x \cdot \ldots \cdot x}_{n \text{ terms}} = x^n,$$
is a continuous function.

Because the product of a real number a and a continuous function is continuous, we see that aI^n, which is defined by
$$aI^n(x) = ax^n,$$
is a continuous function.

The equation that defines a polynomial function P,
$$P(x) = a_n x^n + a_{n-1} x^{n-1} + \ldots + a_1 x + a_0,$$
can be written in the form
$$P(x) = a_n I^n(x) + a_{n-1} I^{n-1}(x) + \ldots + a_1 I(x) + a_0.$$

We know that each term in this equation defines a continuous function, and because the sum of continuous functions is a continuous function, we conclude that a polynomial function is continuous.

The fact that polynomial functions are continuous gives us important information about their graphs. Interpreted geometrically, the fact that polynomial functions are continuous means that there are no "jumps" or "breaks" in the graph of a polynomial function. This property, in turn, leads to a second property that polynomial functions have as a result of their continuity.

Suppose that f is a continuous function and that $(a, f(a))$ and $(b, f(b))$ are points on its graph which are on opposite sides of the horizontal line $y = k$, as in Figure 4-6. What does your intuition tell you about the intersection of the line $y = k$ and the portion of the graph of f that is between the points $(a, f(a))$ and $(b, f(b))$? Before reading any further, try to formulate as careful an answer to this question as you can. You might also think about a related question: If you are given a graph of a continuous function f and you draw a horizontal line so that two points of the graph are on opposite sides of this line, what can you say about the intersection of the graph and the horizontal line?

Figure 4-6

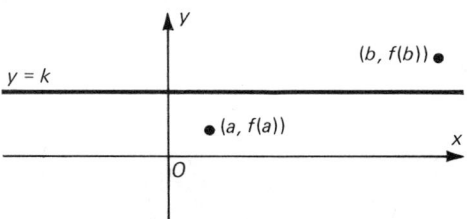

In answer to our first question about the graphs of f and $y = k$, we see that the graph of f must surely intersect the line $y = k$ at least once. It might cross the line $y = k$ several times, as in Figure 4-7, but it clearly must cross the line at least once.

Figure 4-7

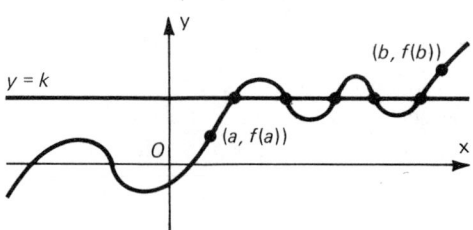

A more formal statement of this property of continuous functions is known as the *Intermediate Value Theorem*.

Theorem 4.3b (***Intermediate Value Theorem***) *Let the function f be continuous for all values of x in the interval $a \leq x \leq b$, with $f(a) \neq f(b)$. If k is a number between $f(a)$ and $f(b)$, then there is at least one number x_0 in the interval $a < x < b$ such that $f(x_0) = k$.*

We are not in a position to prove this useful theorem, so we will accept it without proof. For our purposes, the following informal statement of the theorem will be most useful.

Suppose that f is a continuous function and that $(a, f(a))$ and $(b, f(b))$ are points on the graph of f that are on opposite sides of the horizontal line $y = k$. Then the portion of the graph of f which connects $(a, f(a))$ and $(b, f(b))$ will intersect the line $y = k$ in at least one point.

4.3 EXERCISES

Use the Intermediate Value Theorem to prove each of the following.

1. $P: P(x) = 2x^3 - 3x^2 + 2x - 3$ has a zero between 1 and 2.
2. $P: P(x) = 5x^4 - 2$ has at least one positive zero.
3. $P: P(x) = -3x^3 - 1$ has at least one negative zero.
4. If $P(x)$ is a polynomial such that a_n and a_0 have opposite signs, then $P: x \to P(x)$ has at least one positive zero. (Hint: See Exercises 13–16 on page 168.)
5. If $P(x)$ is a polynomial of odd degree and a_n and a_0 have the same sign, then $P: x \to P(x)$ has at least one negative zero.
6. What does the theorem in Exercise 4 tell you about the graph of $P: x \to P(x)$?
7. What does the theorem in Exercise 5 tell you about the graph of $P: x \to P(x)$?
8. Let f be a continuous function for all x in the interval $a \leq x \leq b$ such that $f(a)$ and $f(b)$ have different signs. If f is a nondecreasing function or a nonincreasing function over this interval, then f has exactly one zero in the interval $a < x < b$.
    a) Show that the equation of the line containing $(a, f(a))$ and $(b, f(b))$ is
    $$y = f(a) + \frac{f(b) - f(a)}{b - a}(x - a).$$
    b) Show that this line intersects the x-axis at the point whose abscissa is
    $$x_1 = \frac{af(b) - bf(a)}{f(b) - f(a)}.$$

c) Let x_0 be the zero of f in the interval $a < x < b$. In each of the following cases draw an appropriate sketch and determine whether $x_0 < x_1$ or $x_0 > x_1$.
   (1) f is nondecreasing and its graph is concave upward.
   (2) f is nondecreasing and its graph is concave downward.
   (3) f is nonincreasing and its graph is concave upward.
   (4) f is nonincreasing and its graph is concave downward.

4.4 Slope Functions of Polynomial Functions

In Section 3.10 of Chapter 3 we derived the slope function $q'_a(x) = 2ax$ for the quadratic function $q_a(x) = ax^2$. The slope function for the general quadratic function $Q_{a,b,c}(x) = ax^2 + bx + c$ is $Q'_{a,b,c}(x) = 2ax + b$. (See Exercise 4, page 156.) It turns out that every polynomial function has a slope function. Since a constant function $P_0(x) = a_0$ has slope 0, it follows that the slope function for such a function is defined by $P'_0(x) = 0$. Similarly, since a linear function $P_1(x) = ax + b$ has constant slope a, it follows that the slope function of a linear function is defined by $P'_1(x) = a$. These remarks are summarized in the chart below, where we have used slightly different notation.

Polynomial function	Slope function
$P_0(x) = a_0$.	$P'_0(x) = 0$.
$P_1(x) = a_1 x + a_0$, with $a_1 \neq 0$.	$P'_1(x) = a_1$.
$P_2(x) = a_2 x^2 + a_1 x + a_0$, with $a_2 \neq 0$.	$P'_2(x) = 2a_2 x + a_1$.

A careful comparison of the two columns above suggests that there is a relationship between the defining equation of a polynomial function and the defining equation of its slope function. This relationship becomes clearer if we add to the list the slope function for the general third-degree polynomial function, commonly called a *cubic function*.

$$P_3(x) = a_3 x^3 + a_2 x^2 + a_1 x + a_0, \quad \text{with } a_3 \neq 0,$$
and
$$P'_3(x) = 3a_3 x^2 + 2a_2 x + a_1.$$

Example 1 If $P(x) = 5x^3 - 2x^2 + 7x - 3$, then
$$P'(x) = (3 \cdot 5)x^2 + (2 \cdot (-2))x + 7 = 15x^2 - 4x + 7.$$

Example 2 If $P(x) = x^3$, then $P'(x) = 3x^2$.

We could derive this formula for $P'_3(x)$ in precisely the same manner, going through precisely the same steps, as we did the formula for the slope function of the general quadratic function (pages 154–156). For the moment, however, we will just accept the formula for $P'_3(x)$, since we intend later to develop the formula for the slope function for a polynomial function P of *any* degree.

We can make the following observations about the slope functions for polynomial functions of degree zero, one, two, and three.

1. The function P'_0 is the zero polynomial.
2. The function P'_1 is a polynomial function of degree 0.
3. The function P'_2 is a polynomial function of degree 1.
4. The function P'_3 is a polynomial function of degree 2.

What would you conjecture about the degree of the slope function P'_n, where P_n is a polynomial function of degree n?

Continuing our comparisons, we make note of the following facts. The constant terms of $P'_1(x)$, $P'_2(x)$, and $P'_3(x)$ are the coefficients of x in $P_1(x)$, $P_2(x)$, and $P_3(x)$, respectively. The coefficients of x in $P'_2(x)$ and $P'_3(x)$ are 2 times the coefficients of x^2 in $P_2(x)$ and $P_3(x)$, respectively. The coefficient of x^2 in $P'_3(x)$ is 3 times the coefficient of x^3 in $P_3(x)$.

By analogy with the earlier cases, it now seems reasonable that, if P_4 is the general fourth-degree polynomial function defined by

$$P_4(x) = a_4x^4 + a_3x^3 + a_2x^2 + a_1x + a_0, \quad \text{with } a_4 \neq 0,$$

then its slope function P'_4 is defined by

$$P'_4(x) = 4a_4x^3 + 3a_3x^2 + 2a_2x + a_1.$$

This is the case, as we could prove by the same technique as we used to find the formula for the slope function to the general quadratic. We could, in fact, employ this same technique to show that the slope function for the general nth-degree polynomial function P_n, defined by

$$P_n(x) = a_nx^n + a_{n-1}x^{n-1} + \ldots + a_2x^2 + a_1x + a_0, \quad \text{with } a_n \neq 0$$

is defined by

$$P'_n(x) = na_nx^{n-1} + (n-1)a_{n-1}x^{n-2} + \ldots + 2a_2x + a_1.$$

This general formula enables us to derive with ease the specific formula which defines the slope function for any given polynomial function.

Example 3 For example, if
$$P(x) = 4x^8 - 3x^5 + 2x^2 - 3x + 7,$$
then
$$P'(x) = 8 \cdot 4x^7 + 5 \cdot (-3)x^4 + 2 \cdot 2x - 3$$
$$= 32x^7 - 15x^4 + 4x - 3.$$

4.4a EXERCISES

Determine the slope function associated with each of the polynomial functions defined below.

1. $P(x) = 2x^3 - x + 7$.
2. $P(x) = -x^3 + x^2$.
3. $P(x) = \frac{1}{2}x^4 - x^3 + 11$.
4. $P(x) = 6x^7 - 22x - 4$.
5. $P(x) = -\frac{3}{4}x^4 + \frac{1}{3}x^3 - \frac{5}{2}x^2$.
6. $P(x) = 0.4x^8 + 2.5x^6 - 1.2x^5 + 7.1$.

Find the slope of the tangent to the graph of each of the following functions at the indicated point. Then write an equation for the tangent line.

7. $f(x) = x^2 - x$; $(3, 6)$.
8. $P(x) = x^4 + x^2 - 10$; $(-2, 10)$.
9. $f(x) = -2x^3 - 9x^2 + 5$; $(-1, -2)$.
10. $P(x) = (x^2 - 2x + 5)^2$; $(0, 25)$.

Determine the coordinates of the points on the graph of each of the following functions at which the slope of the tangent line has the indicated value.

11. $f(x) = x^3 - 3x + 1$; $m = 0$.
12. $f(x) = \frac{1}{4}x^4 + \frac{1}{2}x^2$; $m = 0$.
13. a) Find the equation of the tangent line to the graph of the function p defined by $p(x) = x^3 - 3x^2 + 5$ at the point where $x = 2$.
    b) Find the equation of the tangent line to the graph of $2p$ at the point where $x = 2$.
    c) If l_a and l_b are the linear functions whose graphs are the tangent lines defined in parts a and b, respectively, show that $l_b = 2l_a$.
14. a) Find the equation of the tangent line to the graph of the function p defined by $p(x) = x^4 + 3x - 8$ at the point where $x = -2$.
    b) Find the equation of the tangent line to the graph of $\frac{1}{2}p$ at the point where $x = -2$.
    c) If l_a and l_b are the linear functions whose graphs are the tangent lines defined in parts a and b, respectively, show that $l_b = \frac{1}{2}l_a$.
15. a) Find the equation of the tangent line to the graph of the function f defined by $f(x) = x^4 - 2x^3 - 3x - 10$ at the point where $x = 3$.
    b) Find the equation of the tangent line to the graph of the function g defined by $g(x) = -x^3 + 4x^2 + 4$ at the point where $x = 3$.

c) Find the equation of the tangent line to the graph of $f+g$ at the point where $x = 3$.

d) If l_a, l_b, and l_c are the linear functions whose graphs are the tangent lines defined in parts a, b, and c, respectively, show that $l_c = l_a + l_b$.

16. Repeat Exercise 15 for $f:f(x) = x^5 + 3$ and $g:g(x) = 3x^3 - 2x + 1$ at the points where $x = -1$.

We shall now derive informally the formula for the slope function P'_n for any polynomial function P_n of degree n. We will make use of some geometric properties of the tangents to curves that we will not prove, but that are based upon certain intuitive ideas about tangents.

Some facts that we discussed earlier about the graph of a function will be useful here. For example, from our earlier work on the multiplication of a function f by a positive real number a (see page 118), we know that the graph of af is the graph of f stretched or compressed vertically about the x-axis, depending upon whether a is greater than 1 or less than 1. (See Figure 4-8.) We also know that if f is multiplied by a negative real number a, then the graph of $af = -|a|f = |a|(-f)$ is the graph of f reflected in the x-axis and then stretched or compressed vertically about the x-axis, depending upon whether $|a|$ is greater than or less than 1. (See Figure 4-9.)

Figure 4-8

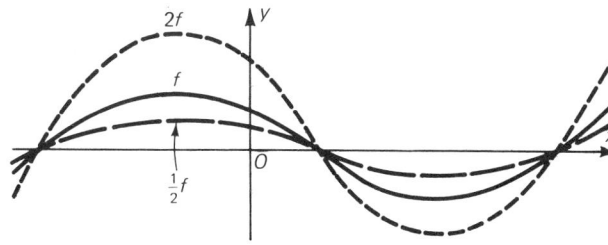

Figure 4-9

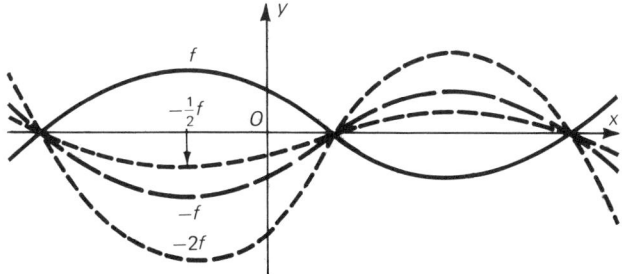

Further, every nonvertical line in the plane is the graph of a constant function C or of a linear function l. If the line is the graph of a constant function, then it is a horizontal line and has slope zero. A constant function multiplied by a real number is again a constant function. (See Figure 4–10.)

Figure 4-10

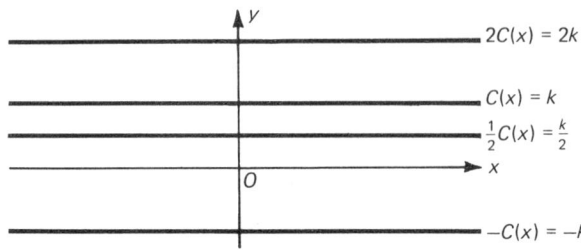

Every nonvertical line which is not horizontal is the graph of a linear function l defined by

$$l(x) = mx + b, \quad \text{with } m \neq 0,$$

where m is the slope of the line. If a is a nonzero real number, then al is a linear function defined by

$$al(x) = (am)x + ab,$$

and has for its graph a line with slope am. If $a = 0$, then al is the zero function and has the x-axis for its graph. (See Figure 4–11.)

Figure 4-11

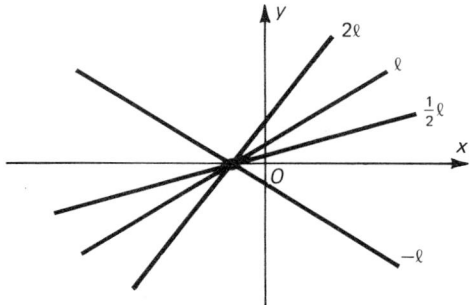

We can summarize the preceding remarks as follows:

If f is any function whose graph is a line with slope m, and a is any real number, then af is a function whose graph is a line with slope am.

Now consider a function f whose graph has a tangent line T at the point $(x_0, f(x_0))$, as shown in Figure 4–12. Because the graph of f has a tangent line at $(x_0, f(x_0))$, the slope function f' is defined at x_0 and $f'(x_0)$ is equal to the slope of the tangent line T. It follows that

$$T(x) = f'(x_0)(x - x_0) + f(x_0).$$

Figure 4-12

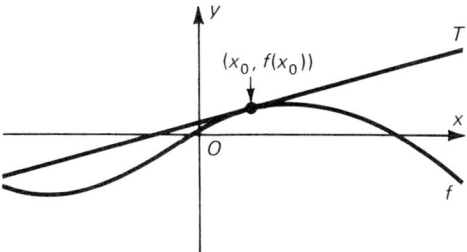

If a is a real number, then the graphs of af and aT are the graphs of f and T after they have been equally stretched or compressed vertically about the x-axis. From our previous discussion, it should be evident that the graph of aT is a line through the point $(x_0, af(x_0))$, which is where the point $(x_0, f(x_0))$ is moved under the stretching or compressing. Because the graph of T is the tangent line to the graph of f at the point $(x_0, f(x_0))$, it seems reasonable that the graph of aT is the tangent line to the graph of af at the point $(x_0, af(x_0))$. (See Figure 4-13.)

Figure 4-13

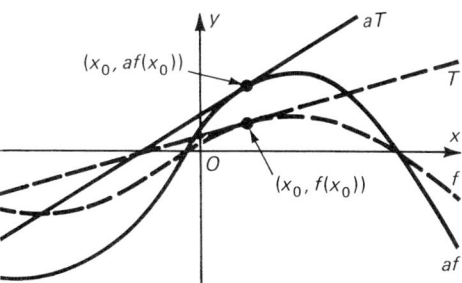

If the graph of aT is the tangent line to the graph of af at the point $(x_0, af(x_0))$, we see that the slope function for the function af is defined at x_0 and is equal to the slope of the graph of aT. That is, $(af)'(x_0)$ is the slope of the graph of aT at the point $(x_0, af(x_0))$. Since the graph of T is a line with slope $f'(x_0)$, it follows that the graph of aT is a line with slope $af'(x_0)$; hence, $(af)'(x_0) = af'(x_0)$. Since x_0 was any real number such that $f'(x_0)$ was defined, we have an informal proof of Theorem 4.4a, which is stated below.

Theorem 4.4a *If a is any real number and the slope function f' exists, then the slope function $(af)'$ exists and is defined by*

$$(af)'(x) = af'(x).$$

4.4 SLOPE FUNCTIONS

What this theorem tells us is that the slope function of the function af is a times the slope function of the function f. Geometrically this means that if a function is multiplied by the real number a, then the slope of its graph at any point is also multiplied by a. These results can be verified by using the formulas given on pages 172–173.

Example 4 If q is defined by $q(x) = x^2$, then $q'(x) = 2x$; and if q_a is defined by $q_a(x) = ax^2$, then $q_a'(x) = 2ax$. But $q_a = aq$, so the earlier equations can be written as
$$(aq)'(x) = q_a'(x) = 2ax = a \cdot 2x = aq'(x).$$

In other words, the slope of $q_a = aq$ at the point $(x_0, aq(x_0))$ is a times the slope of q at the point $(x_0, q(x_0))$.

We know that any nonvertical line is the graph of a constant or a linear function and, conversely, that the graph of a constant or a linear function is a nonvertical line. We also know that the sum of two constant functions is a constant function and that the sum of two linear functions or a linear and a constant function is a linear or a constant function. With these facts in mind, you should be able to justify the following generalization about the sum of two such functions:

If the graphs of functions l_1 and l_2 are nonvertical lines, then the graph of function $l_1 + l_2$ is a nonvertical line and the slope of the graph of $l_1 + l_2$ is the sum of the slopes of the graphs of l_1 and l_2.

Example 5 If $l_1(x) = 2x + 3$ and $l_2(x) = -3x - 2$, then $(l_1 + l_2)(x) = (2x + 3) + (-3x - 2) = -x + 1$. Hence, the slope of l_1 is 2; the slope of l_2 is -3; and the slope of $l_1 + l_2$ is $2 + (-3)$, or -1. (See Figure 4-14.)

Figure 4-14

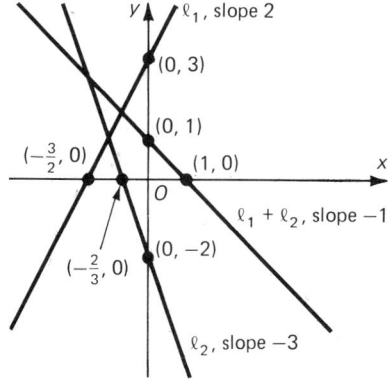

Now suppose that f and g are arbitrary functions with slope functions f' and g' defined at x_0. This means that the graph of f has a tangent line at $(x_0, f(x_0))$ with slope $f'(x_0)$ and the graph of g has a tangent line at $(x_0, g(x_0))$ with slope $g'(x_0)$. Suppose these tangent lines are the graphs of the functions T_1 and T_2, respectively. (See Figure 4–15.) Then T_1 is defined by

$$T_1(x) = f'(x_0)(x - x_0) + f(x_0),$$

and T_2 is defined by

$$T_2(x) = g'(x_0)(x - x_0) + g(x_0).$$

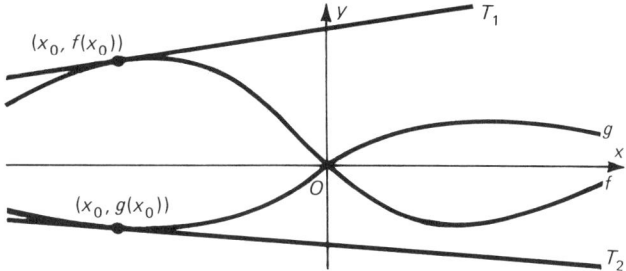

Figure 4-15

Now the graph of $f + g$ contains the point $(x_0, f(x_0) + g(x_0))$, and, since

$$(T_1 + T_2)(x) = (f'(x_0) + g'(x_0))(x - x_0) + (f(x_0) + g(x_0)),$$

the graph of the sum of T_1 and T_2 must be a line through this point. Since, near x_0, T_1 is close to f and T_2 is close to g, it seems reasonable that the graph of $T_1 + T_2$ is the tangent to that of $f + g$ at $(x_0, f(x_0) + g(x_0))$. This is the case, but we will not give a formal proof. (See Figure 4–16.)

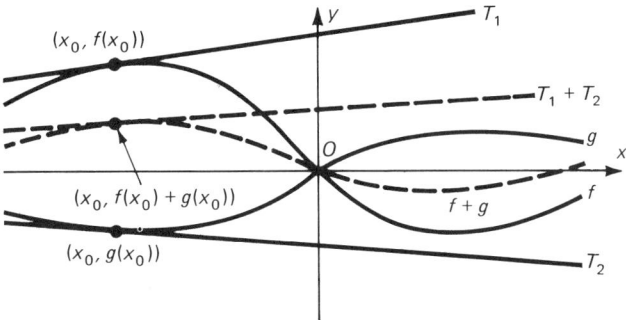

Figure 4-16

The fact that $f + g$ has a tangent at point $(x_0, f(x_0) + g(x_0))$ means that $(f + g)'$ is defined at x_0 and is equal to the slope of the tangent line at that point. That is, $(f + g)'(x_0)$ is the slope of the graph of $T_1 + T_2$ at the point $(x_0, f(x_0) + g(x_0))$. But from the fact that the slope of line $T_1 + T_2$ is the sum of the slopes of lines T_1 and T_2 and the fact that the slopes of lines T_1 and T_2 are $f'(x_0)$ and $g'(x_0)$, respectively, we see that the slope of line $T_1 + T_2$ is $f'(x_0) + g'(x_0)$. Therefore,

$$(f + g)'(x_0) = f'(x_0) + g'(x_0).$$

Since x_0 was chosen arbitrarily, this formula applies for all values of x for which f' and g' are defined. Hence, we have shown that the slope function of the sum of two functions is the sum of their slope functions.

We have just developed an informal proof of the following theorem.

Theorem 4.4b *If the slope functions for functions f and g exist, then the slope function of their sum exists and is defined by*

$$(f + g)'(x) = f'(x) + g'(x).$$

Example 6 Let $f(x) = 2x^2 + 3x$ and $g(x) = -2x + 1$. We then have $(f + g)(x) = f(x) + g(x) = 2x^2 + x + 1$. From the formulas given on page 172, we know that $f'(x) = 4x + 3$, that $g'(x) = -2$, and that $(f + g)'(x) = 4x + 1$. Thus, we see that

$$(f + g)'(x) = 4x + 1 = (4x + 3) + (-2) = f'(x) + g'(x).$$

Example 7 Let $f(x) = 2x^3 + 3x^2 - 2x + 7$ and let $g(x) = x^2 - 5x + 4$. We then have $(f + g)(x) = f(x) + g(x) = 2x^3 + 4x^2 - 7x + 11$. From the formulas given on page 172, we have $f'(x) = 6x^2 + 6x - 2$, $g'(x) = 2x - 5$, and $(f + g)'(x) = 6x^2 + 8x - 7$. Since $6x^2 + 8x - 7 = (6x^2 + 6x - 2) + (2x - 5)$, we see once again that $(f + g)'(x) = f'(x) + g'(x)$.

4.4b EXERCISES

For each of the functions defined below, verify that $(af)' = af'$.

1. $f(x) = 15x^3 - 15x^2$; $a = \frac{1}{5}$.
2. $f(x) = -6x^5 + 3x^3 - 6x$; $a = -3$.

For each exercise below, verify that $(f + g)' = f' + g'$.

3. $f(x) = 3x + 5$; $g(x) = 7x^2 - 3x + 6$.
4. $f(x) = 3x^3 - 6x^2 + 11$; $g(x) = 2x^4 + 7x - 7$.
5. If $f(x) = 6x^4 + x^3 - 4x^2 + 8x - 32$, find an equation of the tangent line to the graph of $\frac{1}{4}f$ at the point where $x = -2$.
6. If $f(x) = 3x^3 + 1$ and $g(x) = x^3 - 3x^2$, find an equation of the tangent line to the graph of $f + g$ at the point where $x = 1$.
7. If $f(x) = 3x^3 + 1$ and $g(x) = 7x^2 - 4$, write a defining equation for each of the following slope functions.
   a) $(2f)'$ b) $(3g)'$ c) $2f' + 3g'$ d) $(2f + 3g)'$
8. Repeat Exercise 7 for $f(x) = 2x^4 + x^2 - 1$ and $g(x) = x^3 - x^2 - 2$.

In each of the following exercises, p is a polynomial function. Use the information given to determine an equation that defines p.

9. $p(0) = 3$ and $p'(x) = 2x + 5$.
10. $p(1) = 1$ and $p'(x) = 3x^2$.
11. $p(0) = 2$ and $p'(x) = 6x^2 + 4x$.
12. $p(1) = 1$ and $p'(x) = -12x^3 + 5$.

Theorems 4.4a and 4.4b can be combined into a single theorem, and, in fact, they are special cases of the following theorem.

Theorem 4.4c *If a and $b \in R$ and the slope functions for f and g both exist, then the slope function for $af + bg$ exists and is defined by*

$$(af + bg)'(x) = af'(x) + bg'(x).$$

Proof Since the slope functions for f and g are defined, it follows from Theorem 4.4a that the slope functions for af and bg are defined and that $(af)'(x) = af'(x)$ and $(bg)'(x) = bg'(x)$. Now that we know that the slope functions for af and bg are defined, it follows from Theorem 4.4b that the slope function for $af + bg$ is defined and that

$$(af + bg)'(x) = (af)'(x) + (bg)'(x) = af'(x) + bg'(x).$$

This completes our proof. □

If a_1, a_2, and a_3 are real numbers and the slope functions for f_1, f_2, and f_3 exist, then we can apply Theorem 4.4c twice to show that

$$(a_1 f_1 + a_2 f_2 + a_3 f_3)'(x) = (a_1 f_1 + a_2 f_2)'(x) + a_3 f_3'(x)$$
$$= a_1 f_1'(x) + a_2 f_2'(x) + a_3 f_3'(x).$$

To generalize, suppose that n is a positive integer, that a_1, a_2, ..., a_n are real numbers, and that the slope functions for f_1, f_2, ..., f_n are defined. Then Theorem 4.4c can be applied n times to show that

$$(a_1 f_1 + a_2 f_2 + \ldots + a_n f_n)'(x) = a_1 f_1'(x) + a_2 f_2'(x) + \ldots + a_n f_n'(x).$$

Theorem 4.4c can now be used to carry our derivation of P_n' one step further. We know, from the discussion on page 170, that the polynomial function P_n, defined by

$$P_n(x) = a_n x^n + a_{n-1} x^{n-1} + \ldots + a_2 x^2 + a_1 x + a_0,$$

can be written in the form

$$P_n(x) = a_n I^n(x) + a_{n-1} I^{n-1}(x) + \ldots + a_2 I^2(x) + a_1 I(x) + a_0.$$

By Theorem 4.4c, it follows that, if the slope functions exist for I, $I^2, \ldots, I^{n-1}, I^n$, where $I^n(x) = x^n$, then

$$P'_n(x) = a_n(I^n)'(x) + a_{n-1}(I^{n-1})'(x) + \ldots + a_2(I^2)'(x) + a_1 I'(x).$$

We already know that $I'(x) = 1$ and $(I^2)'(x) = 2x$. In the exercises you will show that $(I^n)'$ is defined for all x, and

$$(I^n)'(x) = nx^{n-1}.$$

On page 173, we stated that, if P_n is the polynomial function defined by $P_n(x) = a_n x^n + a_{n-1} x^{n-1} + \ldots + a_2 x^2 + a_1 x + a_0$, then the slope function P'_n is defined by

$$P'_n(x) = a_n(I^n)'(x) + a_{n-1}(I^{n-1})'(x) + \ldots + a_2(I^2)'(x) + a_1 I'(x).$$

Since we now have $(I^n)'(x) = nx^{n-1}$, we conclude that *the slope function P'_n is defined by*

$$P'_n(x) = na_n x^{n-1} + (n-1)a_{n-1} x^{n-2} + \ldots + 2a_2 x + a_1.$$

Notice that this is the formula we assumed would be a reasonable one for the slope function of P_n earlier in this section on page 173.

Example 8 As another example of how this formula can be used to obtain the equation for a slope function, suppose that you are given the polynomial function defined by $P(x) = x^{17} + 3x^5 - 4x^2 + 7$. Using the formula, you can immediately write the following equation defining the slope function P': $P'(x) = 17x^{16} + 15x^4 - 8x$.

4.4c EXERCISES

In each of the following exercises, verify that $(af + bg)' = af' + bg'$.

1. $f(x) = 3x^2 + 6x - 7$; $g(x) = -4x^2 - 2x + 3$; $a = -2$; $b = 3$.
2. $f(x) = x^3 - 6x^2$; $g(x) = -2x^2 + 4x - 6$; $a = 2$; $b = -\frac{1}{2}$.
3. $f(x) = 3x^4 + 3x^2 + 15$; $g(x) = -x^3 - 16x$; $a = \frac{1}{3}$; $b = -3$.
4. $f(x) = 6x^5 - 3x^3 + 7x$; $g(x) = 5x^4 - 11$; $a = -\frac{1}{4}$; $b = \frac{1}{2}$.

Let $f: f(x) = x^n$, $n \in I^+$. Prove each of the following theorems.

5. If $n > 1$, the x-axis is the tangent line to the graph of f at $(0, 0)$.
6. If n is even, then $(0, 0)$ is the minimum point on the graph of f.
7. If n is odd, then $(0, 0)$ is neither a minimum point nor a maximum point on the graph of f.

The slope function is useful in approximating functional values. Exercises 8–11 illustrate this method of approximation.

8. Let $P(x_0, f(x_0))$ be a point on the graph of a continuous function f. Let h be a small positive number and P_1 and P_2 be the respective

points of intersection of the lines $x = x_0 - h$ and $x = x_0 + h$ with the graph of f.

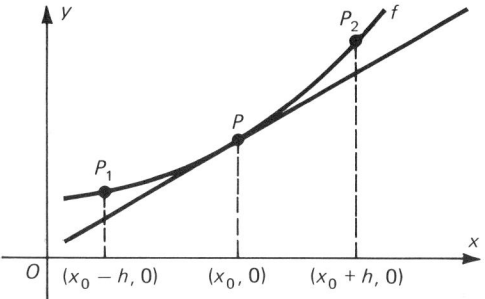

a) Show that the equation of the tangent to the graph of f at P is
$l(x) = f'(x_0)(x - x_0) + f(x_0)$.
b) Since the graph of f is close to the tangent line near P, the value of $l(x_0 - h)$ is close to the value of $f(x_0 - h)$ and the value of $l(x_0 + h)$ is close to the value of $f(x_0 + h)$. Show that
$$l(x_0 - h) = -f'(x_0)(h) + f(x_0);$$
$$l(x_0 + h) = f'(x_0)(h) + f(x_0).$$

9. a) Use $f:f(x) = x^2$, $x_0 = 10$, and $h = 0.1$ in one of the equations from Exercise 8b to find an approximate value for $(10.1)^2$.
b) Calculate $(10.1)^2$ and compare your answer with that in part a to find the error in the approximation.

10. a) Use $f:f(x) = x^3$, $x_0 = 10$, and $h = 0.1$ in one of the equations from Exercise 8b to find an approximate value for $(9.9)^3$.
b) Calculate the error in this approximation.

11. The volume, V, of a sphere may be considered as a function of its radius r, and this function is defined by $V(r) = \frac{4}{3}\pi r^3$.
a) Find an approximation for the volume of gas which may be contained in a spherical plastic balloon whose outer diameter is 200 cm. if the thickness of the balloon's plastic "skin" is 0.001 cm. (Leave your answer in terms of π.)
b) Find the approximate volume of the plastic needed to make the balloon described in part a. (Assume no stretching.)

12. Show that $(I^n)'$ is defined for all x, and $(I^n)'(x) = nx^{n-1}$.

4.5 Some Properties of Slope Functions

In the last section, we discussed the slope function of a polynomial function in some detail. We can now use the slope function to help us learn something about the graph of a polynomial function.

We know that the value of the slope function for a given x_0 is the slope of the tangent to the graph of a function f at the point $(x_0, f(x_0))$. Hence, if $f'(x_0)$ is positive for some x_0, then it would seem from geometry that the function f must be an increasing function near x_0. By contrast, if $f'(x_0)$ is negative at x_0, then it would seem that f must be a decreasing function near x_0. Stated in another way, if the slope of the tangent to the graph of f at the point $(x_0, f(x_0))$ is positive, then we would expect the graph to be rising near $(x_0, f(x_0))$; and if the slope of the tangent at this point is negative, we would expect the graph to be falling near $(x_0, f(x_0))$. Figure 4-17 shows that these conjectures are reasonable.

Figure 4-17

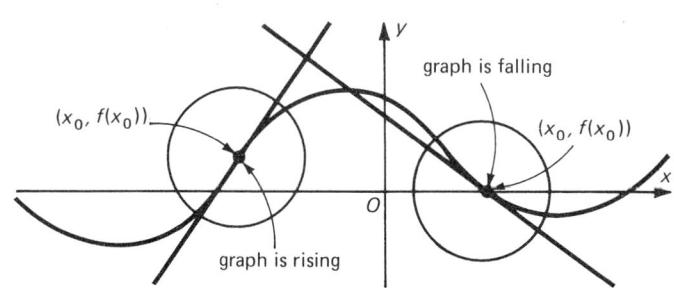

Suppose that the assertion just made is true. As a consequence, if the slope function f' takes on positive values for all x in some interval $a \leq x \leq b$, then the function f is an increasing function over that interval. On the other hand, if f' is negative for all values of x in some interval $c \leq x \leq d$, then f is a decreasing function over that interval. These results are in fact valid, but we will not try to prove them now. Instead, we will accept the following theorem, which agrees with our geometric intuition.

Theorem 4.5a *If the function f has a slope function f' and $f'(x) > 0$ for all x on an interval or a half-line, then f is an increasing function on the interval or half-line. If $f'(x) < 0$ for all x on an interval or a half-line, then f is a decreasing function on the interval or half-line.*

Example 1 If l is the linear function defined by

$$l(x) = mx + b, \quad x \in R,$$

we know that the graph of l is a line with slope m. Since $l'(x) = m$, Theorem 4.5a tells us that l is an increasing function if $m > 0$ and is a decreasing function if $m < 0$. This is consistent with what we already know about linear functions (see page 65 of Chapter 2).

Example 2 Now consider the quadratic function q, defined by $q(x) = x^2$. From the formula for the slope function of a polynomial function, we see that

$$q'(x) = 2x.$$

Because $2x$ is negative when x is negative and is positive when x is positive, we see that q is a decreasing function for negative values of x and an increasing function for positive values of x. That is, the graph of q is falling when $x < 0$ and is rising when $x > 0$. We already know that the graph of q has these characteristics by other means, but it is interesting to see how easily we can obtain this information from the slope function q'.

In Figure 4–18, we show the graphs of q and q' together. Note that when the graph of q' lies below the x-axis, then the graph of q is falling; and when the graph of q' lies above the x-axis, then the graph of q is rising. Also note that at the origin, where the graph of q' crosses from below the x-axis to above the x-axis, the graph of q stops falling and starts rising.

Figure 4-18

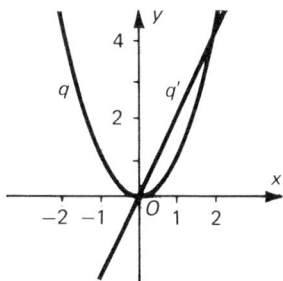

Example 3 For a third example, consider the cubic function f defined by

$$f(x) = 32x^3 - 6x - 1.$$

(We first discussed this function on pages 165–166. There, this polynomial function was referred to as p.) From the formula for the slope function of a polynomial function, we see that

$$\begin{aligned} f'(x) &= 3 \cdot 32x^2 - 6 \\ &= 96x^2 - 6. \end{aligned}$$

This equation can be written as $f'(x) = 6(16x^2 - 1)$, and, by factoring the binomial $16x^2 - 1$, we obtain

$$f'(x) = 6(4x + 1)(4x - 1).$$

We can use this equation to determine for which values of x the slope function $f'(x)$ is positive and for which values of x it is negative. Notice that, because 6 is positive, $f'(x)$ will be positive when both the factors $(4x + 1)$ and $(4x - 1)$ are positive or when both of them are negative. Furthermore, $f'(x)$ will be negative when just one of the factors $(4x + 1)$ and $(4x - 1)$ is positive.

We see that $4x + 1 = 0$ when $x = -\frac{1}{4}$; that $4x + 1 < 0$ when $x < -\frac{1}{4}$; and that $4x + 1 > 0$ when $x > -\frac{1}{4}$. These results are illustrated in Figure 4-19.

Figure 4-19

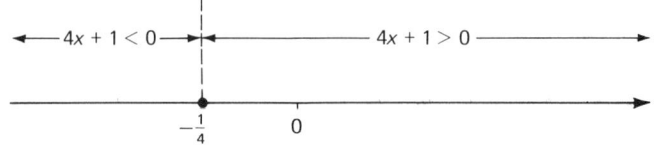

Similarly, we see that $4x - 1 = 0$ when $x = \frac{1}{4}$; that $4x - 1 < 0$ when $x < \frac{1}{4}$; and that $4x - 1 > 0$ when $x > \frac{1}{4}$. Figure 4-20 shows the results for both factors, $(4x + 1)$ and $(4x - 1)$.

Figure 4-20

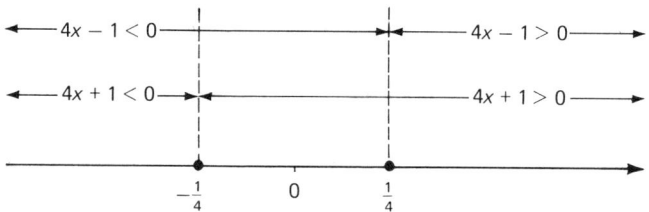

The diagram shows clearly that both $(4x + 1)$ and $(4x - 1)$ are negative when $x < -\frac{1}{4}$ and that both of these factors are positive when $x > \frac{1}{4}$. These facts imply that $f'(x)$ is positive when $x < -\frac{1}{4}$ or when $x > \frac{1}{4}$. Thus, by Theorem 4.5a, f is an increasing function and its graph is rising if $x < -\frac{1}{4}$ or if $x > \frac{1}{4}$.

Notice that if x is in the interval $-\frac{1}{4} < x < \frac{1}{4}$, then $4x - 1 < 0$ and $4x + 1 > 0$, which implies that $f'(x)$ is negative over this interval. By Theorem 4.5a, this means that f is a decreasing function and that its graph is falling in the interval $-\frac{1}{4} < x < \frac{1}{4}$. These facts are all consistent with what we learned earlier about the graph of f, which is shown again in Figure 4-21.

Figure 4-21

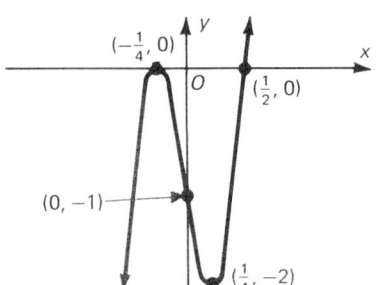

We could have determined when $f'(x)$ was positive and when it was negative in another way. We first observe that f' is a quadratic function, and, since we know a great deal about quadratic functions, we can easily draw its graph. Those values of x for which the graph of f' is above the x-axis are those values for which $f'(x)$ is positive; and those values of x for which the graph of f' is below the x-axis are those values for which $f'(x)$ is negative.

The results just obtained can be seen immediately when we graph both f and f' on the same set of coordinate axes, as shown in Figure 4-22. We observe that when the graph of f' is below the x-axis, the graph of f is falling; and when the graph of f' is above the x-axis, the graph of f is rising. This is, of course, just a geometrical interpretation of Theorem 4.5a for this particular example.

Figure 4-22

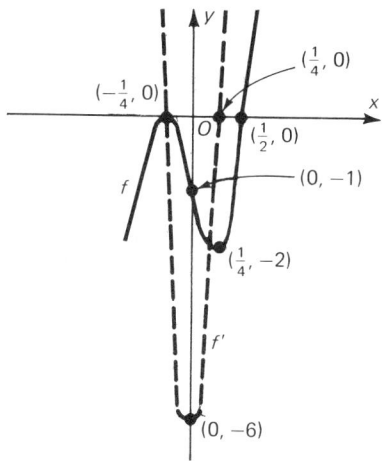

Now consider the point $(-\frac{1}{4}, 0)$ on the graph of f. This point is higher than all nearby points on the graph of f, since for all values of x sufficiently near $-\frac{1}{4}$, $f(x) < f(-\frac{1}{4})$. Nevertheless, this point is not the highest or "maximum" point on the graph, since if $x > \frac{1}{2}$, then $f(x) > 0$, which means that $f(x) > f(-\frac{1}{4})$. We say that a point such as $(-\frac{1}{4}, 0)$ is a *local* or *relative maximum point* of the graph, and we say that the function f has a *relative maximum* of $f(-\frac{1}{4}) = 0$ at $x = -\frac{1}{4}$. Similarly, we say that a point such as $(\frac{1}{4}, -2)$ is a *relative minimum point* of the graph of f and that the function f has a *relative minimum* of $f(\frac{1}{4}) = -2$ at $x = \frac{1}{4}$. We now give a general definition.

Definition 4.5a If x_0 is a point in the domain of a function f and if there is an open interval, $a < x < b$, containing x_0 such that, for every x in this interval, $f(x_0) \geq f(x)$, we say that f has a **relative maximum**

$f(x_0)$ at the point x_0. We also say that the point $(x_0, f(x_0))$ is a **relative maximum point** of the graph of f.

If, for every x in the interval $a < x < b$, $f(x_0) \leq f(x)$, then we say that f has a **relative minimum** $f(x_0)$ at the point x_0 and that the point $(x_0, f(x_0))$ is a **relative minimum point** of the graph of f.

Notice that the function f has a relative maximum or minimum at a point in its domain, not at a point of its graph.

For function f, whose graph is shown in Figure 4-22, the tangents to the graph of f at $(-\frac{1}{4}, 0)$ and $(\frac{1}{4}, -2)$ are horizontal lines. In general, if a function f has a relative maximum or minimum at a point x_0 and f' is defined at x_0, then $f'(x_0) = 0$, and the tangent to the graph of f at $(x_0, f(x_0))$ is horizontal. We state this as a theorem.

Theorem 4.5b *If a function f has a relative maximum or minimum at a point x_0 where f' is defined, then $f'(x_0) = 0$.*

Proof For our proof we will also assume that f' is continuous. Although the theorem is true without this assumption, the functions that we will consider do have continuous slope functions. We will use an indirect proof. Suppose that f has a relative maximum at x_0 and that $f'(x_0) > 0$. Then, since f' is continuous, $f'(x) > 0$ in some interval containing x_0. (Why?) But then, by Theorem 4.5a, f is an increasing function on this interval. This contradicts the assumption that f has a relative maximum at x_0. A similar contradiction results if $f'(x_0) < 0$. Thus, if f has a relative maximum at x_0, then $f'(x_0)$ must equal zero. In precisely the same way, the assumption that f has a relative minimum at x_0 leads to the conclusion that $f'(x_0) = 0$. $\square$

Note that Theorem 4.5b does *not* say that if $f'(x_0) = 0$, then f has a relative maximum or minimum at x_0. The function f defined by $f(x) = x^3$ serves as a counterexample, since $f'(x) = 3x^2$ so that $f'(0) = 0$, but it is easy to show that f is an increasing function at $x = 0$.

4.5 EXERCISES

For each function f defined below, find the regions in the domain of f for which the graph of f is rising and the regions for which the graph of f is falling. Then, if possible, sketch the graphs of f and f' on the same set of axes.

1. $f(x) = x^2 + 6x$.
2. $f(x) = -x^2 + 8x - 1$.
3. $f(x) = x^n$, n an even positive integer.
4. $f(x) = x^n$, n an odd positive integer.

For each function g defined below, use the graphical technique described on page 186 to find the regions in the domain of g for which the graph of g is rising and the regions for which the graph is falling.

5. $g(x) = 3x^3 - 4x$.
6. $g(x) = 2x^3 - 15x^2 + 12$.
7. $g(x) = -x^3 - 3x^2 + 24x - 8$.
8. $g(x) = x^3 - 4x^2 - 3x$.
9. Let f be the function defined by $f(x) = x^3 + 3x^2 + 3x + 1$.
   a) For what values of x, if any, is $f'(x) = 0$?
   b) For what values of x, if any, is $f'(x) > 0$?
   c) For what values of x, if any, is $f'(x) < 0$?
   d) Is $(-1, 0)$ a relative maximum or a relative minimum point on the graph of f? Explain your answer.
10. We have seen that if $(x_0, f(x_0))$ is a relative maximum or a relative minimum point on the graph of a polynomial function f, then $f'(x_0) = 0$. The results of Exercise 9 are clear proof that the converse of this statement is not true. However, if $f'(x_0) = 0$, then the values of $f'(x)$ for values of x close to x_0 may be used to determine whether or not $(x_0, f(x_0))$ is a relative maximum or a relative minimum point on the graph.
    a) Let the point with coordinates $(x_0, f(x_0))$ be a relative maximum point on the graph of a polynomial function f. Consider values of x close to, but less than, x_0. As x increases, do the values of $f(x)$ increase or decrease? What does this tell you about the value of $f'(x_0 - h)$, where $0 < h < c$ and c is a small positive number?
    b) Consider values of x close to, but greater than, x_0. As x increases, do the values of $f(x)$ increase or decrease? What does this tell you about the value of $f'(x_0 + h)$, where $0 < h < c$ and c is a small positive number?
    c) Follow the pattern of parts a and b to investigate the signs of $f'(x_0 - h)$ and $f'(x_0 + h)$ when $(x_0, f(x_0))$ is a relative minimum point and $0 < h < c$, where c is a small positive number.
    d) The figures shown illustrate the two remaining cases in which the tangent to the graph of a polynomial function f at a point P is a horizontal line. What will be true in these cases about the values of $f'(x_0 - h)$ and $f'(x_0 + h)$, where $0 < h < c$ and c is a small positive number?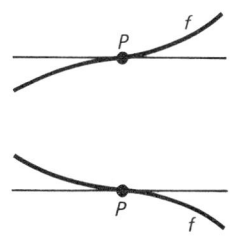
    e) Let f be a polynomial function with $f'(x_0) = 0$, and let c be a positive number. Complete the following statements about the point $(x_0, f(x_0))$.
        (1) If $f'(x_0 - h) > 0$ and $f'(x_0 + h) < 0$ for all h such that $0 < h < c$, then ⁓.

(2) If $f'(x_0 - h) < 0$ and $f'(x_0 + h) > 0$ for all h such that $0 < h < c$, then ~~~.

(3) If $f'(x_0 - h)$ and $f'(x_0 + h)$ agree in sign for all h such that $0 < h < c$, then ~~~.

In each of the following exercises, find the coordinates of each point on the graph of f at which $f'(x) = 0$. Then use the theorems from Exercise 10e to determine if the point is a relative maximum or a relative minimum point, or neither.

11. $f(x) = 18x - 3x^2$.
12. $f(x) = -x^3 + 12x$.
13. $f(x) = x^3 - 3x^2 + 5$.
14. $f(x) = \frac{1}{3}x^3 + \frac{1}{2}x^2 - 6x$.

4.6 Concavity

Figure 4-23 shows the graph of q, defined by $q(x) = x^2$, with tangent lines drawn at various points on the graph.

Figure 4-23

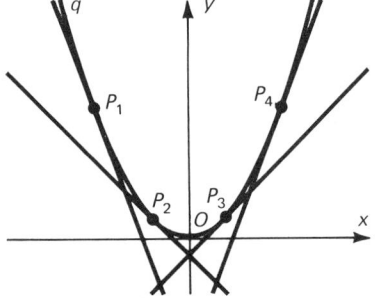

If we think of a point P moving along the graph of q from left to right and of the line through P as always tangent to the graph of q, we can interpret the diagram as showing the tangent line to the graph of q at the point P at various times during its movement. We see that this tangent line will rotate about point P in a counterclockwise direction as P moves from left to right. A good way to visualize this is to slide a straightedge along the graph of q while trying to keep it tangent to the graph at every point. (See Figure 4-24.) At points on the graph of q with negative abscissas, the tangent line has a negative slope, and, as the point P approaches the vertex of the parabola, the tangent line rotates counterclockwise until it is horizontal at the vertex $(0, 0)$. As the point P continues moving to the right through points with positive abscissas, the tangent line through P takes on a positive slope and the slope of this tangent line keeps increasing as P moves further upward to the right.

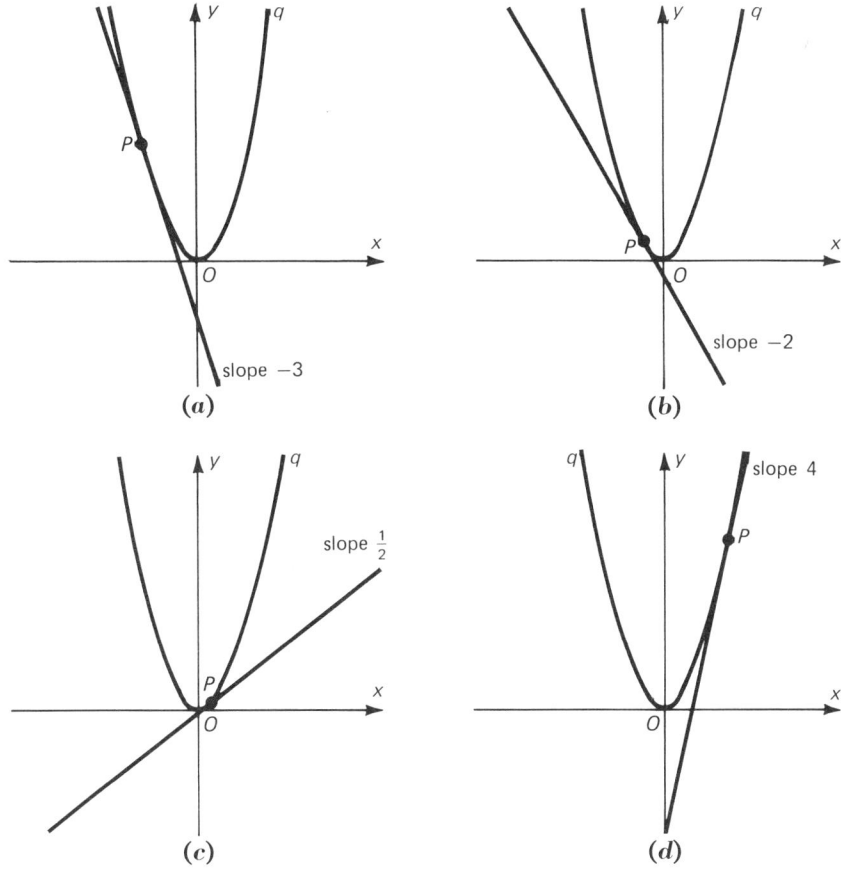

Figure 4-24

Thus, as P moves along the graph of q from left to right, the slope of the tangent line through P increases. This is just another way of saying that the slope function q' is an increasing function. We already knew this, since the graph of q' is a line with slope 2.

Notice that the slope function of q, which is defined by $q'(x) = 2x$, is a linear function that also has a slope function. The slope function of q', which we will denote by $(q')'$ or, more simply, by q'', is defined by $q''(x) = 2$. We again see that, by Theorem 4.5a, q' is an increasing function. This follows from the fact that q'', the slope function of q', is greater than zero for all real numbers.

You already know, of course, that the graph of q is concave upward, and, as you can see from Figure 4–24, the graph is above each of the tangent lines. Perhaps you can see the relationship between this property of the graph of q and the following property:

4.6 CONCAVITY 191

As a point P moves along the graph of q from left to right, the line through P which is tangent to the graph of q rotates in a counterclockwise direction.

Presently, we will explain, in mathematical terms, the connection between these two properties.

If we were to analyze the graph of the function $-q$, which is defined by $-q(x) = -x^2$, in the same manner that we have analyzed the graph of the function q, we would observe that this graph has the following properties:

1. The graph of $-q$ is below each of its tangent lines and is concave downward. (See Figure 4–25.)
2. As a point P moves along the graph of $-q$ from left to right, the line through P which is tangent to the graph of $-q$ rotates in a clockwise direction.

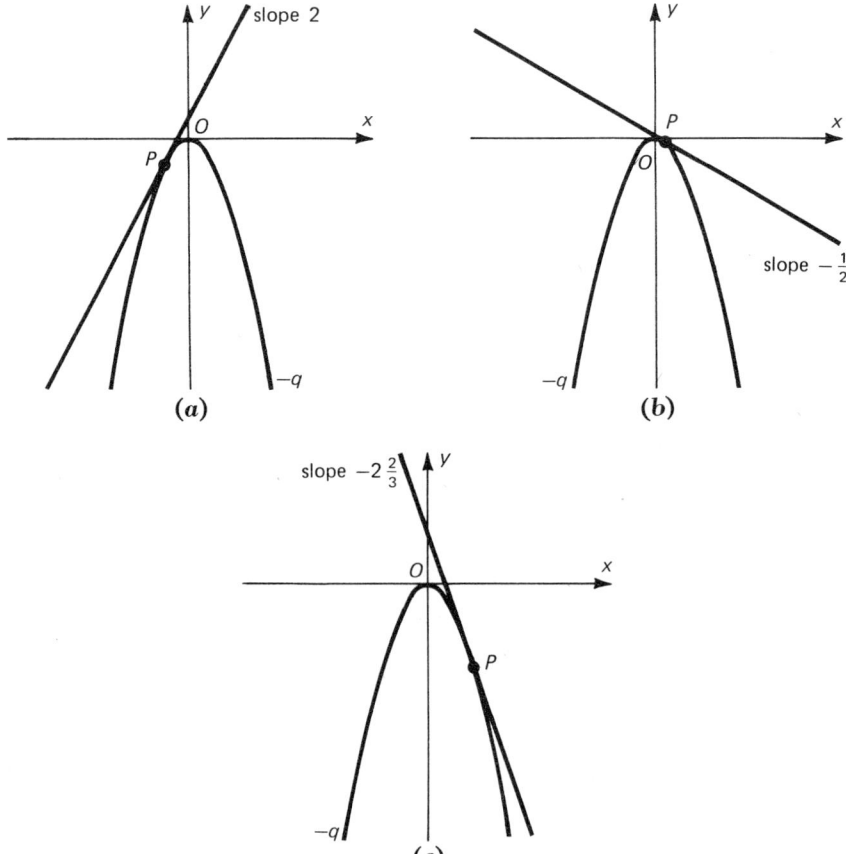

Figure 4-25

192 POLYNOMIAL FUNCTIONS

Let us now see what observations we can make about the smooth and continuous curve shown in Figure 4-26, which we will assume to be the graph of some function f.

Figure 4-26

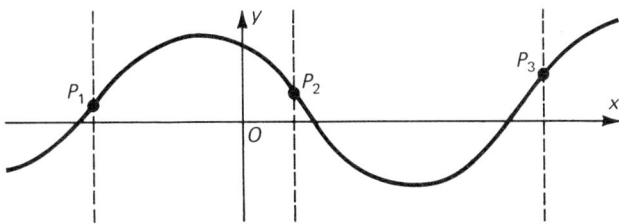

1. As a point P moves along the graph of f from left to right, the tangent line through P behaves as follows:
   a) The tangent line rotates in a counterclockwise direction when P is on the portion of the graph to the left of the vertical line through P_1 and also when P is on the section of the graph between the vertical lines through P_2 and P_3.
   b) The tangent line rotates in a clockwise direction when P is on the portion of the graph between the vertical lines through P_1 and P_2 and also when P is on the portion of the graph to the right of the vertical line through P_3.
2. The graph of f has the following properties:
   a) The graph is concave upward in the region to the left of the vertical line through P_1 and in the region between the vertical lines through P_2 and P_3.
   b) The graph is concave downward in the region between the vertical lines through P_1 and P_2 and in the region to the right of the vertical line through P_3.

If you were to experiment with other graphs of this kind, you should be convinced that the following generalization holds for all such curves: If, as a point P moves along a portion of a smooth, continuous curve from left to right, the tangent line to the curve through P rotates in a counterclockwise direction, then that portion of the graph is concave upward. If the tangent line rotates in a clockwise direction, then that portion of the graph is concave downward.

Now, suppose that the tangent line to the graph of a function f at a point P is rotating in a counterclockwise direction as P moves from left to right along a portion of the graph. This means that the slope of the tangent line is increasing and, therefore, that the slope function f' is an increasing function over that interval in the domain of the function f. If the slope function f' itself has a slope function, f'', and if $f''(x)$ is positive for $a \leq x \leq b$, then, by Theorem 4.5a, f' is an increasing function over this interval. Similarly, if $f''(x)$ is negative for $c \leq x \leq d$, then f' is a decreasing function over this interval. (See Figure 4–27 on page 194.)

Figure 4-27

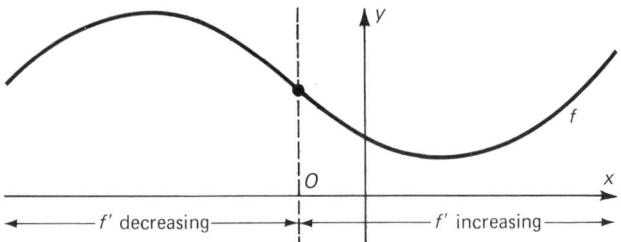

On the basis of the preceding discussion, we now state the following theorem. We will not give a rigorous proof of this theorem, since such a proof would require properties of functions that we have not yet established. However, the theorem should seem reasonable to you in light of our discussion.

Theorem 4.6a *Suppose that the function f has a slope function f' on the interval $a < x < b$. Then, if the portion of the graph of f between the vertical lines $x = a$ and $x = b$ is concave upward, f' is an increasing function on the interval $a < x < b$, and conversely. If this portion of the graph is concave downward, f' is a decreasing function on $a < x < b$, and conversely.*

If, in addition, f' has a slope function f'' and $f''(x) > 0$ for $a < x < b$, then the portion of the graph of f between $x = a$ and $x = b$ is concave upward. If $f''(x) < 0$ for all x in the interval $a < x < b$, then the portion of the graph of f between $x = a$ and $x = b$ is concave downward.

This theorem will be an extremely valuable aid to us as we try to produce good approximations to the graphs of various functions.

Example 1 Consider again the cubic function f defined by $f(x) = 32x^3 - 6x - 1$. As we have seen, f' is defined by $f'(x) = 96x^2 - 6$. Using the formula for the slope function of a quadratic function, we see that f'' is defined by $f''(x) = 2 \cdot 96x = 192x$. Now $f''(x)$ is clearly negative for negative x and positive for positive x. Thus, by Theorem 4.6a, it follows that the graph of f is concave downward to the left of the y-axis and concave upward to the right of the y-axis. (You should turn back to the graph of f given in Figure 4–21 on page 186 to see that these observations agree with the graph given there.) Since the graph of f is concave downward to the left of the y-axis and concave upward to the right of the y-axis, the graph changes concavity at point $(0, -1)$.

If $(x_0, f(x_0))$ is a point on the graph of a continuous function f and if there are numbers a and b in the domain of f, with $a < x_0 < b$, such that the portions of the graph of f between the vertical lines $x = a$ and $x = x_0$ and between the vertical lines $x = x_0$ and $x = b$ have different concavity (that is, if one portion is concave upward and the other concave downward), then $(x_0, f(x_0))$ is said to be a **point of inflection** of the graph of f.

If $(x_0, P(x_0))$ is a point of inflection on the graph of any polynomial function P, then $P''(x_0) = 0$. This is a special case of the following theorem.

Theorem 4.6b *If $(x_0, f(x_0))$ is a point of inflection on the graph of a function f having a slope function f' and if $f''(x_0)$ is defined, then $f''(x_0) = 0$.*

Proof Since $(x_0, f(x_0))$ is a point of inflection, there are numbers a and b in the domain of f such that the two portions of the graph between the vertical lines $x = a$ and $x = x_0$ and between the vertical lines $x = x_0$ and $x = b$, respectively, have different concavity. From Theorem 4.6a, we infer that $f'(x)$ must be increasing on one of the intervals, $a < x < x_0$ or $x_0 < x < b$, and decreasing on the other. But this means that $f'(x)$ must have a relative maximum or a relative minimum at x_0. (Why?) From Theorem 4.5b, we can now conclude that $(f')'(x_0) = f''(x_0) = 0$, which proves the theorem. □

You should note that Theorem 4.6b does *not* state that if $f''(x_0) = 0$, then the graph of f has an inflection point at $(x_0, f(x_0))$. The function defined by $f(x) = x^4$ serves as a counterexample, since $f''(x) = 12x^2$ so that $f''(0) = 0$, but it is easy to show that the entire graph of f is concave upward.

4.6 EXERCISES

For each function f defined in Exercises 1–6, find the values of x (intervals, half-lines, or the entire x-axis) for which the graph of f is concave upward and the values of x for which the graph is concave downward.

1. $f(x) = 6x^2 - 13x + 49$.
2. $f(x) = x^3 - 2x^2 + 5x - 6$.
3. $f(x) = -\frac{5}{6}x^3 + \frac{3}{2}x^2 - \frac{11}{3}$.
4. $f(x) = x^4 - 6x^2 + 10$.
5. $f(x) = x^4 - \frac{7}{6}x^3 - 6x^2 + 3x - 5$.
6. $f(x) = x^n$, n an even positive integer.

For each of the functions defined below, find the coordinates of each point on the graph of f at which $f''(x) = 0$. Then, for each point $(x_0, f(x_0))$, investigate the values of $f''(x_0 - h)$ and $f''(x_0 + h)$, where $0 < h < c$ and c is a small positive number, and state whether the point is a point where the graph of f changes concavity.

7. $f(x) = -2x^3 + 5x + 11$.
8. $f(x) = x^3 - 9x^2 + 22$.
9. $f(x) = 2x^4 - 4x^3 + 3x^2 + 10$.
10. $f(x) = -x^4 + 6x^3 - 12x^2$.

4.7 Basic Properties of Graphs of Polynomial Functions

Let us begin this section by summarizing some important properties of polynomial functions that have already been discussed.

1. A polynomial function P has a slope function P' which is also a polynomial function.
2. The slope function P' has a slope function P'', which is also a polynomial function.
3. If we can determine when $P(x)$ is negative, zero, and positive, then we can determine some of the basic properties of the function P and its graph. For example, if $P(x) < 0$ for $x_0 < x < x_1$, we know that the portion of the graph of P between the vertical lines $x = x_0$ and $x = x_1$ is below the x-axis. When $P(x) = 0$ for some x, we know that the graph of P intersects the x-axis at $(x, P(x))$. For $x_2 < x < x_3$, if $P(x) > 0$, then the portion of the graph of P between the vertical lines $x = x_2$ and $x = x_3$ is above the x-axis.
4. If we can determine when $P'(x)$ is negative, zero, and positive, we can use Theorem 4.5a to decide when the graph of P is falling, when it has a horizontal tangent, and when it is rising.
5. If we can determine when $P''(x)$ is negative, zero, and positive, we can use Theorems 4.6a and 4.6b to decide where the graph of P is concave downward, where the concavity of the graph of P *may* change—that is, where there *may be* inflection points, and where the graph of P is concave upward.

Example 1 For a specific example, consider the function P defined by

$$P(x) = x^3 - 4x, \quad x \in R.$$

The equation for P can be written in the following factored form:

$$P(x) = x(x - 2)(x + 2).$$

We first observe that $P(x) = 0$ if and only if at least one of the three factors x, $(x - 2)$, and $(x + 2)$ is zero; that is, when $x = 0$, $x = 2$, or $x = -2$. From this, we see that the graph of P intersects the x-axis at the points $(0, P(0)) = (0, 0)$, $(2, P(2)) = (2, 0)$, and $(-2, P(-2)) = (-2, 0)$.

Next, we will determine when $P(x)$ is positive and when it is negative. First of all, you will recall that the product of two or more nonzero real numbers is positive if and only if an even number of the factors are negative.

Now, for every value of x other than 0, 2, and -2, the factors x, $(x - 2)$, and $(x + 2)$ are nonzero real numbers. Therefore, $P(x)$ will be positive for those values of x, other than 0, 2, -2, for which an even number of these factors are negative. Since there are only three factors, this means $P(x)$ will be positive only when none or two of the three factors are negative. To help us determine those values of x for which $P(x)$ is positive, we can use illustrations like those shown for Example 3 on page 186.

Since the factor x is negative for all $x < 0$, we begin with the illustration in Figure 4–28. Figure 4–29 also shows that the factor $x - 2$ is negative for all $x < 2$ and that the factor $x + 2$ is negative for all $x < -2$. As we have already noted, $P(x)$ is positive when either two or none of the factors are negative. So, Figure 4–29 shows clearly that $P(x)$ is positive only when $-2 < x < 0$ or when $x > 2$. Since we have seen that $P(x) = 0$ when $x = -2$, $x = 0$, or $x = 2$, we further see that $P(x)$ is negative when $x < -2$ or $0 < x < 2$.

Figure 4-28

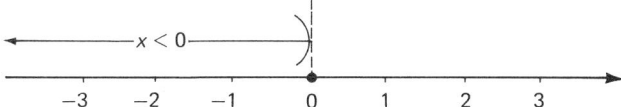

Figure 4-29

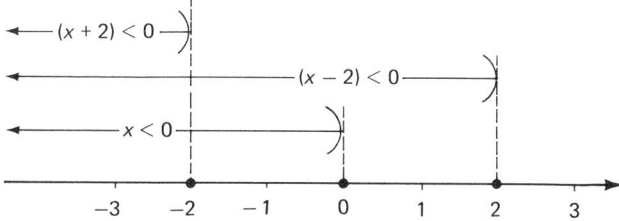

Now any point $(x, P(x))$ on the graph of P for which $P(x) > 0$ lies above the x-axis, and any point for which $P(x) < 0$ lies below the x-axis. Hence, the graph of P must be contained in the shaded portions of the plane indicated in Figure 4–30 on page 198.

Figure 4-30

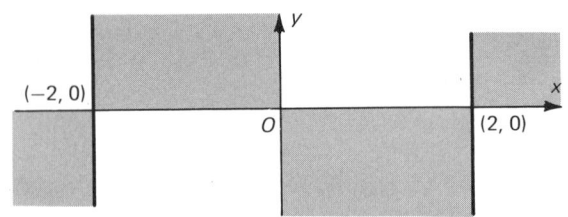

If we can determine when the slope function $P'(x)$ is negative, zero, and positive, we can gain further valuable information about the behavior of the function P and its graph. From Theorem 4.5a, we know that P will be an increasing function and that its graph will be rising to the right over any interval in which the value of its slope function $P'(x)$ is positive. Further, we know that P will be a decreasing function with a graph that is falling to the right for those intervals in which $P'(x)$ is negative. We also know that the graph of P will have a horizontal tangent when $P'(x) = 0$.

Since function P is defined by $P(x) = x^3 - 4x$, its slope function is defined by

$$P'(x) = 3x^2 - 4.$$

This equation for P' can be factored as follows:

$$P'(x) = 3\left(x - \frac{2}{\sqrt{3}}\right)\left(x + \frac{2}{\sqrt{3}}\right).$$

It is immediately apparent that $P'(x) = 0$ only when $x = \frac{\pm 2}{\sqrt{3}}$. Since the constant factor 3 is positive, $P'(x) > 0$ whenever the other two factors are both positive or both negative. $P'(x) < 0$ whenever only one of these two factors is negative. The factor $\left(x - \frac{2}{\sqrt{3}}\right)$ is negative only for $x < \frac{2}{\sqrt{3}}$, and the factor $\left(x + \frac{2}{\sqrt{3}}\right)$ is negative only for $x < \frac{-2}{\sqrt{3}}$, as Figure 4–31 illustrates.

Figure 4-31

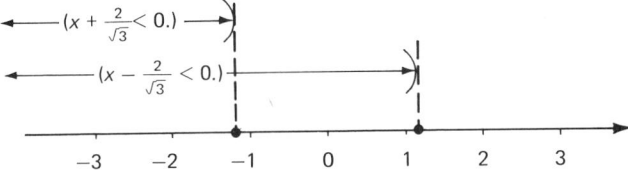

From this figure, we see that $P'(x)$ is positive for $x < \frac{-2}{\sqrt{3}}$ or $x > \frac{2}{\sqrt{3}}$ and is negative for $\frac{-2}{\sqrt{3}} < x < \frac{2}{\sqrt{3}}$. Thus, from Theorem 4.5a, we know that P is an increasing function with a rising graph for $x < \dfrac{-2}{\sqrt{3}}$ or for $x > \dfrac{2}{\sqrt{3}}$ and that P is a decreasing function with a falling graph for $\dfrac{-2}{\sqrt{3}} < x < \dfrac{2}{\sqrt{3}}$. Since $P'(x)$ is zero for $x = \dfrac{\pm 2}{\sqrt{3}}$, we also know that the graph of P has horizontal tangents at the points $\left(\dfrac{-2}{\sqrt{3}}, P\left(\dfrac{-2}{\sqrt{3}}\right)\right) = \left(\dfrac{-2}{\sqrt{3}}, \dfrac{16}{3\sqrt{3}}\right)$ and $\left(\dfrac{2}{\sqrt{3}}, P\left(\dfrac{2}{\sqrt{3}}\right)\right) = \left(\dfrac{2}{\sqrt{3}}, \dfrac{-16}{3\sqrt{3}}\right)$.

We now use all that we have learned about the graph of P to construct the illustration in Figure 4–32.

Figure 4-32

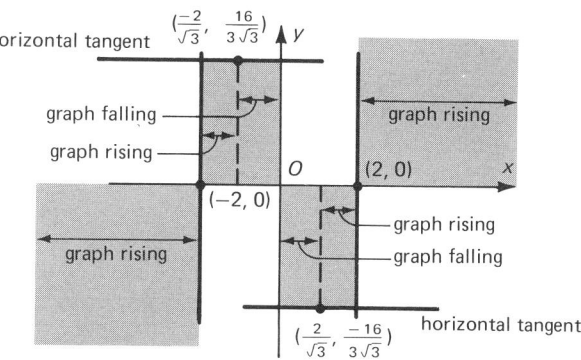

Even with the information given in this figure, we are still not able to give a good characterization of the graph of P. (By a *characterization* of the graph of P we mean a graph that is not necessarily an exact graph of P, but one that has the same intercepts, intervals of increase and decrease, high and low points, concavity, horizontal tangents, symmetry, behavior for large values of x, and so on. It should be obvious that any graph of P which we will be able to draw will be only a characterization of the actual graph.) As you see in Figure 4–33, we can construct at least two different graphs that have the properties noted in Figure 4–32.

Figure 4-33

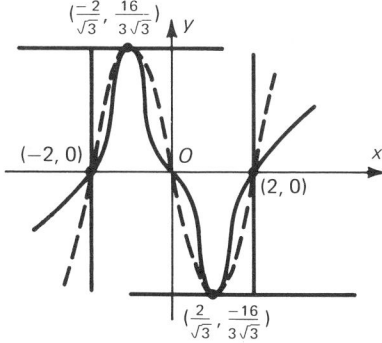

What we still need to know is for what regions of the domain of P the graph of P is concave upward and for what regions it is concave downward. We can get this information by using the slope function of the slope function and Theorem 4.6a. Remember that this theorem establishes that the graph of P will be concave upward over any interval for which $P''(x) > 0$ and that it will be concave downward over any interval for which $P''(x) < 0$. Since P' is defined by $P'(x) = 3x^2 - 4$, P'' is defined by

$$P''(x) = 6x.$$

Clearly, $P''(x)$ is positive for $x > 0$ and negative for $x < 0$. Therefore, the graph of P is concave upward for positive x and concave downward for negative x.

With this added information, we are now able to give a reasonably good characterization of the graph of P, as shown in Figure 4-34.

Figure 4-34

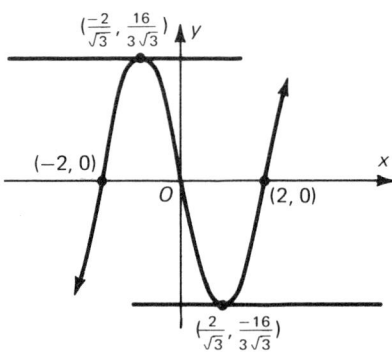

We could improve on this characterization in several ways. One way would be to compute the coordinates of a few more points and then plot them. For example, we might compute $(\frac{1}{2}, P(\frac{1}{2}))$, $(\frac{3}{2}, P(\frac{3}{2}))$ and $(\frac{5}{2}, P(\frac{5}{2}))$. Since $P(x) = -P(-x)$, P is an odd function, and, hence, its graph is symmetric with respect to the origin (see Definition 2.15a, page 100, and Theorem 2.15b, page 102). This means that for every pair of coordinates we compute, such as $(\frac{1}{2}, P(\frac{1}{2})) = (\frac{1}{2}, -\frac{15}{8})$, we know immediately another pair of coordinates, $(-\frac{1}{2}, -(-\frac{15}{8})) = (-\frac{1}{2}, \frac{15}{8})$.

We could also compute the slope of the tangent lines, $P'(x)$, at certain points $(x, P(x))$ on the graph of P and then use this knowledge to improve our characterization of the graph. We often gain useful information by determining the slope of the tangent lines at the points where the graph crosses the coordinate axes (intercepts) and at the points where the concavity of the graph changes (inflection points).

The procedure we have just used for gaining information about the function P can be used for many polynomial functions. We now summarize the information that can be obtained about the graph of a function by the procedure used in this example. If we are given a polynomial function P such that $P(x)$, $P'(x)$, and $P''(x)$ can all be factored into the product of a constant and one or more linear factors, then we can determine the regions over which each of these functions is positive, zero, and negative. This information about P, P', and P'' will tell us the following:

1. Where the graph of P is above the x-axis, where the graph intersects the coordinate axes, and where it is below the x-axis.
2. Where the graph of P is rising, where the graph has horizontal tangents, and where it is falling.
3. Where the graph of P is concave upward, where it may have inflection points, and where it is concave downward.

With such knowledge, we can draw a good characterization of the graph of P, which illustrates not only the properties just enumerated, but also others deriving from these, such as points at which the graph has relative maximum or minimum points.

To summarize the discussion so far, if, for a polynomial function P, we can factor $P(x)$, $P'(x)$, and $P''(x)$ each into the product of a constant and one or more linear factors, then we will be able to determine the basic properties of P and of its graph. Since the functions P' and P'' are polynomial functions whenever P is a polynomial, the important question becomes, when and how can we factor a polynomial into the product of a constant and linear factors? The answers to this question will be the topic of the next section.

4.7 EXERCISES

For each of the following products, use the graphical technique illustrated on pages 197–198 to determine the regions over which the product is positive and the regions over which it is negative.

1. $P(x) = x(x - 3)(x + 2)$.
2. $P(x) = (x - 1)(2x + 5)(x - 7)$.
3. $P(x) = x(x + 5)(3x - 8)$.
4. $P(x) = 3x(5x - 12)(5x + 12)$.
5. $P(x) = x(2x + 7)(x - 5)(x + 4)$.
6. $P(x) = (x - 1)(x - 2)(x + 1)(x + 2)$.

In each of Exercises 7–14, express $P(x)$, $P'(x)$, and $P''(x)$ each as a product of linear factors. For each product, determine the values of x for which the product is negative, zero, and positive, and use this information to sketch the graph of P.

7. $P(x) = x^3 - 3x$.
8. $P(x) = 12x - x^3$.

9. $P(x) = x^3 - 3x^2$.
10. $P(x) = x^3 + 6x^2$.
11. $P(x) = x^3 - 6x^2 + 9x$.
12. $P(x) = -x^3 - 6x^2 - 9x$.
13. $P(x) = x^4 - 4x^3$.
14. $P(x) = x^4 - 6x^2$.

4.8 The Remainder and Factor Theorems

It is quite clear that if a polynomial $P(x)$ has $(x - c)$ as a factor, that is, if $P(x) = (x - c)Q(x)$, where $Q(x)$ is a polynomial; then $P(c) = 0$. This follows from the fact that $P(c) = (c - c)Q(c) = 0 \cdot Q(c) = 0$. Now suppose that $P(x)$ is a polynomial and that $P(c) = 0$. Does it follow that $(x - c)$ is a factor of $P(x)$? The answer to this question is yes, but before proving this property of polynomials, we will first prove the following theorem, which is called the *Remainder Theorem*.

Theorem 4.8a (Remainder Theorem) If $P(x)$ is a polynomial of positive degree n, then the remainder upon dividing $P(x)$ by $x - c$ is $P(c)$.

Proof To prove the theorem, we need to show that

$$P(x) = (x - c)Q(x) + P(c),$$

where $Q(x)$ is a polynomial of degree $n - 1$. Now if

$$P(x) = a_n x^n + a_{n-1} x^{n-1} + \ldots + a_0,$$

then

$$P(c) = a_n c^n + a_{n-1} c^{n-1} + \ldots + a_0.$$

Therefore,

$$(*) \quad P(x) - P(c) = a_n(x^n - c^n) + a_{n-1}(x^{n-1} - c^{n-1}) + \ldots + a_1(x - c).$$

Notice that each of the terms on the right side of this last equation has a factor of the form $x^k - c^k$. If we now use the identity

$$x^k - c^k = (x - c)(x^{k-1} + cx^{k-2} + \ldots + c^{k-2}x + c^{k-1}),$$

it follows that every term on the right side of equation (*) has $(x - c)$ as a factor. It is now a simple matter to determine the polynomial $Q(x)$ of degree $n - 1$ and to complete the proof. These steps are left as an exercise. ☐

Example 1 If we divide $P(x) = x^5 - 4x^3 + x^2 - 5$ by $x + 2$, we obtain $Q(x) = x^4 - 2x^3 + x - 2$, with remainder -1. $P(-2) = (-2)^5 - 4(-2)^3 +$

$(-2)^2 - 5 = -1$, which verifies that $P(-2)$ is equal to the remainder upon division by $x - (-2)$. Notice also that the quotient $Q(x)$ is of degree $5 - 1$, or 4.

Theorem 4.8a makes the proof of the following theorem, called the *Factor Theorem*, a simple matter.

Theorem 4.8b (Factor Theorem) If $P(x)$ is a polynomial of positive degree and $P(c) = 0$, then $(x - c)$ is a factor of $P(x)$. Conversely, if $(x - c)$ is a factor of $P(x)$, then $P(c) = 0$.

Proof We must first show that, if $P(c) = 0$, then $P(x) = (x - c)Q(x)$. From the Remainder Theorem, we know that

$$P(x) = (x - c)Q(x) + P(c),$$

where $Q(x)$ is a polynomial of degree $n - 1$. It follows immediately that if $P(c) = 0$ then $(x - c)$ is a factor of $P(x)$.

On the other hand, if $(x - c)$ is a factor of $P(x)$, that is, if

$$P(x) = (x - c)R(x),$$

where R is a polynomial in x, then

$$P(c) = (c - c)R(c) = 0 \cdot R(c) = 0.$$

This completes the proof of the theorem. $\square$

The Factor Theorem can be very helpful in determining the zeros of a polynomial function. For example, if we are given a third degree polynomial $P(x)$ and we can see by inspection that $P(c) = 0$ for some real number c, then we know that we can factor $P(x)$ in the following way:

$$P(x) = (x - c)Q(x).$$

From the proof of the theorem, we know also that $Q(x)$ is a polynomial of degree two. Any zeros of $P(x)$ other than c, if there are any, must be the zeros of $Q(x)$, whose linear factors, if any, can be determined by factoring or by using the quadratic formula.

Example 2 As a specific example, consider

$$P(x) = x^3 - x^2 + x - 1.$$

By inspection, we see that $P(1) = 1 - 1 + 1 - 1 = 0$ and therefore that 1 is a zero of $P(x)$. From the Factor Theorem, we know that

$$P(x) = (x - 1)Q(x),$$

where $Q(x)$ is a polynomial of degree 2. Since we know that $(x-1)$ must be a factor of $x^3 - x^2 + x - 1$, we divide by $(x-1)$, as follows:

$$\begin{array}{r}
x^2 + 1 \\
x - 1 \overline{\smash{)}\,x^3 - x^2 + x - 1} \\
\underline{x^3 - x^2 } \\
0 + x - 1 \\
\underline{x - 1} \\
0
\end{array}$$

Hence, $P(x) = (x-1)(x^2+1)$.

We see that $(x^2 + 1) = Q(x)$ does not have any linear factors over the set of real numbers. Therefore, we conclude that $P(x)$ has $x = 1$ as its only zero in the set of real numbers. This means, of course, that the graph of function P crosses the x-axis only at the point $(1, 0)$.

Example 3 The polynomial

$$P(x) = x^4 - 3x^3 + 2x^2 + x - 1,$$

has 1 as a zero. It can therefore be factored into the form

$$P(x) = (x-1)R(x).$$

Where $R(x)$ is a polynomial of degree three, we now divide $P(x)$ by $(x-1)$ to obtain $R(x)$, as shown below.

$$\begin{array}{r}
x^3 - 2x^2 + 1 \\
x - 1 \overline{\smash{)}\,x^4 - 3x^3 + 2x^2 + x - 1} \\
\underline{x^4 - x^3 } \\
-2x^3 + 2x^2 \\
\underline{-2x^3 + 2x^2 } \\
0 + x - 1 \\
\underline{x - 1} \\
0
\end{array}$$

Thus,
$$P(x) = (x-1)R(x) = (x-1)(x^3 - 2x^2 + 1).$$

Again, it is easy to see that 1 is a zero of $R(x) = x^3 - 2x^2 + 1$. Using the same division process, we see that $x^3 - 2x^2 + 1$ factors into $(x-1)(x^2 - x - 1)$. Finally, using the quadratic formula, we factor $(x^2 - x - 1)$ into its linear factors, $\left(x - \dfrac{1 + \sqrt{5}}{2}\right)$ and $\left(x - \dfrac{1 - \sqrt{5}}{2}\right)$. The result is that $P(x)$ can be written as the product of four linear factors, two of which have the same constant term:

$$P(x) = (x-1)(x-1)\left(x - \dfrac{1 + \sqrt{5}}{2}\right)\left(x - \dfrac{1 - \sqrt{5}}{2}\right).$$

Hence, $P(x)$ has three zeros, 1, $\dfrac{1+\sqrt{5}}{2}$, and $\dfrac{1-\sqrt{5}}{2}$. This means that the graph of function P has x-intercepts at the three points $(1, 0)$, $\left(\dfrac{1+\sqrt{5}}{2}, 0\right)$, and $\left(\dfrac{1-\sqrt{5}}{2}, 0\right)$.

In Examples 2 and 3, if we could next determine the linear factors of $P'(x)$ and $P''(x)$, we then could use the results to learn more about the graphs of the two functions.

4.8 EXERCISES

For each of the following exercises, write $P(x)$ in the form $P(x) = (x - c)Q(x) + P(c)$.

1. $P(x) = 3x^3 - 4x^2 + 2x - 15$; $c = 3$.
2. $P(x) = x^4 + 16$; $c = -2$.
3. $P(x) = 4x^3 - 3x^2 + 5x - 2$; $c = \frac{1}{2}$.

In each of the following exercises, determine the value of k that will make the statement true.

4. When $P(x) = 2x^3 + 6x^2 + 2x$ is divided by $x + 3$, the remainder is k.
5. When $P(x) = 3x^3 + kx^2 - 8x + 4$ is divided by $x - 2$, the remainder is -4.
6. When $P(x) = 2x^3 + 3x^2 + kx + 2$ is divided by $x + \frac{1}{2}$, the remainder is 5.

In each of the following exercises, use the given information to express $P(x)$ as a product of linear factors. Then tabulate the zeros of $P(x)$.

7. $P(x) = x^3 + 2x^2 - 23x - 60$ and $P(5) = 0$.
8. $P(x) = 2x^3 - 3x^2 - 23x + 12$ and $P(-3) = 0$.
9. $P(x) = 6x^3 + 2x^2 - \frac{25}{2}x - 6$ and $P(-\frac{1}{2}) = 0$.

In each of the following exercises, express $P(x)$ as a product of linear factors and tabulate the zeros of $P(x)$.

10. $P(x) = x^3 - x^2 - 9x + 9$.
11. $P(x) = 2x^3 + 5x^2 - 17x - 20$.
12. $P(x) = x^3 + 2x^2 - 3x - 6$.
13. Determine whether each of the following is a true statement or a false statement, and justify your answer.
    a) If n is a positive integer, then $x - c$ is a factor of $x^n + c^n$.
    b) If n is a positive even integer, then $x + c$ is a factor of $x^n - c^n$.
    c) If n is a positive odd integer, then $x + c$ is a factor of $x^n - c^n$.
    d) If n is a positive even integer, then $x + c$ is a factor of $x^n + c^n$.
    e) If n is a positive odd integer, then $x + c$ is a factor of $x^n + c^n$.

14. Suppose that
$$P(x) - P(c) = a_n(x^n - c^n) + a_{n-1}(x^{n-1} - c^{n-1}) + \ldots + a_1(x - c).$$
Use the fact that
$$x^k - c^k = (x - c)(x^{k-1} + cx^{k-2} + \ldots + c^{k-2}x + c^{k-1}),$$
where k is a positive integer, to prove that
$$P(x) = (x - c)Q(x) + P(c),$$
where $Q(x)$ is a polynomial of degree $n - 1$.

5

Zeros of Polynomial Functions and Complex Numbers

Introduction

In our work with functions we will be concerned for the most part only with the real numbers. There will be occasions, however, when the real numbers will not be adequate for our needs. For example, in solving polynomial equations, we find that there is no real number which satisfies the equation $x^2 + 1 = 0$.

In fact, it is interesting to establish a kind of hierarchy among various sets of numbers in terms of their usefulness in solving equations. A great variety of equations do have integral solutions, and the set of integers is adequate for solving such equations. But an equation like $2x + 1 = 0$ has no integral solutions. This equation does, of course, have a rational solution, but the set of rational numbers is not rich enough in numbers to supply solutions to the equation $x^2 - 2 = 0$, for example. The real numbers do contain solutions to the equation $x^2 - 2 = 0$, but do not, as noted above, have any solutions of $x^2 + 1 = 0$. In fact, any polynomial equation of the form $x^2 + a = 0$, with $a > 0$, has no real-number solutions.

It is obviously desirable to have a system that is rich enough in numbers to allow us to assert that any polynomial equation will have solutions in that system. To continue with the example discussed above, the system should include a number that is a solution to the equation $x^2 + 1 = 0$. It happens that if we add to the set of real numbers a solution of $x^2 + 1 = 0$, we have a set of numbers called the *complex numbers*. The set of complex numbers under the operations of addition and multiplication forms the field of complex numbers. Most importantly, this field is rich enough in numbers to supply solutions to every polynomial equation.

In the eighteenth century, mathematicians took the bold step of introducing the symbol i to represent the solution of the equation $x^2 + 1 = 0$.[1] Since i is the solution of $x^2 + 1 = 0$, it follows that $i^2 + 1 = 0$, or $i^2 = -1$. As a solution for this equation, i was introduced for purely theoretical reasons, since science and technology had not yet advanced to the point of raising problems that required solutions for such equations. Because i did not belong to the set of numbers thought of as the "real" numbers (the square of any "real" number is nonnegative), it was called an **imaginary number**.[2] In fact, any multiple of i and a real number was called a "pure imaginary number." It was assumed that the number i could be combined with real numbers by the various algebraic operations. For example, if the product of i and the real number b is added to the real number a, the result $a + bi$ was called a **complex number**. Thus, $3 + 2i$, $-1 + i$, and $4 + \sqrt{2}i$ are examples of complex numbers. Historically, the real number a in $a + bi$ was called the *real part* of the complex number, and bi was called the *imaginary part*.

5.1 Operations on Complex Numbers

If a and b are any real numbers, then $a + bi$ is called a *complex number*. The number a is called the "real component" and b is called the "imaginary component."

Two complex numbers, $a + bi$ and $c + di$, are equal if and only if $a = c$ and $b = d$, that is, if and only if their real and imaginary components are respectively equal.

By assuming that i satisfies all the usual algebraic rules together with the added property that $i^2 = -1$, we can perform the operations of addition and subtraction for complex numbers. To add two complex numbers, the real parts and the imaginary parts are added independently, and the distributive property is used to simplify the results. For example, the sum of $2 + 3i$ and $3 - 2i$ is obtained as follows:

$$(2 + 3i) + (3 - 2i) = (2 + 3) + (3i - 2i)$$
$$= 5 + (3 - 2)i$$
$$= 5 + i.$$

---

[1] The idea of solutions outside the set of real numbers for equations such as $x^2 + 1 = 0$ had been troubling mathematicians for several centuries prior to the introduction of i.

[2] The adjective "imaginary" applied to this number is indicative of the reluctance to accept i as a number having the same status as real numbers. Earlier, there had been a similar feeling about negative numbers when they were first introduced.

More generally, if $a + bi$ and $c + di$ represent any two complex numbers, then

$$(a + bi) + (c + di) = (a + c) + (bi + di)$$
$$= (a + c) + (b + d)i.$$

The product of two complex numbers is obtained in the same way that the product of two binomials is found; that is, by formally carrying out the algebraic operations. In simplifying the results, the fact that $i^2 = -1$ is used. For example,

$$(2 + 3i) \cdot (3 - 2i) = 2 \cdot 3 + 2 \cdot (-2i) + (3i) \cdot 3 + (3i) \cdot (-2i)$$
$$= 6 - 4i + 9i - 6i^2$$
$$= 6 + 5i - 6(-1)$$
$$= 12 + 5i.$$

In general, the product of any two complex numbers $a + bi$ and $c + di$ may be found as follows:

$$(a + bi)(c + di) = ac + a(di) + (bi)c + (bi)(di)$$
$$= ac + (ad)i + (bc)i + (bd)i^2$$
$$= ac + (ad + bc)i + bd(-1)$$
$$= (ac - bd) + (ad + bc)i.$$

The difference of two complex numbers is obtained as follows:

$$(a + bi) - (c + di) = a + bi + (-c - di) = (a - c) + (b - d)i.$$

The quotient of two complex numbers $\dfrac{a + bi}{c + di}$ in the form $e + fi$ is not as readily obtained. We shall discuss this in the next section after introducing the concept of the conjugate of a complex number.

We close this section with the observation that the real numbers are a subset of the complex numbers since any complex number $a + bi$, where $b = 0$, is the real number a.

5.1 EXERCISES

In each of Exercises 1–16, first determine the sum and then the product of the complex numbers, expressing each result in the form $a + bi$.

1. $2i, 3i$.
2. $-3i, -4i$.
3. $8i, 2 + 3i$.
4. $-7i, -1 + 3i$.
5. $9 - 2i, -8 + 3i$.
6. $8 - 7i, 6 - 5i$.
7. $-2 + 5i, 4 - 7i$.

8. $1 - i, 1 + i$.
9. $3 - 2i, 2 - 3i$.
10. $3 - 2i, 3 + 2i$.
11. $5 + 5i, 5 + 5i$.
12. $\frac{3}{5} + \frac{4}{5}i, -\frac{4}{5} + \frac{3}{5}i$.
13. $\frac{1}{3} + \frac{1}{2}i, -\frac{5}{3} + \frac{1}{9}i$.
14. $0.1 - 0.6i, -10 + 100i$.
15. $\sqrt{2} + 3i, 2\sqrt{2} + i$.
16. $\frac{\sqrt{2}}{2} + \frac{\sqrt{2}}{2}i, \frac{\sqrt{2}}{2} - \frac{\sqrt{2}}{2}i$.

In each of Exercises 17–22, perform the indicated operations and express the result in the form $a + bi$.

17. $(-i) - (1 + 3i)$
18. $(-6) - (-5 + 8i)$
19. $(6 + 3i) + (1 - 4i) - (2 - 7i)$
20. $(-5 + 3i) - (5 + 7i) + (8 - i)$
21. $-(4 - 5i) + (7 - i) - (-3)$
22. $-(6 + 4i) - (1 - 5i) - 5i$

In Exercises 23 and 24, determine the real values of x and y that satisfy the given equation.

23. $(x + y) + (x - y)i = 7 + i$.
24. $(x - y) + (2x + 3y)i = 4 + 13i$.
25. a) Complete the following table.
    $i^1 = i$ $\qquad$ $i^5 = i^4 \cdot i = \underline{\qquad}$
    $i^2 = i \cdot i = -1$ $\qquad$ $i^6 = i^5 \cdot i = \underline{\qquad}$
    $i^3 = i^2 \cdot i = -1 \cdot i = -i$ $\qquad$ $i^7 = i^6 \cdot i = \underline{\qquad}$
    $i^4 = i^3 \cdot i = \underline{\qquad}$ $\qquad$ $i^8 = i^7 \cdot i = \underline{\qquad}$

    b) Study your answers in part a. What is the value of i^{17}? Of i^{100}? Of i^{103}? Explain.

5.2 Complex Conjugates and the Quotient of Complex Numbers

The complex numbers $a + bi$ and $a - bi$ are said to be **conjugates** of one another. We shall adopt the convention of denoting the conjugate of a complex number c by $\bar{c}$. Thus, since $a + bi$ and $a - bi$ are conjugates of each other, $\overline{a + bi} = a - bi$.

Example 1 The complex numbers $1 + i$ and $1 - i$ are conjugates of one another; that is, $\overline{1 + i} = 1 - i$. The numbers $\frac{3}{4} - \sqrt{2}i$ and $\frac{3}{4} + \sqrt{2}i$ are conjugates; that is, $\overline{\frac{3}{4} - \sqrt{2}i} = \frac{3}{4} + \sqrt{2}i$.

Two observations that we can make immediately about conjugates are the following:

1. Since every real number can be written in the form $a + 0i$, it follows immediately that $\overline{a + 0i} = a - 0i$, so that every real number is its own *conjugate*.

2. Since $\overline{a + bi} = a - bi$, it follows that $\overline{\overline{a + bi}} = \overline{a - bi} = a + bi$, which means that every complex number is the conjugate of its conjugate.

There are some other useful properties of conjugates we shall need.

1. The sum of a complex number and its conjugate is a real number.
$$(a + bi) + \overline{(a + bi)} = (a + bi) + (a - bi)$$
$$= 2a.$$

2. The product of a complex number and its conjugate is a real number.
$$(a + bi)\overline{(a + bi)} = (a + bi)(a - bi)$$
$$= a^2 + b^2 + (ba - ab)i$$
$$= a^2 + b^2.$$

3. The conjugate of the sum of two complex numbers is the sum of their conjugates.
$$\overline{(a + bi) + (c + di)} = \overline{(a + c) + (b + d)i}$$
$$= (a + c) - (b + d)i$$
$$= (a - bi) + (c - di)$$
$$= \overline{(a + bi)} + \overline{(c + di)}.$$

4. The conjugate of the product of two complex numbers is the product of their conjugates.
$$\overline{(a + bi)(c + di)} = \overline{(ac - bd) + (ad + bc)i}$$
$$= (ac - bd) - (ad + bc)i$$
$$= (a - bi) \cdot (c - di)$$
$$= \overline{(a + bi)} \cdot \overline{(c + di)}.$$

5. The square roots of any negative number exist and are conjugates of one another. If a is positive, the square roots of $-a$ are $\sqrt{a}\, i$ and $-\sqrt{a}\, i$.

Now consider the quotient of two complex numbers, $\dfrac{(a+bi)}{(c+di)}$. If we multiply the numerator and denominator by the conjugate of the denominator we obtain

$$\frac{a+bi}{c+di} = \frac{(a+bi)(c-di)}{(c+di)(c-di)} = \frac{(ac+bd)+(bc-ad)i}{c^2+d^2}$$

$$= \left(\frac{ac+bd}{c^2+d^2}\right) + \left(\frac{bc-ad}{c^2+d^2}\right)i.$$

This gives us the quotient of two complex numbers as a complex number of the form $e + fi$ and we could think of the above equation as a formula for this quotient. The formula is much too complicated to memorize, however, and in practice you should just use the procedure described above. For example let us express the quotient of $1 + \sqrt{2}\,i$ and $1 - \sqrt{2}\,i$ as a complex number of the form $e + fi$. Since the conjugate of $(1 - \sqrt{2}\,i)$ is $(1 + \sqrt{2}\,i)$, we have

$$\frac{1+\sqrt{2}\,i}{1-\sqrt{2}\,i} = \frac{(1+\sqrt{2}\,i)(1+\sqrt{2}\,i)}{(1-\sqrt{2}\,i)(1+\sqrt{2}\,i)} = \frac{1-(\sqrt{2})^2+2\sqrt{2}\,i}{1^2+(\sqrt{2})^2}$$

$$= \frac{-1+2\sqrt{2}\,i}{3} = -\frac{1}{3} + \frac{2\sqrt{2}}{3}i.$$

5.2 EXERCISES

Express each of the following quotients in the form $a + bi$.

1. $\dfrac{1}{i}$
2. $\dfrac{5-2i}{i}$
3. $\dfrac{i}{4-2i}$
4. $\dfrac{3}{1-i}$
5. $\dfrac{i}{6+3i}$
6. $\dfrac{6+i}{i-3}$
7. $\dfrac{1-i}{1+i}$
8. $\dfrac{16+11i}{2+5i}$
9. $\dfrac{7-5i}{i-3}$
10. $\dfrac{-2+3i}{i-1}$

Use property 5 on page 211 to find the complex solutions of each of the following equations.

11. $2x^2 + 50 = 0.$
12. $\frac{1}{3}x^2 + 12 = 0.$
13. $\frac{2}{5}x^2 = -250.$

14. $x^2 + 20 = 0$.
15. $5x^2 = -320$.
16. $\frac{1}{4}x^2 = -18$.
17. Prove that the complex conjugate of the sum of m complex numbers is the sum of the conjugates of these numbers.
18. Prove that the complex conjugate of the product of m complex numbers is the product of the m complex conjugates of these numbers. [Hint: use the associative property of multiplication of complex numbers.]

5.3 Complex Solutions of Quadratic Equations

One of the motivations for developing the complex number system was to provide solutions for all quadratic equations with real coefficients. As we have seen i is a solution for the quadratic equation $x^2 + 1 = 0$ which has no real solutions. It is easy to see that $-i$ is also a solution to this equation. If a is a positive real number, then $x^2 + a = 0$, which has no real solutions, has the two solutions $\sqrt{a}\,i$ and $-\sqrt{a}\,i$. It is, in fact, true that any quadratic equation with real coefficients that has no real solutions has two complex solutions which are conjugates. To verify this we recall how the quadratic formula was developed.

Let a, b and c be real numbers, $a \neq 0$, and consider the equation

$$ax^2 + bx + c = 0.$$

This equation can be written as

$$a\left(x^2 + \frac{b}{a}x\right) = -c$$

or

$$x^2 + \frac{b}{a}x = -\frac{c}{a}.$$

If we now "complete the square" on the left-hand side by adding the square of $\dfrac{b}{2a}$ to both sides of the equation, we have

(1) $$x^2 + \frac{b}{a}x + \frac{b^2}{4a^2} = \frac{b^2}{4a^2} - \frac{c}{a}$$

or

(2) $$\left(x + \frac{b}{2a}\right)^2 = \frac{b^2 - 4ac}{4a^2}.$$

If $b^2 - 4ac$ is a positive, we can take the square root of both sides and with a little algebraic manipulation obtain the formula for the two real roots:

$$x = \frac{-b \pm \sqrt{b^2 - 4ac}}{2a}.$$

If $b^2 - 4ac$ is negative, then $4ac - b^2$ is positive and the square roots of the right-hand side of equation (2) are

$$\pm \frac{\sqrt{4ac - b^2}}{2a} i,$$

and the two roots of the equation are

$$x = \frac{-b \pm \sqrt{4ac - b^2}\, i}{2a}.$$

These two roots are clearly conjugates. Thus, we see that if any second degree polynomial with real coefficients has a complex zero C, then its conjugate $\bar{C}$ is also a zero of the polynomial. From the properties of conjugates we have noted on page 211, we can prove a far more general statement.

Theorem 5.3a *If P is a polynomial with real coefficients and C is a complex number such that $P(C) = 0$ and if $\bar{C}$ is the conjugate of C, then $P(\bar{C}) = 0$.*

Before proving Theorem 5.3a we state the following auxiliary theorem or lemma.

Lemma 5.3b *If $P(x)$ is a polynomial with real coefficients (that is, if $P(x)$ is a real polynomial) and C is any complex number, then $\overline{P(C)} = P(\bar{C})$.*

The proof of this lemma is an exercise in the use of properties of conjugates and we leave it for the exercises.

The proof of Theorem 5.3a now follows easily.

Proof Assume that $P(C) = 0$. Then $\overline{P(C)} = \bar{0}$. We know that $\bar{0} = 0$ since the real number 0 is its own conjugate. But by Lemma 5.3b $\overline{P(C)} = P(\bar{C})$; therefore, $P(\bar{C}) = 0$. □

Note that if C is a real number, Theorem 5.3a is trivial, since every real number is its own complex conjugate. In case C is a real number, the theorem simply says that if C is a zero of P, then $\bar{C} = C$ is a zero of P.

If, however, C is a nonreal complex number, the theorem does tell us something important about the zeros of the function. It tells us that nonreal zeros of a real polynomial function occur in pairs that are conjugates of each other.

5.3 EXERCISES

For each of Exercises 1–10, determine whether the roots of the given equation are real numbers or nonreal complex numbers. Then determine the roots of the equation.

1. $x^2 - 6x + 10 = 0$.
2. $x^2 + 8x + 20 = 0$.
3. $x^2 - 4x + 1 = 0$.
4. $x^2 + 10x + 28 = 0$.
5. $x^2 - 12x + 3 = 0$.
6. $x^2 - 4x + 12 = 0$.
7. $2x^2 - 2x + 1 = 0$.
8. $4x^2 + 4x = 1$.
9. $4x^2 - 4x + 3 = 0$.
10. $3x^2 + 6x = -4$.

11. Let r_1 and r_2 denote the roots of $ax^2 + bx + c = 0$, where a, b, and c are real numbers and $a \neq 0$. Prove that

   a) $r_1 + r_2 = -\dfrac{b}{a}$,   b) $r_1 \cdot r_2 = \dfrac{c}{a}$.

Use the theorem in Exercise 11 to determine whether each of the following statements is true or false.

12. The roots of $x^2 - 6x + 13 = 0$ are $-3 + 2i$ and $-3 - 2i$.
13. The roots of $x^2 - 10x - 24 = 0$ are $5 - 7i$ and $5 + 7i$.
14. The roots of $x^2 - 10x + 22 = 0$ are $5 - \sqrt{3}$ and $5 + \sqrt{3}$.
15. The roots of $9x^2 - 12x + 5 = 0$ are $\tfrac{2}{3} + \tfrac{1}{3}i$ and $\tfrac{2}{3} - \tfrac{1}{3}i$.
16. Prove Lemma 5.3b on page 214. [Hint: see Exercise 17 and 18 on page 213.]

5.4 The Fundamental Theorem of Algebra

We have seen that in certain cases it is possible to express a polynomial $P(x)$ as the product of constant and linear factors, which provide us with useful information about the graph of the function P. It is not always possible, however, to express a polynomial as the product of linear factors over the set of real numbers. Consider, for example, the polynomial equation $P(x) = x^2 + 1$. From our work with the complex number field, we see that $x^2 + 1$ can be factored as follows:

$$P(x) = (x + i)(x - i) = x^2 + 1,$$

where i is the pure imaginary number having the property that $i^2 = -1$.

When we have referred to *linear factors* in our discussion, we have actually meant a **linear factor in the real-number field,** which is an expression of the form $ax + b$, where a and b are *real* numbers. (In fact, in most of the work up to this point, we have been concerned mainly with linear factors of the form $x - c$, that is, with factors where $a = 1$.) A **linear factor in the complex-number field** is an expression of the form $\alpha x + \beta$, where α and β are complex numbers. Although the polynomial $P(x) = x^2 + 1$ cannot be expressed in linear factors in the real-number field, we see that it can be expressed in linear factors in the complex-number field.

An extremely important property of any polynomial of degree n is that it is the product of n linear factors in the complex-number field. Specifically, if P is a polynomial with real coefficients defined by

$$P(x) = a_n x^n + a_{n-1} x^{n-1} + \ldots + a_1 x + a_0,$$

then there exist n pairs of complex numbers (possibly real) (α_k, β_k), for $k = 1, 2, \ldots, n$, such that

$$P(x) = (\alpha_1 x + \beta_1)(\alpha_2 x + \beta_2) \ldots (\alpha_n x + \beta_n).$$

The pairs of complex numbers (α_k, β_k) need not be distinct; indeed, they may all be the same.

Example 1 $P(x) = x^4 + 4x^3 + 6x^2 + 4x + 1$
$= (x + 1)(x + 1)(x + 1)(x + 1) = (x + 1)^4.$
Hence, in this case, $(\alpha_k, \beta_k) = (1, 1)$ for $k = 1, 2, 3, 4$.

It is often desirable to factor the polynomial in such a way that all the linear factors have the number 1 as the coefficient of x. This is always possible provided that we admit a constant factor as well as the linear factors. To see this, we need only observe that if

$$P(x) = (\alpha_1 x + \beta_1)(\alpha_2 x + \beta_2) \ldots (\alpha_n x + \beta_n),$$

then we can divide each linear factor by its leading coefficient, α_k, and then multiply the whole expression by the product of all these coefficients. Thus,

$$P(x) = (\alpha_1 x + \beta_1)(\alpha_2 x + \beta_2) \ldots (\alpha_n x + \beta_n)$$

$$= \left[\alpha_1\left(x + \frac{\beta_1}{\alpha_1}\right)\right]\left[\alpha_2\left(x + \frac{\beta_2}{\alpha_2}\right)\right] \ldots \left[\alpha_n\left(x + \frac{\beta_n}{\alpha_n}\right)\right]$$

$$= (\alpha_1 \alpha_2 \ldots \alpha_n)\left[\left(x + \frac{\beta_1}{\alpha_1}\right)\left(x + \frac{\beta_2}{\alpha_2}\right) \ldots \left(x + \frac{\beta_n}{\alpha_n}\right)\right]$$

$$= K(x - r_1)(x - r_2) \ldots (x - r_n),$$

where $K = \alpha_1 \cdot \alpha_2 \cdot \ldots \cdot \alpha_n$ and $r_k = \dfrac{-\beta_k}{\alpha_k}$, for $k = 1, 2, \ldots, n$.

Example 2 The polynomial
$$P(x) = 6x^3 - 19x^2 + x + 6$$
can be factored as follows:
$$P(x) = (x - 3)(2x + 1)(3x - 2).$$
This product, in turn, can be written as
$$P(x) = 6(x - 3)(x - (-\tfrac{1}{2}))(x - \tfrac{2}{3}).$$

The fact that any polynomial expression is the product of linear factors in the complex-number field is a consequence of two theorems, the Factor Theorem and the *Fundamental Theorem of Algebra*. The proof of the latter theorem is beyond the scope of this book, and we state the theorem without proof.

Theorem 5.4a (Fundamental Theorem of Algebra) *Every polynomial function of degree n where n is a positive integer, has at least one zero that is a complex number.*

Since the set of complex numbers contains the set of real numbers, this zero of the function may be a real number. (In fact, the fundamental theorem of algebra also holds if the polynomial used in defining the function has *complex* coefficients, but keep in mind that we are concerned only with polynomials with *real* coefficients.) Notice that, if the zero of the function is a real number a, we know that the graph of the function intersects the x-axis at the point $(a, 0)$.

In proving our next theorem about the linear factors of a polynomial, we need to point out that the Remainder Theorem, page 202, and the Factor Theorem, page 203, hold even if c in the linear factor $x - c$ represents a complex number and even if the coefficients of the polynomials in question are complex numbers. You should read the proofs of these two theorems again to see that they would be valid even under these conditions. The proofs would still be valid because all the algebraic operations that were used and the algebraic identity that was used,
$$(x^k - c^k) = (x - c)(x^{k-1} + cx^{k-2} + \ldots + c^{k-2}x + c^{k-1})$$
are also valid if c and the coefficients of the given polynomial are arbitrary complex numbers.

Of special interest to us is the case in which the coefficients of the terms of the polynomial are complex numbers that happen to be real numbers and c in the linear factor $x - c$ is a nonreal complex number. In this case, even though the coefficients of $P(x)$ are all real numbers, *some* of the coefficients of $Q(x)$ in the factorization
$$P(x) = (x - c)Q(x)$$

must be nonreal. The following specific example should help convince you that this must be so.

Example 3 If $P(x) = x^4 - 1$, then

$$P(i) = i^4 - 1 = (i^2)^2 - 1 = (-1)^2 - 1 = 0.$$

Thus, $x - i$ is a linear factor of $P(x)$, so that

$$P(x) = (x - i)Q(x),$$

where $Q(x)$ is a polynomial of degree three. Dividing by $x - i$, we get the following results.

$$\begin{array}{r}
x^3 + ix^2 - x - i \\
x - i \overline{)x^4 + 0 \cdot x^3 + 0 \cdot x^2 + 0 \cdot x - 1} \\
\underline{x^4 - ix^3} \\
ix^3 \\
\underline{ix^3 + x^2} \\
-x^2 \\
\underline{-x^2 + ix} \\
-ix - 1 \\
\underline{-ix - 1} \\
0
\end{array}$$

Therefore,

$$P(x) = (x - i)(x^3 + ix^2 - x - i),$$

and $Q(x)$ does have complex coefficients that are not real.

We can now prove the following theorem concerning the linear factors of a polynomial with real coefficients. This theorem, which will be useful to us in our analysis of a polynomial function and its graph, asserts that $P(x)$ is the product of the constant factor a_n and n linear factors, where the constant terms in the linear factors are not necessarily real.

Theorem 5.4b *If $P(x)$ is a polynomial of degree n with real coefficients,*

$$P(x) = a_n x^n + a_{n-1} x^{n-1} + \ldots + a_0,$$

then $P(x)$ is of the form

$$P(x) = a_n(x - r_1)(x - r_2) \ldots (x - r_n),$$

where $r_1, r_2, \ldots, r_n$ are complex (possibly real) numbers.

Proof By the Fundamental Theorem of Algebra, $P(x)$ has at least one zero that is a complex number. In other words, there is some complex number r_1 such that $P(r_1) = 0$. By the Factor Theorem, this means that

$$P(x) = (x - r_1)Q_{n-1}(x),$$

where $Q_{n-1}(x)$ is a polynomial of degree $n - 1$. If $n - 1 > 0$, that is, if the degree of $P(x)$ is greater than 1, we can again use the Fundamental Theorem of Algebra with respect to $Q_{n-1}(x)$ to get

$$Q_{n-1}(x) = (x - r_2)Q_{n-2}(x),$$

where $Q_{n-2}(x)$ is a polynomial of degree $n - 2$. Therefore,

$$P(x) = (x - r_1)(x - r_2)Q_{n-2}(x).$$

After repeating this process n times, $Q_{n-n}(x) = Q_0(x)$ is a polynomial of degree zero, that is, a constant. Because the product of the n factors $(x - r_1)$, $(x - r_2)$, . . . , $(x - r_n)$ is a polynomial of degree n, where the coefficient of x^n is 1, and because the product of this polynomial and $Q_0(x)$ must be equal to $P(x) = a_n x^n + \ldots + a_0$, the constant factor must be a_n. □

We use Theorem 5.4b, together with the fact that nonreal zeros of a real polynomial function occur in pairs that are conjugates, (a consequence of Theorem 5.3a) to obtain the following result.

Theorem 5.4c *A real polynomial function of odd degree has at least one real zero.*

Now let us review some of the properties of polynomials that we have been discussing. First, any real polynomial of degree n is the product of a constant and n linear factors, where the constant terms in the linear factors are complex numbers. That is, if P is a polynomial of degree n, then

$$P(x) = a(x - r_1)(x - r_2) \ldots (x - r_n),$$

where a is the coefficient of x^n in $P(x)$ and the r's are complex numbers.

There is no reason to expect that the n numbers $r_1, r_2, \ldots, r_n$ in this factorization will be distinct. They can in fact all be equal, all be different, or satisfy a condition between these extremes. If the constant term for k of the n linear factors, $1 \leq k \leq n$, does have the same value r, then r is said to be a **zero of multiplicity k of the polynomial** or a **root of multiplicity k** of the equation $P(x) = 0$.

Example 4 In the following case,

$$P(x) = (x - 1)(x - 3)^2(x - 4)^5,$$

1 is a zero of function P of multiplicity 1, 3 is a zero of multiplicity 2, and 4 is a zero of multiplicity 5.

A second property of polynomials already established is that if one of the linear factors has a nonreal constant term, then there is another linear factor which has the conjugate of this constant term for its constant term. Using the commutative property of multiplication, we see that the equation for any polynomial of degree n has the following form:

$$P(x) = a(x - r_1) \cdots (x - r_j)(x - C_1)(x - \bar{C}_1) \cdots (x - C_k)(x - \bar{C}_k),$$

where the r's denote the real zeros of P and the C's denote the nonreal zeros. In this representation, some of the r's with different subscripts may be equal, and the same may be said of the C's. It should also be clear, since P was assumed to be of degree n, that the sum of the number of real and complex zeros must equal n; that is, $j + 2k = n$.

As you will note, we have grouped the linear factors corresponding to complex conjugate zeros in pairs. The reason for this will be made apparent in the next section.

We have noted that the sum and product of a complex number and its complex conjugate are real numbers. Given two linear expressions of the form $(x - C_1)$ and $(x - \bar{C}_1)$, where C_1 and $\bar{C}_1$ are complex conjugates, then the product of these two factors is a second-degree polynomial with real coefficients. To see this, we note that

$$(x - C_1)(x - \bar{C}_1) = x^2 - (C_1 + \bar{C}_1)x + C_1\bar{C}_1.$$

Since the coefficient of x, $C_1 + \bar{C}_1$, and the constant term, $C_1\bar{C}_1$, are both real numbers (why?) we see that $(x - C_1)(x - \bar{C}_1)$ is equal to a second-degree polynomial with real coefficients.

Example 5 As a specific example, consider the linear expressions $(x - i)$ and $\overline{(x - i)} = (x + i)$. The product of these two expressions is $(x - i)(x + i) = x^2 + 1$, which is a second-degree polynomial with real coefficients.

The polynomial $x^2 + 1$ that we obtained in this example is called an *irreducible quadratic expression* or an *irreducible polynomial of degree 2*. We now give a general definition of this kind of polynomial.

Definition 5.4a A polynomial of degree 2 with real coefficients is said to be **irreducible** (over the reals) if it cannot be factored into the product of two linear factors with real coefficients.

You already know that a real polynomial, $ax^2 + bx + c$, has real zeros if and only if its discriminant, $b^2 - 4ac$, is greater than or equal to zero. It follows that a polynomial of degree 2 is irreducible if and only if its discriminant is less than zero. Thus, we can easily decide whether or not $ax^2 + bx + c$ is irreducible simply by determining whether or not $b^2 - 4ac < 0$.

5.4 EXERCISES

In each of Exercises 1–3, express $P(x)$ in the form $P(x) = (x - c)Q(x)$.

1. $P(x) = x^3 + 4x$; $c = 2i$.
2. $P(x) = x^4 - 16$; $c = -2i$.
3. $P(x) = x^3 + x^2 + x + 1$; $c = i$.

In each of Exercises 4–9, use the information given to express $P(x)$ as a product of linear factors in the complex field. Then tabulate the zeros of $P(x)$.

4. $P(x) = x^3 - 2x + 4$ and $P(-2) = 0$.
5. One factor of $P(x) = x^3 + 3x^2 + 16x + 48$ is $x + 3$.
6. $P(x) = x^4 - 2x^3 + 2x^2 - 2x + 1$ and $P(i) = P(-i) = 0$.
7. $P(x) = x^3 - 3x^2 + x - 3$ and $P(i) = 0$.
8. $P(x) = x^3 + x^2 + 4x + 4$ and $P(-2i) = 0$.
9. $P(x) = x^3 - 5x^2 + 9x - 5$ and $P(2 - i) = 0$.

For each of Exercises 10–15, write a polynomial $P(x)$ of lowest possible degree and with real coefficients that has the given numbers as zeros.

10. $2 + i$.
11. -2 and $3i$.
12. 3 and $5i$.
13. 2 and $1 - i$.
14. 1, -1, and $-2i$.
15. -2, 3, and $i - 4$.

In each of Exercises 16–18, express $P(x)$ as a product of linear factors in the complex-number field and tabulate the zeros of $P(x)$.

16. $P(x) = x^3 + 1$.
17. $P(x) = x^3 - 1$.
18. $P(x) = x^3 - 5x^2 + 9x - 5$.

19. Suppose that $P(x)$ is a polynomial with real coefficients and c is a nonreal complex number such that $P(c) = 0$. Prove that at least one of the coefficients of $Q(x)$ in the factorization
$$P(x) = (x - c)Q(x)$$
must be nonreal.

20. Let c_1, c_2, and c_3 be the zeros of $P(x) = x^3 + px^2 + qx + r$. Prove each of the following.
    a) $c_1 + c_2 + c_3 = -p$.
    b) $c_1c_2 + c_1c_3 + c_2c_3 = q$.
    c) $c_1c_2c_3 = -r$.

5.5 Properties of Irreducible Quadratic Expressions

The information given in the last section can be most helpful in analyzing a polynomial function and in drawing its graph. In this section, we will show how this information can be used.

First of all, several of the ideas discussed in the last section can be combined into the following important generalization:

If $P(x)$ is a real polynomial of positive degree, then $P(x)$ is the product of linear factors and irreducible quadratic expressions. That is,

$$P(x) = (x - r_1) \ldots (x - r_j)(a_1 x^2 + b_1 x + c_1) \ldots (a_k x^2 + b_k x + c_k),$$

where the r's, a's, b's, and c's are all real numbers,

and $j + 2k = n$, with $0 \leq j \leq n$ and $0 \leq k \leq \dfrac{n}{2}$.

We remarked earlier, in section 5.4, that if we know when $P(x)$ is negative, zero, and positive, then we know a great deal about the function P and its graph. In that section, we saw how the sign of $P(x)$ can be determined when P has only linear factors whose constant terms are real numbers. We will now see how we can determine the sign of $P(x)$ when some of the factors of $P(x)$ are irreducible quadratic expressions. Then the generalization expressed above assures us that, given any polynomial function P of degree n with known linear and irreducible quadratic factors, we can determine where $P(x)$ is negative, zero, and positive.

Suppose that an irreducible quadratic expression $q(x) = ax^2 + bx + c$, is a factor of a polynomial $P(x)$. We can easily show that q has no real zeros; that is, we can show that there is no real number r such that $q(r) = 0$. For, if there were such a real number r, then a second zero of q would also have to be real, since nonreal zeros of polynomials occur as pairs of complex conjugates. But by the Factor Theorem (page 203), this means that $q(x)$ is factorable into the product of a real constant and two real linear factors. This, in turn, implies that $q(x)$ is reducible rather than irreducible, which contradicts our assumption that $q(x)$ is irreducible. Hence, $q(x)$ has no real zeros.

From the property that q has no real zeros, it follows that $q(x)$ must be positive for all real values of x, or $q(x)$ must be negative for all real values of x. To see this, suppose that $q(x)$ is positive for $x = x_1$ and negative for $x = x_2$. It then follows by the Intermediate Value Theorem (page 171), that there must be a value of x between x_1 and x_2 for which $q(x) = 0$. But this contradicts the fact that $q(x)$ has no real zeros and, hence, $q(x)$ is never zero. Therefore, $q(x)$ has the same sign for all real values of x.

Since an irreducible quadratic expression has the same sign for all values of x, we can determine that sign by evaluating the expression for any value of x. For example, $q(x) = x^2 + 1$ defines an irreducible quadratic expression. Since $q(0) = 1$, we can infer that $q(x) > 0$ for all $x \in R$. This leads to the observation that, if $q(x) = ax^2 + bx + c$ defines an irreducible quadratic expression (that is, if $b^2 - 4ac < 0$), then the sign of $q(x)$ must be that of c, because $q(0) = c$.

Now suppose we are given a polynomial $P(x)$ of degree n, and that we can determine in some manner all of the real zeros of P and the multiplicity of each of these zeros. In this case, the function P is defined as follows:

$$P(x) = (x - r_1)^{k_1}(x - r_2)^{k_2} \ldots (x - r_j)^{k_j}R(x);$$

where $r_1, \ldots, r_j$ are the distinct real zeros of P; $k_1, \ldots, k_j$ are their respective multiplicities; and $R(x)$ is a constant, an irreducible quadratic expression, or the product of a constant and one or more irreducible quadratic expressions.

Example 1 The zeros of the polynomial function P, defined by

$$P(x) = x^7 - x^6 - 14x^5 + 54x^4 - 103x^3 + 115x^2 - 88x + 60,$$

are 2, with multiplicity 2, and -5, with multiplicity 1. The equation for P can then be written as

$$P(x) = (x - 2)^2(x + 5)(x^2 + 1)(x^2 - 2x + 3).$$

In the example just given, $R(x)$ is the product of the two irreducible quadratic expressions $x^2 + 1$ and $x^2 - 2x + 3$. In every case, $R(x)$ will be a single factor or the product of factors, each of which has the same sign for all values of x. Hence, $R(x)$ will have the same sign for all values of x. It is clear that, to decide what that sign is, all we need do is determine the sign of $R(x)$ for an arbitrary value of x. Thus, in the preceding example, where $R(x) = (x^2 + 1)(x^2 - 2x + 3)$, since $R(0) = 1 \cdot 3 = 3$ it follows that $R(x) > 0$ for all $x \in R$.

Combining all of the information we have so far, we see that we can determine the sign of $P(x)$ for any polynomial P as follows.

1. Draw a horizontal real-number line.
2. If $R(x) < 0$, draw a parallel line above the real-number line. If the sign of $R(x)$ is positive, do nothing. [Remember that in determining whether $P(x)$ is negative or positive, we need only determine whether an odd or even number of negative factors are contained in P. Hence, we need not be concerned about positive factors.]
3. Next determine if an odd power of a linear factor of $P(x)$ with a real constant term is negative for some values of x. (Why do you need consider only odd powers of these factors?) Then draw a half-line above that portion of the real-number line for which this factor is negative; otherwise, do nothing. For example, corre-

sponding to the factor $(x - 1)^3$, which is negative for all $x < 1$, we would draw a left half-line, open on the right, as shown in Figure 5-1.

Figure 5-1

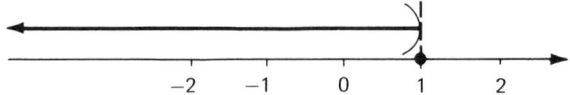

We will now apply the procedure just outlined to a specific example.

Example 2 Let the polynomial function P be defined by
$$P(x) = (x - 1)^3(x + 1)(x - 2)^2(x^2 + 2x + 3)(-x^2 + x - 1).$$
The quadratic expressions $(x^2 + 2x + 3)$ and $(-x^2 + x - 1)$ are both irreducible, as can be verified by computing their discriminants. Thus, in the general formula given on page 223,
$$R(x) = (x^2 + 2x + 3)(-x^2 + x - 1).$$
Since we know that $R(x)$ has the same sign for all x and since $R(0) = -3$, $R(x)$ is negative for all x. We therefore draw a parallel line above the number line, as in Figure 5-2, to show that this factor of $P(x)$ is negative.

Figure 5-2

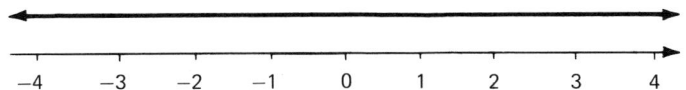

Now the linear factor $(x - 1)$ appears to the third power, which is an odd power. Therefore, since $(x - 1) < 0$ for $x < 1$, $(x - 1)^3 < 0$ for $x < 1$, and we add to our diagram the left half-line, as shown in Figure 5-3. We also note that 1, which is circled in the figure, is a zero of P.

Figure 5-3

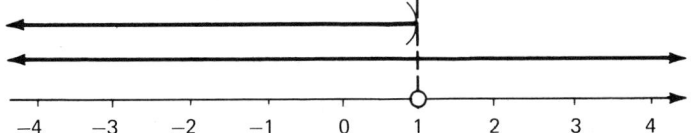

The linear factor $(x + 1)$ appears to the first power, which is also an odd power. Since $x + 1 < 0$ for all $x < -1$, we add another half-line to our figure, as shown in Figure 5-4. We also note that -1 is a zero of P.

Figure 5-4

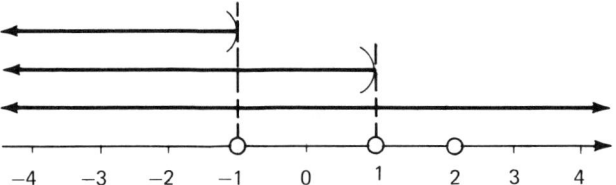

Because the linear factor $(x - 2)$ appears to the second power, it is nonnegative for all x; so we do nothing more, except to note that 2 is a zero of P.

We have now considered all the factors of P and we can use Figure 5-5 to determine the sign of $P(x)$ for all x. We see that this diagram has the following properties:

1. Corresponding to each of the factors which are odd powers of a linear factor, there is a half-line above that portion of the real-number line for which this power is negative.
2. Corresponding to $R(x)$, which is the product of all those factors that are not powers of linear factors with real constant terms, there is a line drawn over the entire number line to indicate that $R(x)$ is negative for all x.

Since each line drawn above the number line, or a portion of it, represents a negative factor, it should be clear that any portion of the real-number line with an odd number of lines above it represents an interval over which $P(x)$ is negative. Of course, any portion of the real-number line which has an even number of these lines above it represents an interval over which $P(x)$ is positive. These results are summarized in Figure 5-5.

Figure 5-5

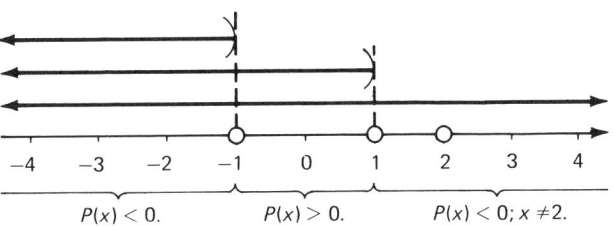

The diagram shows clearly that $P(x) > 0$ for all x in the interval $-1 < x < 1$ and that $P(x) < 0$ for all x in the interval $x < -1$ or $x > 1$ and $x \neq 2$. From the linear factors of P, we have noted that the real zeros of P are 1, -1, and 2. That is, $P(x) = 0$ if and only if x is 1, -1, or 2. Combining all this information, we can now draw the illustration in Figure 5-6 as a first step in graphing function P. The graph of P must lie

Figure 5-6

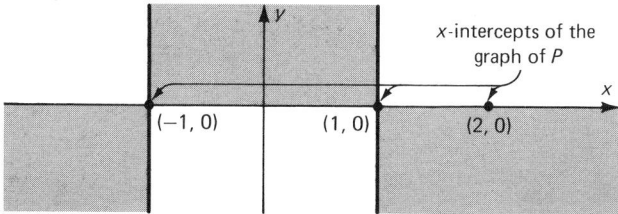

in those portions of the plane indicated by the shaded regions in the figure.

If we were to carry this example further, we would next consider P', the slope function of P. If $P'(x)$ is factorable, we could determine where the graph of P is rising, where it is falling, and where it has horizontal tangents. We could gain this information by analyzing P' in much the same way as we analyzed P; that is, by determining when $P'(x) > 0$, when $P'(x) < 0$, and when $P'(x) = 0$.

After considering P', we could consider P'', the slope function of the slope function of P. Again, if $P''(x)$ is factorable, we could analyze this function to determine when the graph of P is concave upward and when it is concave downward. With all this information, we would then be able to make a rough sketch of the graph of P.

5.5 EXERCISES

In each of Exercises 1–6, determine the values of x for which $P(x)$ is negative, zero, or positive. Then make a sketch similar to the one in Figure 5–6 to show the regions of the plane in which the graph of P must be contained.

1. $P(x) = (x - 3)(x^2 + 4)$.
2. $P(x) = (x + 1)^2(x^2 + 3)$.
3. $P(x) = (x - 2)^3(x^2 + x + 1)$.
4. $P(x) = (x - 1)(x + 1)(x^2 - 2x + 2)$.
5. $P(x) = x^2(x + 5)(-x^2 + x - 3)$.
6. $P(x) = x(x + 2)^2(x - 3)^3(x^2 + 5)$.

In each of Exercises 7 and 8, determine a and b so that the statement will be true.

7. $P(x) = x^4 - 9x^3 - (a + b)x^2 + bx + 1$ has 1 as a zero of multiplicity 2.
8. $P(x) = x^4 + (2 - b)x^3 + ax^2 - (a + b)x + 4$ has -2 as a zero of multiplicity 2.

In each of Exercises 9–12, find the values of x for which $P(x)$, $P'(x)$, and $P''(x)$ are respectively negative, zero, and positive. Then use this information to sketch the graph of P.

9. $P(x) = x^4 - 6x^2 - 7$.
10. $P(x) = x^3 + 4x^2 + 5x$.
11. $P(x) = -\frac{1}{3}x^4 - \frac{1}{3}x^3$.
12. $P(x) = \frac{1}{20}x^5 - x^3$.
13. If P is a polynomial function and r is a real zero of multiplicity k of P, it may be proved that r is a zero of multiplicity $k - 1$ of P'. To consider the geometric implications of this statement, let
$$P(x) = (x - r_1)^{k_1}(x - r_2)^{k_2} \cdots (x - r_j)^{k_j} R(x),$$
where $r_1, r_2, \ldots, r_j$ are the distinct real zeros of P; $k_1, k_2, \ldots, k_j$ are their respective multiplicities; and $R(x)$ is a constant, an irre-

ducible quadratic expression, or the product of a constant and one or more such expressions.

a) Suppose that $k_1 = 1$. Prove that the graph of P crosses the x-axis at $(r_1, 0)$. That is, for $0 < h < c$, where c is a small positive number, show that $P(r_1 - h)$ and $P(r_1 + h)$ differ in sign.

b) Suppose that $k_2 = 2$. Prove that the graph of P has a relative maximum or minimum at $(r_2, 0)$.

c) Suppose that $k_3 = 3$. Prove that the graph of P crosses the x-axis at $(r_3, 0)$, but that the x-axis is the tangent line to the graph of P at $(r_3, 0)$.

14. Let P be a polynomial function with r as a real zero of multiplicity $k > 2$. Describe the characteristics of the graph of P in the vicinity of the point $(r, 0)$ in each of the following cases:
    a) k is even.   b) k is odd.

In each of Exercises 15–18, use the theorem from Exercise 13 to sketch each portion of the graph of P that is near a point of intersection with the x-axis.

15. $P(x) = (x - 1)(x - 3)^2(x^2 + 1)$.
16. $P(x) = (x - 1)^2(x - 3)(x^2 + 1)$.
17. $P(x) = (x - 1)^3(x - 3)(x^2 + 1)$.
18. $P(x) = (x - 1)^2(x - 3)^3(x^2 + 1)$.

5.6 Zeros of Polynomials

In the preceding sections of this chapter, we have seen that determining all the real zeros, and their multiplicities, of a real polynomial enables us to determine the values for which the polynomial is negative, zero, or positive. We have also learned that if we are given a polynomial $P(x)$, then we can determine the polynomials $P'(x)$ and $P''(x)$. Further, we saw that if we could decide when $P'(x)$ and $P''(x)$ are negative, zero, and positive, we could then discover the basis properties of the function P, determined by the polynomial $P(x)$, and draw a good characterization of its graph.

With these facts in mind, we see that it is very important for us to be able to find the real zeros of real polynomials. The purpose of the rest of this chapter is to develop some techniques for determining these zeros. In the following discussion, we will use the word polynomial to mean a real polynomial (a polynomial with real coefficients), unless we specifically state otherwise.

In our work with polynomials so far, most of the specific examples have had integral coefficients. This need not be the case, of course, since a polynomial is defined to be an expression of the form

$$a_n x^n + a_{n-1} x^{n-1} + \ldots + a_1 x + a_0,$$

where n is a nonnegative integer and $a_0, a_1, \ldots, a_n$ are arbitrary real numbers. The expression

$$\pi x^5 - \sqrt{2}x^3 + x^2 - \sqrt{5},$$

for example, is a polynomial with real coefficients that are not integers. The reason for using polynomials with integral coefficients in this book (as in most textbooks) is to present the important ideas without having to carry out needlessly complicated calculations. Moreover, many polynomials used in solving practical problems have integral or rational coefficients, so that a study of such polynomials is useful as well as convenient.

Finding the zeros of a specific polynomial can be a very simple task or an extremely complicated one. In fact, we will find that for certain polynomials we will not be able to determine the exact values of some of their zeros, but will be forced to approximate their values. It is usually, but not always, the case that the higher the degree of the polynomial the more difficult it is to determine all of its zeros.

We begin by summarizing what we already know about determining the zeros of a polynomial. A polynomial of degree zero has no zeros. (Why?) To find the zero of a polynomial of degree one, of the form $ax + b$, is a trivial matter; the zero is found by using the formula $x = \dfrac{-b}{a}$.

To determine the zeros of a polynomial of degree two, we have the quadratic formula,

$$x = \frac{-b \pm \sqrt{b^2 - 4ac}}{2a}.$$

Remember that this formula gives the real zeros of the polynomial $ax^2 + bx + c$ when the discriminant $b^2 - 4ac$ is not negative. When the discriminant is negative, the polynomial has no real zeros.

The formula for finding the zero of a polynomial of degree one and the quadratic formula both require only a finite number of algebraic operations on the coefficients of the given polynomial. In the case of the quadratic equation, these operations include addition, multiplication, subtraction, division, and the extraction of a root. In view of these facts, it might seem reasonable to expect that, given a polynomial of degree n, where n is a positive integer, there should be a formula involving a finite number of these basic algebraic operations which would give the real roots of the polynomial.

When such formulas do exist, we say that the polynomial can be *solved by radicals*. It is the case that any polynomial of degree three or four can also be solved by radicals, but the required formulas are much too complicated for general use. From the first half of the sixteenth century, when these formulas were discovered, until the early part of the nineteenth century, many mathematicians tried to derive similar formulas for polynomials of degree five and greater. None of these attempts were successful because, as we mentioned early in the chapter, no such formulas exist. The fact that no formulas exist for polynomials of degree greater than four was proved by the Norwegian mathematician Niels Henrik Abel in 1824.

To emphasize the point we made earlier, Abel did *not* prove that it is impossible to determine the zeros of any polynomial of degree five or greater. In many instances, it is a simple matter to find the zeros of such polynomials. For example, the polynomial

$$x^6 - 12x^5 + 58x^4 - 144x^3 + 193x^2 - 132x + 36,$$

which looks so complicated, can be factored rather easily into

$$(x-1)^2(x-2)^2(x-3)^2.$$

This factorization shows immediately that the zeros of the polynomial are 1, 2, 3, each being of multiplicity 2.

What Abel did prove is that there is no *general* formula involving only a finite number of algebraic operations on the coefficients of any polynomial of degree five or greater that will give the zeros of the polynomial. Even though no general formulas exist for finding the zeros of a polynomial of degree greater than four, there are techniques which make it possible to approximate the zeros of a polynomial to any desired degree of accuracy. We will discuss one of these techniques in the last section of this chapter. In the following section, we will discuss some techniques for determining the *rational* zeros of polynomials with *rational* coefficients.

5.7 Polynomials with Rational Coefficients

As we have already noted, most of the polynomials that we have considered have had integral coefficients. In this section, we will discuss methods for finding all the rational zeros of such polynomials and of polynomials with rational coefficients.

Actually, the method is developed for polynomials with integral coefficients and then extended to apply to polynomials with rational coefficients. The method can be extended because it is always possible to multiply a polynomial with rational coefficients by an appropriate nonzero integer to obtain a polynomial with integral coefficients having the same zeros as the original polynomial.

Example 1 Suppose that

$$P(x) = \tfrac{1}{2}x^3 + \tfrac{6}{5}x^2 - \tfrac{7}{10}x - 1.$$

Then the polynomial obtained by multiplying $P(x)$ by 10,

$$Q(x) = 10P(x) = 5x^3 + 12x^2 - 7x - 10,$$

has exactly the same zeros as $P(x)$.

We can, therefore, limit ourselves to the problem of determining all the rational roots of a polynomial of degree n with integral coefficients, $a_0, \ldots, a_n$,

$$P(x) = a_n x^n + a_{n-1} x^{n-1} + \ldots + a_1 x + a_0.$$

Now if $\dfrac{p}{q}$ is a rational number expressed in lowest terms (that is, if p and q have no common integral divisor greater than 1) and if $\dfrac{p}{q}$ is a zero of $P(x)$, then

$$a_n \left(\frac{p}{q}\right)^n + a_{n-1} \left(\frac{p}{q}\right)^{n-1} + \ldots + a_1 \frac{p}{q} + a_0 = 0.$$

On multiplying through by q^n, we have

$$a_n p^n + a_{n-1} p^{n-1} q + \ldots + a_1 p q^{n-1} + a_0 q^n = 0, \text{ or}$$
$$a_n p^n = -(a_{n-1} p^{n-1} q + \ldots + a_1 p q^{n-1} + a_0 q^n).$$

Since the right side of this last equation contains q in every term, it must be divisible by q. But then the left side of the equation $a_n p^n$ must also be divisible by q. Since p^n cannot be divisible by q (remember that p and q have no common factors), a_n must be divisible by q.

But we can also rearrange the equation

$$a_n p^n + a_{n-1} p^{n-1} q + \ldots + a_1 p q^{n-1} + a_0 q^n = 0$$

to obtain

$$a_0 q^n = -(a_n p^n + a_{n-1} p^{n-1} q + \ldots + a_1 p q^{n-1}).$$

After inspecting this form of the equation, you should be able to carry out an argument similar to the one just given to show that a_0 must be divisible by p.

The results of the preceding discussion are stated in the following theorem.

Theorem 5.7a *If $P(x)$ is a polynomial with integral coefficients,*

$$P(x) = a_n x^n + a_{n-1} x^{n-1} + \ldots + a_1 x + a_0,$$

and if $\dfrac{p}{q}$ is a rational number in lowest terms that is a zero of $P(x)$, then q is a factor of a_n and p is a factor of a_0.

This theorem gives us a direct method of determining the rational zeros of $P(x)$. Suppose that we let S_{a_0} denote the set of all integral factors of a_0 (we are assuming that $a_0 \neq 0$) and S_{a_n} denote the set of all integral factors of a_n. Then Theorem 5.7a says that *if* there are any rational zeros of $P(x)$, then they must belong to the set

$$S = \left\{ \frac{p}{q} \mid p \in S_{a_0} \text{ and } q \in S_{a_n} \right\}.$$

Thus, to determine if there are any rational zeros of $P(x)$ we could test every member (in lowest terms) of S.

Example 2 Let $P(x) = 6x^3 - 19x^2 + 16x - 4$. Using the notation introduced above, we see that
$$S_{a_0} = S_{-4} = \{1, 2, 4, -1, -2, -4\} \text{ and that}$$
$$S_{a_n} = S_6 = \{1, 2, 3, 6, -1, -2, -3, -6\}.$$

We next list the rational numbers in lowest terms that can be obtained from the set S:

$$\pm 1, \ \pm 2, \ \pm 4, \ \pm \tfrac{1}{2}, \ \pm \tfrac{1}{3}, \ \pm \tfrac{1}{6}, \ \pm \tfrac{2}{3}, \text{ and } \pm \tfrac{4}{3}.$$

From Theorem 5.7a, we conclude that the rational zeros of $P(x)$, if any exist, must be included in the preceding list.

By direct computation, we can determine that 2 is an integral zero of $P(x)$. Having discovered that 2 is a zero of $P(x)$, we can proceed in one of two ways: The first way would be to continue to check the other possible rational zeros of $P(x)$ given in the list. The second would be to use the Factor Theorem to obtain $P(x) = (x - 2)(6x^2 - 7x + 2)$. We could then use the quadratic formula to determine the zeros of $6x^2 - 7x + 2$.

In using this theorem about rational roots, there are some important observations you can make that will shorten the amount of computing and testing you need to do. First, if a polynomial of degree n has integral coefficients and $a_n = 1$, then the only possible rational zeros are the *integers* which divide a_0.

Example 3 If $P(x) = x^4 - 5x^2 + 2x - 2$, then the only *possible* rational zeros of $P(x)$ are ± 1 and ± 2.

Second, no rational zero of a polynomial can be greater than $|a_0|$, if a_0 is different from zero.

Example 4 If $P(x) = 3x^3 - 2x^2 + 5x - 2$, then any rational zeros of $P(x)$ must lie in the interval $-2 \leq x \leq 2$.

Of course, if the constant term of a polynomial is zero, then, before using any of the procedures just described, we would first factor out the highest possible power of x.

Example 5 If $P(x) = x^6 - 4x^4 + 3x^3$, we would first factor $P(x)$ into $x^3(x^3 - 4x + 3)$ and then apply the procedures to the polynomial $Q(x) = x^3 - 4x + 3$.

5.7a EXERCISES

For each of Exercises 1–10, list all the possible rational zeros of $P(x)$. Then determine which of these are actually zeros of $P(x)$.

1. $P(x) = x^3 - x^2 - x + 1$.
2. $P(x) = x^3 - 3x - 2$.
3. $P(x) = x^3 - 4x^2 - 11x - 6$.
4. $P(x) = x^3 + 3x^2 - 4x - 12$.
5. $P(x) = x^4 + 5x^3 + 9x^2 - x - 14$.
6. $P(x) = x^4 + 3x^3 + x^2 + x - 6$.
7. $P(x) = 2x^3 + 3x^2 - x - 1$.
8. $P(x) = 3x^3 - 10x^2 + 5x + 4$.
9. $P(x) = 2x^3 - x^2 - 2x + 1$.
10. $P(x) = 3x^4 - 8x^3 - 28x^2 + 64x - 15$.
11. Prove: If $P(x) = a_n x^n + a_{n-1} x^{n-1} + \ldots + a_1 x + a_0$ is a polynomial of degree n with integral coefficients and $\frac{p}{q}$ is a rational number in lowest terms that is a zero of $P(x)$, then a_0 must be divisible by p.
12. a) Prove: If P is a polynomial function of degree n, where $n > 1$, and $P(a) = P(b) = 0$, where $a < b$, then there is a number x_0 between a and b (that is, $a < x_0 < b$) such that $P'(x_0) = 0$.
    b) Complete the following statement: "Between any two consecutive points of intersection with the x-axis on the graph of a polynomial function of degree n, $n > 1$, there is a _____."
13. Let $(a, P(a))$ and $(b, P(b))$ be two points on the graph of a polynomial function P of degree n, $n > 1$.
    a) Find the linear function l, whose graph is the line through the points $(a, P(a))$ and $(b, P(b))$.
    b) Prove that the function $P - l$ satisfies the hypothesis of the theorem to be proved in Exercise 12a.
    c) Use the results of part b of this exercise to prove that there is a number x_0 between a and b such that the tangent to the graph of P at $(x_0, P(x_0))$ is parallel to the line through $(a, P(a))$ and $(b, P(b))$.

We next have the task of deciding which, if any, of these rational numbers actually are zeros of P. What is the most efficient way to proceed at this point? We could simply start with the first number listed and then proceed to test each number in order. This would be time-consuming and also quite unnecessary. What we need are *tests* that will give us estimates of where the zeros of P are located. For example, there is a simple test, described in a subsequent theorem (Theorem 5.7c), which tells us immediately that this particular polynomial has no negative zeros. This one test would reduce the number of candidates for rational zeros from 16 to 8.

There are many and varied tests for deciding where the zeros of a polynomial might be located. We will limit ourselves to describing several that are relatively simple to apply. As a first example, Theorem 5.7b provides an easy way for determining that a certain class of polynomials cannot have any positive zeros.

Theorem 5.7b *If the coefficients of a polynomial are all of the same sign, either positive or negative, then the polynomial has no positive zeros.*

Proof If the coefficients of $P(x)$ are all positive and $x_0 > 0$, then $P(x_0) > 0$. The other part of the proof is analogous and is left as an exercise. □

Theorem 5.7c *If a polynomial has the property that all the coefficients of its even powers are of one sign and all the coefficients of its odd powers are of the opposite sign, then the polynomial has no negative zeros.*

Proof This theorem can be proved either directly by considering cases or in the following way, which makes it a corollary of Theorem 5.7b. If $P(x)$ is a polynomial in x, then $P(-x) = R(x)$ is a polynomial in x whose zeros are the negatives (additive inverses) of the zeros of $P(x)$. If the coefficients of $P(x)$ satisfy the conditions of this theorem, then it follows that the coefficients of $P(-x) = R(x)$ are all of one sign. (Explain why this is so.) This means by Theorem 5.7b, that the zeros of $R(x) = P(-x)$ cannot be positive. Hence, the zeros of $P(x)$ cannot be negative. □

Theorem 5.7c could have simplified our task of determining any rational zeros of the polynomial $P(x) = 6x^3 - 19x^2 + 16x - 4$ in Example 2. Since this polynomial satisfies the condition of the theorem, it cannot have any negative zeros. In the final section of this chapter, we will prove another theorem that is most useful in determining intervals in which the zeros of a polynomial function will occur.

5.7b EXERCISES

In each of Exercises 1–8, list all the possible rational zeros of $P(x)$. Then use Theorems 5.7b and 5.7c to reduce this list. Finally, determine the rational zeros of $P(x)$.

1. $P(x) = x^3 - 2x^2 + 3x - 4$.
2. $P(x) = x^4 + 2x^3 + 2x^2 + 2x + 1$.
3. $P(x) = -x^3 - 6x^2 - 11x - 6$.

4. $P(x) = 2x^3 - 7x^2 + 7x - 2$.
5. $P(x) = 2x^3 - 5x^2 + 8x - 3$.
6. $P(x) = 2x^4 - 5x^3 + 4x^2 - 5x + 2$.
7. $P(x) = 4x^4 + 4x^3 + 13x^2 + 12x + 3$.
8. $P(x) = 6x^4 + 23x^3 + 29x^2 + 16x + 4$.
9. Prove: If $P(x)$ is a polynomial whose coefficients are all negative, then $P(x)$ has no positive zeros.

5.8 Isolating Zeros of Polynomials

It is often helpful to be able to locate zeros of a function in the sense that we can say a zero must lie between two numbers, a and b. If $P(x)$ is a polynomial such that $P(a)$ and $P(b)$ are of opposite sign, that is, such that $P(a)P(b) < 0$, then the Intermediate Value Theorem tells us that there must be at least one number x_0 between a and b such that $P(x_0) = 0$.

Example 1 Let $P(x) = x^5 - 4x^3 + x + 1$. It is a simple matter to compute $P(0) = 1$, $P(1) = -1$ and $P(2) = 3$. Since $P(0)$ and $P(1)$ are of opposite sign, there must be at least one zero of $P(x)$ between 0 and 1. Also, since $P(1)$ and $P(2)$ are of opposite sign, there must be at least one zero of $P(x)$ between 1 and 2. Interpreted geometrically, this means that the graph of function P intersects the x-axis at least once between (0, 0) and (1, 0) and at least once between (1, 0) and (2, 0).

The following theorem establishes this useful property of polynomial functions. The property is easily understood when you interpret it geometrically. It simply says that, given a polynomial function such that one point of its graph is above the x-axis and the other is below the x-axis, then there must be at least one point of the graph that is on the x-axis.

Theorem 5.8a *If $P(x)$ is a polynomial and $P(a)P(b) < 0$, where $a < b$, then there is at least one number x_0 in the interval $a < x_0 < b$ such that $P(x_0) = 0$.*

Since P is a continuous function, this theorem is a corollary of the Intermediate Value Theorem, Theorem 4.3b. As we remarked earlier, this theorem can often be used in a systematic way to determine the approximate location of the zeros of a polynomial.

Example 2 Consider the polynomial $P(x) = 8x^3 - 36x^2 + 46x - 15$. First of all, Theorem 5.4a tells us that this polynomial has no negative zeros. Hence,

we need consider only nonnegative domain elements. To simplify comdutation, we choose integers and compute the functional values corresponping to 0, 1, 2, and 3. By direct computation, we have $P(0) = -15$, $P(1) = 1$, $P(2) = -3$, and $P(3) = 15$. From these results, we know that $P(x)$ changes sign between $x = 0$ and $x = 1$, between $x = 1$ and $x = 2$, and between $x = 2$ and $x = 3$. Therefore, by Theorem 5.8a, zeros of $P(x)$ are between 0 and 1, between 1 and 2, and between 2 and 3. Since $P(x)$ is a polynomial of degree three, we now have determined certain intervals in which the zeros of $P(x)$ must be. (Why?) We sometimes say that, when we have located each of the zeros of a polynomial between two specific numbers, we have *isolated* the zeros of the polynomial. By Theorem 5.7a, we know that possible rational-number zeros of $P(x)$ are contained in
$$\{1, 3, 5, 15, \tfrac{1}{2}, \tfrac{3}{2}, \tfrac{5}{2}, \tfrac{15}{2}, \tfrac{1}{4}, \tfrac{3}{4}, \tfrac{5}{4}, \tfrac{15}{4}, \tfrac{1}{8}, \tfrac{3}{8}, \tfrac{5}{8}, \tfrac{15}{8}\}.$$

(We have included only the possible rational zeros that are positive, since we already know that there are no negative zeros.) What we now know about the approximate locations of the zeros—between 0 and 1, 1 and 2, and 2 and 3—quickly reduces the possibilities to

$$\{\tfrac{1}{2}, \tfrac{3}{2}, \tfrac{5}{2}, \tfrac{1}{4}, \tfrac{3}{4}, \tfrac{5}{4}, \tfrac{1}{8}, \tfrac{3}{8}, \tfrac{5}{8}, \tfrac{15}{8}\}.$$

As it turns out, the first three members of this set are the zeros of $P(x)$.

This example outlines a definite procedure for determining all of the rational roots of polynomials with rational coefficients. Usually we will, on discovering the first rational zero r_1, divide the given polynomial of degree n, $P_n(x)$, by $(x - r_1)$ to obtain

$$P_n(x) = (x - r_1)P_{n-1}(x).$$

We then apply the same procedure to the polynomial $P_{n-1}(x)$, which is of degree $n - 1$. Continuing in this manner, we will eventually factor out all linear factors corresponding to rational roots and obtain

$$P_n(x) = (x - r_1)(x - r_2) \cdots (x - r_k)P_{n-k}(x).$$

How do we now proceed with the polynomial $P_{n-k}(x)$, which has no rational zeros? Furthermore, what do we do with a polynomial $P(x)$ which does not have rational coefficients? In either case, if the polynomial is of second degree or less, there is no problem, since we have formulas for first and second degree polynomials. But if $P(x)$ is of the third degree or higher, how do we proceed?

Given any polynomial, we can always use the procedure of Theorem 5.8a to isolate zeros. In fact, once we have isolated a zero, we can "refine" this isolation again and again to any desired degree of accuracy. What we mean by "refining" the isolation is taking smaller and smaller intervals within the original interval, $a < x_0 < b$, described in the theorem. For example, if we have isolated a zero of $P(x)$ between 0 and 1, we can then divide the interval from 0 to 1 in 10 subintervals,

$$0 \leq x \leq .1, .1 \leq x \leq .2, \ldots, .9 \leq x \leq 1.$$

Either $P(x)$ must be zero at some endpoint of one of these intervals, or there must be a change of sign for $P(x)$ for the endpoints of one of these intervals. Suppose, for instance, that $P(.1) \cdot P(.2) < 0$. Then we know by Theorem 5.8a that a zero of $P(x)$ lies between .1 and .2. We could then subdivide the interval $.1 \leq x \leq .2$ into subintervals and repeat the process to any degree of accuracy.

Such a process would, in general, be extremely laborious and time-consuming, and there are other techniques which often prove simpler. Unfortunately, we do not have time to present any further methods for approximating zeros of polynomials which cannot be factored. In your future mathematics courses, you will learn such techniques.

5.8 EXERCISES

1. Verify that $P(x) = -x^3 - 3x^2 + 2x + 5$ has a real zero between -4 and -3, and prove that this zero is irrational.
2. Verify that $P(x) = x^3 - x^2 + x - 3$ has a real zero between 1 and 2, and prove that this zero is irrational.

Each of the following polynomials has exactly one real zero. In each case, determine the consecutive integers between which this zero lies. Then determine if this zero is rational or irrational.

3. $P(x) = 2x^3 - x^2 - 3$.
4. $P(x) = 2x^3 + x^2 + x - 1$.
5. $P(x) = -2x^3 + 7x^2 - 7x + 5$.
6. $P(x) = 2x^3 - x^2 + x + 1$.

All of the real zeros of each of the following polynomials are between -4 and 4. For each zero, determine the consecutive integers between which that zero lies. Then determine which of these two integers is closer to the zero.

7. $P(x) = -x^3 + 3x + 1$.
8. $P(x) = x^3 + 2x - 13$.
9. $P(x) = 2x^3 - x^2 + 5x - 3$.
10. $P(x) = x^4 - x^2 - 3$.

6

Trigonometric Functions; Sine and Cosine

Introduction

In this chapter we will consider the sine and cosine functions which were developed in the study of triangles, more specifically, right triangles. These functions belong to an important class of functions called **trigonometric functions**. (The word *trigonometry* comes from three Greek words, *tri*—three, *gonia*—angle, and *metria*—measurement.)

Originally the trigonometric functions were defined as functions of angles. For mathematical and scientific purposes these functions were generalized to functions of a real variable. An understanding of this generalization requires some knowledge of the notion of an angle and the measurement of angles. We shall briefly investigate these topics in the next section.

In this chapter we will analyze the trigonometric functions, as in the study of polynomial functions, by determining their zeros, the regions of their domains where they are positive and negative, and certain properties of their slope functions. This analysis will enable us to draw good approximations of their graphs.

6.1 Measures of Angles

In geometry, an angle is defined as the union of two noncollinear closed rays that have a common endpoint (see Figure 6-1), and $\angle R_1VR_2$ is identical with $\angle R_2VR_1$. In trigonometry, however, $\angle R_1VR_2$ and

∡R_2VR_1 are considered as different directed angles; that is, the angles defined by the ordered pairs of rays (R_1, R_2) and (R_2, R_1) are different.

Figure 6-1

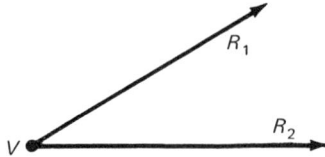

If we construct a circle of radius r with its center at the common endpoint of the closed rays R_1 and R_2, we define the angle ∡(R_1, R_2) as the angle that includes that part of the circumference of the circle "covered" in moving in a counterclockwise direction around the circle from ray R_1 to R_2. (See Figure 6-2.)

Figure 6-2

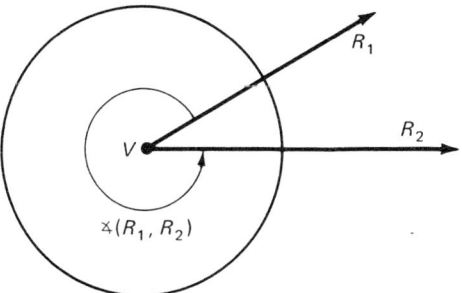

Similarly, ∡(R_2, R_1) is the angle that includes that part of the circumference of the circle "covered" in moving in a counterclockwise direction around the circle from ray R_2 to R_1. (See Figure 6-3.) Given an angle, ∡(R_1, R_2), it is conventional to refer to ray R_1 as the **initial side** of the angle and ray R_2 as the **terminal side** of the angle.

Figure 6-3

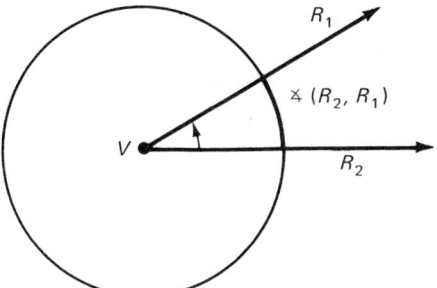

To obtain the degree measure of a directed angle, the circumference of the circle is divided into 360 equal units of arc, called **degrees**, and the angle is measured by the number of degrees it contains. (Technically, there are no angles corresponding to measures of 180° and 0°, because the rays R_1 and R_2 are collinear in these cases. But we will often be imprecise and speak of such "angles.") Of course, one could divide the circumference into any number of equal units of arc and produce an angle measure. The degree measure which divides the circumference into 360 equal units is Babylonian in origin and may have resulted originally from the fact that, by their reckoning, there were approximately 360 days in a year.

A directed angle is said to be in **standard position** on a set of coordinate axes if its vertex is at the origin and its initial side coincides with the positive x-axis. The angle is said to terminate in a particular quadrant if its terminal side is included in that quadrant.

For example, as illustrated in Figure 6–4, an angle of 155° terminates in quadrant II.

Figure 6-4

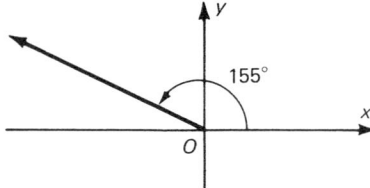

Another useful way of measuring directed angles is to measure them on the basis of the ratio of the length of the arc of the circle which they intercept to the radius of the circle. The unit of measure is the *radian*, which is the measure of the angle that intercepts an arc of the circle equal in length to the radius of the circle. You may recall from geometry that this definition is independent of the radius of the circle used; that is, an angle whose measure is $\frac{\pi}{2}$ radians in a circle of radius 1 is congruent to an angle whose measure is $\frac{\pi}{2}$ radians in any circle of radius r, $r > 0$. If the circle is the unit circle of radius 1, then an angle of 1 radian is the angle (with vertex at the center of the circle) that intercepts an arc of the circle which is 1 unit long. (See Figure 6–5.)

Figure 6-5

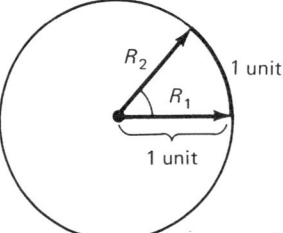

An angle of 1 radian is approximately the same as an angle of 57.3°. This is easy to see, since the circumference of a circle of unit radius is 2π and an angle of 1 radian intercepts an arc of length 1. Hence, we have

$$\frac{1 \text{ radian}}{360°} = \frac{1}{2\pi}.$$

6.1 MEASURES OF ANGLES

Therefore,
$$1 \text{ radian} = \frac{360°}{2\pi} = \frac{180°}{\pi},$$

which is approximately 57.3°. From the ratio $\frac{1 \text{ radian}}{360°} = \frac{1}{2\pi}$, we also see that $1° = \frac{\pi}{180}$ radians.

The familiar degree measures of 45°, 60°, 90°, and 180° correspond to $\frac{\pi}{4}, \frac{\pi}{3}, \frac{\pi}{2}$, and π radians, respectively. (You should verify these for yourself.) In working with radian measure, it is customary to give the numerical measure of the angle without using the word "radian." For example, we speak of "an angle of $\frac{\pi}{2}$," instead of "an angle of $\frac{\pi}{2}$ radians."

From the way that we have defined radian measure of directed angles, every angle has a unique measure in radians. This measure will be greater than or equal to zero and less than 2π.

We can, however, extend the definition of radian measure. First, we can extend radian measures to values greater than or equal to 2π by thinking of the terminal side R_2 of the angle as having been rotated a complete revolution or more from its initial position—coinciding with the initial side R_1—to its final position. Thus, we can assign to the angle, $\measuredangle(R_1, R_2)$ in Figure 6-6, not only the value $\frac{\pi}{3}$, but also the value $\frac{\pi}{3} + 2\pi$. In a similar way, we can also assign to $\measuredangle(R_1, R_2)$ the values $\frac{\pi}{3} + 4\pi$, $\frac{\pi}{3} + 6\pi$, and, indeed, any value in the form $\frac{\pi}{3} + 2n\pi$, where n is a nonnegative integer.

Figure 6-6

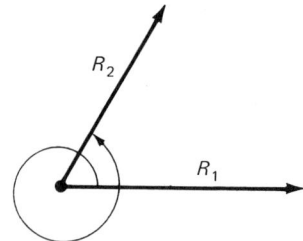

To extend radian measure to negative values, we adopt the convention of thinking of the terminal side of the angle as forming positive angles as it is rotated in a counterclockwise direction and negative angles as it is rotated in a clockwise direction.

On this basis, the angle $\measuredangle(R_1, R_2)$ in our example may be assigned the value $-\frac{5}{3}\pi$. (See Figure 6-7.) However, $-\frac{5}{3}\pi = \frac{\pi}{3} + 2(-1)\pi$. Therefore, using the notion of rotation for more than a complete revolution (but in a clockwise direction), we can assign to this same angle the values $-\frac{5}{3}\pi - 2\pi = \frac{\pi}{3} + 2(-2)\pi$, $-\frac{5}{3}\pi - 4\pi = \frac{\pi}{3} + 2(-3)\pi$, and, in fact, any value in the form $\frac{\pi}{3} + 2n\pi$, where n is a negative integer.

Figure 6-7

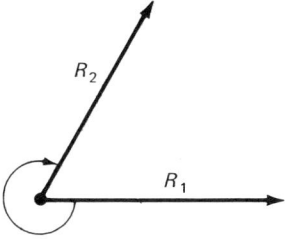

Combining these results with our earlier extension, we see that we can assign to the angle $\measuredangle(R_1, R_2)$ any radian measure of the form

$$\frac{\pi}{3} + 2n\pi, \quad \text{where } n \text{ is an integer.}$$

Now, any real number t must be in precisely one interval of the form

$$2n\pi \leq t < 2n\pi + 2\pi.$$

Hence, corresponding to any real number t, there will be a unique angle measured in radians.

6.1a EXERCISES

Express each of the following directed angles in radian measure. (Answers may be left in terms of π.) Then draw the angle in standard position and name the quadrant in which the angle terminates.

1. 30°
2. 200°
3. 300°
4. 135°
5. −48°
6. 120°

Express each of the following directed angles in degree measure. Then draw the angle in standard position and name the quadrant in which it terminates.

7. $\dfrac{\pi}{12}$

8. $-\dfrac{5\pi}{6}$

9. $-\dfrac{5\pi}{2}$

10. $\dfrac{13\pi}{6}$

11. $\dfrac{5\pi}{18}$

12. $-\dfrac{7\pi}{4}$

Draw each of the following directed angles in standard position. Then determine the coordinates of the point P where the terminal side of the angle intersects the circle with radius 1 whose center is at the origin. [Hint: draw the segment $\overline{PM}$ that is perpendicular to the x-axis and consider the right triangle OPM. Answers may be left in radical form.]

13. $\dfrac{\pi}{6}$

14. $\dfrac{4\pi}{3}$

15. $\dfrac{3\pi}{4}$

16. $-\dfrac{\pi}{3}$

17. $-\dfrac{7\pi}{6}$

18. $\dfrac{7\pi}{4}$

19. Angles AOB and COD are central angles in circle O, whose radius is r. The measure of angle AOB is 1 radian and that of angle COD is t radians. Use the fact that the measures of central angles are proportional to the measures of their intercepted arcs to prove that the length of $\overset{\frown}{CD}$ is rt.

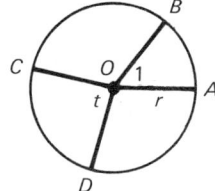

20. In the accompanying diagram, the measure of central angle AOB is t radians. Show that the area of sector AOB is $\tfrac{1}{2}r^2t$. [Hint: the ratio of the area of a sector to the area of a circle is equal to the ratio of its central angle to the measure of a complete revolution.]

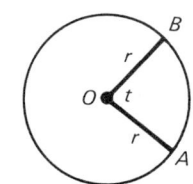

We have seen that corresponding to each real number t there is a unique angle measured in radians. Corresponding to every such angle there is a unique point $P(x, y)$ on the unit circle, C, with center at the origin.

$$C = \{(x, y) \mid x^2 + y^2 = 1\}$$

These two correspondences combine to give us a single correspondence that associates with every real number t a unique point $P(x, y)$ on the circle C. We can describe this correspondence in an interesting way. You know from geometry that the circumference of the unit circle is equal to 2π. Suppose that we imagine the nonnegative real-number line wrapped around the unit circle in a counterclockwise direction, with point 0, or the origin, of the number line coinciding with point $(1, 0)$ on the unit circle. Corresponding to any point on the nonnegative real-number line is a unique point on the unit circle. Because the circumference of the circle is 2π, the point 2π on the number line will correspond to $(1, 0)$, as does the point 0. Figure 6-8 indicates some of the corresponding points.

Figure 6-8

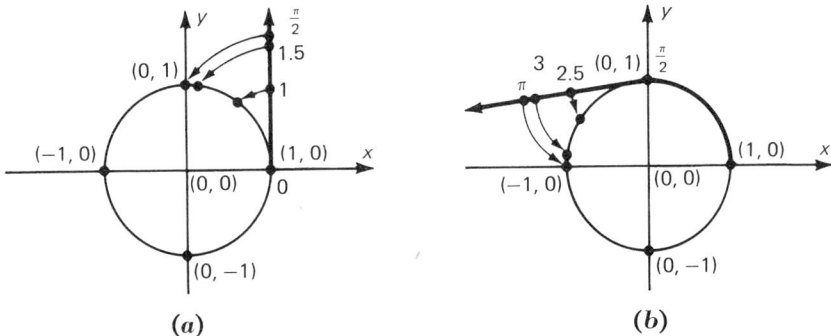

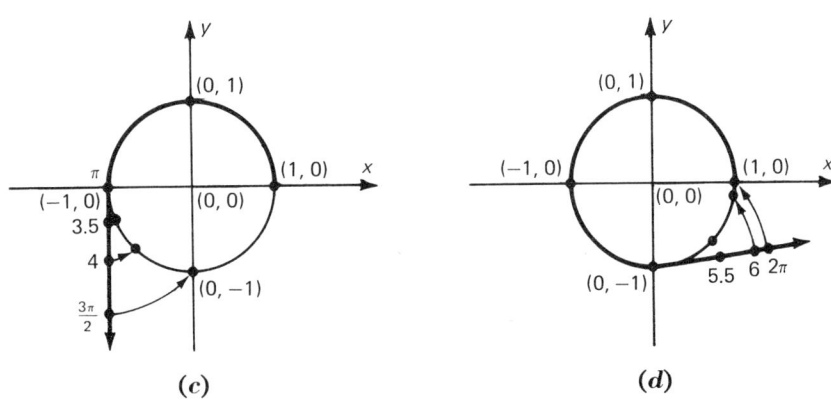

6.1 MEASURES OF ANGLES

It is clear that, if we continue to wrap the number line around the unit circle, many points on the line will correspond to the same point on the circle. For example, $1 + 2\pi$ on the number line will correspond to the same point as does 1; 3π, or $\pi + 2\pi$, will correspond to the same point as does π; 4π, or $2\pi + 2\pi$, will correspond to the same point as do 0 and 2π, and so on. In general, given any point t on the nonnegative real line, the point $t + 2\pi$ will correspond to the same point as t does on the unit circle. As a matter of fact, the points on the real line,

$$t, t + 2\pi, t + 4\pi, \ldots, t + n(2\pi), \ldots, \quad \text{for } t \geq 0,$$

will all correspond to the same point on the circle. Thus, any one point on the circle is the image of infinitely many points on the nonnegative real line, with this set of points on the line being equally spaced at intervals of 2π. On the other hand, there is precisely one point on the unit circle that is the image of any point of the nonnegative real-number line.

If we again let the origin correspond to the point (1, 0) and if we wrap the negative real-number line around the unit circle in a clockwise direction, we get a correspondence between negative real numbers and points on the unit circle as shown in Figure 6-9. Notice that $-\dfrac{\pi}{2}$ corresponds to the same point as did $\dfrac{3\pi}{2}$ in Figure 6-8 on page 243. Similarly, $-\pi$ corresponds to the same point as π, -2π corresponds to the same point as 0 and 2π, and so on.

Figure 6-9

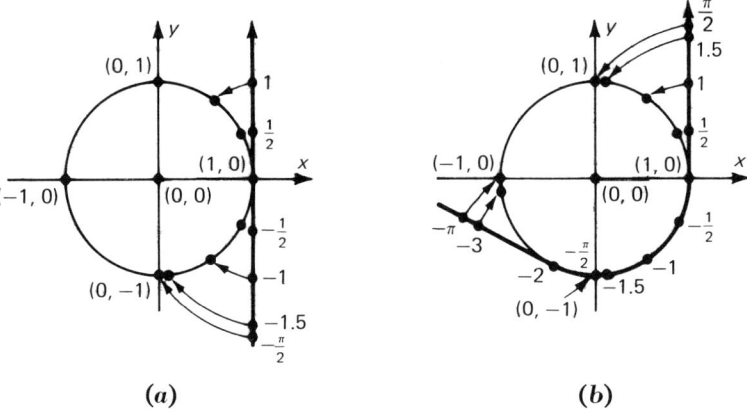

(a) (b)

If we combine the wrapping of the negative real line around the unit circle with the wrapping of the nonnegative real line, we see that we have a correspondence which relates each real number t to a unique point on the unit circle.

6.1b EXERCISES

In each of Exercises 1–16, determine the coordinates of the point on the unit circle about the origin that corresponds to the given real number.

1. $\dfrac{\pi}{3}$
2. $\dfrac{3\pi}{2}$
3. $\dfrac{5\pi}{6}$
4. $\dfrac{\pi}{4}$
5. 7π
6. $\dfrac{3\pi}{4}$
7. $-\dfrac{\pi}{3}$
8. $\dfrac{2\pi}{3}$
9. 4π
10. $-\dfrac{5\pi}{4}$
11. -7π
12. $\dfrac{11\pi}{6}$
13. $\dfrac{4\pi}{3}$
14. $-\dfrac{7\pi}{6}$
15. $\dfrac{5\pi}{3}$
16. $-\dfrac{3\pi}{4}$

For each of the following exercises indicate whether the points on the unit circle corresponding to the given real numbers are symmetric with respect to the x-axis, the y-axis, or the origin.

17. $\dfrac{\pi}{4}$ and $-\dfrac{\pi}{4}$
18. $\dfrac{\pi}{6}$ and $\dfrac{7\pi}{6}$
19. $\dfrac{\pi}{3}$ and $\dfrac{2\pi}{3}$
20. $\dfrac{\pi}{4}$ and $\dfrac{3\pi}{4}$

21. $\dfrac{2\pi}{3}$ and $\dfrac{4\pi}{3}$

22. $\dfrac{7\pi}{6}$ and $\dfrac{11\pi}{6}$

23. t and $-t$

24. t and $\pi + t$

25. t and $\pi - t$

26. $\dfrac{\pi}{2} + t$ and $\dfrac{\pi}{2} - t$

27. $\dfrac{3\pi}{2} - t$ and $\dfrac{3\pi}{2} + t$

28. $\dfrac{\pi}{2} - t$ and $\dfrac{3\pi}{2} + t$

6.2 Sine and Cosine Functions Defined

In the last section we described a correspondence which related to every real number t a unique point $P(x, y)$ on the unit circle C.

$$t \to P(x(t), y(t))$$

We write $x(t)$ and $y(t)$, since the coordinates x and y are uniquely determined by t and are, therefore, functions of t. These functions x and y are the trigonometric functions **cosine** (pronounced cō′sīn) and **sine** (pronounced sīn) respectively. When used in equations and formulas, "sine" is abbreviated to "sin" and "cosine" is abbreviated to "cos". It is customary to write cos t and sin t instead of cos(t) and sin(t), except in those cases where the parentheses prevent ambiguity, as, for example, in sin($t - \pi$).

The values of trigonometric functions for certain values of t are immediately apparent from the nature of the correspondence $t \to P(\cos t, \sin t)$. Since the number 0 on the real-number line corresponds to the point (1, 0) on the circle, we have cos 0 = 1, and sin 0 = 0. Other values, listed below, can be read off from Figure 6-10.

t	$P(\cos t, \sin t)$	$\cos t$	$\sin t$
$-\pi$	$(-1, 0)$	-1	0
$-\pi/2$	$(0, -1)$	0	-1
0	$(1, 0)$	1	0
$\pi/2$	$(0, 1)$	0	1
π	$(-1, 0)$	-1	0

Figure 6-10

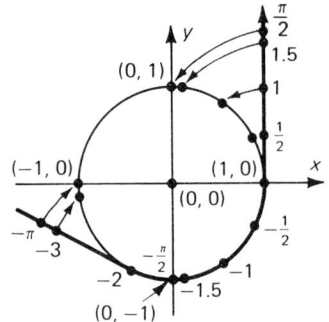

The sine function is the set of ordered pairs each of whose first components is a real number corresponding to a point on the real-number line and each of whose second components is the *ordinate* of the corresponding point on the unit circle. The cosine function is the set of ordered pairs each of whose first components is a real number corresponding to a point on the real line and each of whose second components is the *abscissa* of the associated point on the unit circle.

In Section 6.1 we saw that corresponding to each real number t there is a unique angle (directed angle in standard position) with radian measure t. The terminal side of this angle intersects the unit circle C in the point $P(\cos t, \sin t)$. It is therefore possible to consider the cosine and sine functions as functions of angles instead of real numbers. With this interpretation we would consider the cosine function as assigning to an angle with radian measure t the numerical value $\cos t$. In elementary courses in trigonometry this is done. For our purposes, and for most scientific purposes, it is more convenient to think of the trigonometric functions as functions of real numbers. There are occasions, however, when it is useful to consider them as functions of angles.

6.2 EXERCISES

For each of the following exercises, locate the point on the unit circle corresponding to the real number t. Then give the values of $\sin t$ and $\cos t$.

1. $t = 0$.
2. $t = \dfrac{-\pi}{3}$.
3. $t = \dfrac{\pi}{4}$.
4. $t = \dfrac{-\pi}{2}$.
5. $t = \dfrac{11\pi}{4}$.
6. $t = \dfrac{-4\pi}{3}$.
7. $t = \dfrac{-8\pi}{3}$.
8. $t = 7\pi$.
9. $t = \dfrac{-3\pi}{2}$.
10. $t = \dfrac{11\pi}{6}$.
11. $t = \dfrac{-5\pi}{6}$.
12. $t = \dfrac{13\pi}{6}$.

Recalling that (cos t, sin t) are the coordinates of a point on the unit circle, find the required value in each of the following exercises.

13. If $0 \leq t \leq \dfrac{\pi}{2}$ and $\cos t = \dfrac{4}{5}$, find $\sin t$.

14. If $0 \leq t \leq \dfrac{\pi}{2}$ and $\sin t = \dfrac{5}{13}$, find $\cos t$.

15. If $\dfrac{\pi}{2} \leq t \leq \pi$ and $\sin t = \dfrac{15}{17}$, find $\cos t$.

16. If $\dfrac{\pi}{2} \leq t \leq \pi$ and $\cos t = -\dfrac{1}{3}$, find $\sin t$.

17. If $\pi \leq t \leq \dfrac{3\pi}{2}$ and $\cos t = -\dfrac{3}{5}$, find $\sin t$.

18. If $\dfrac{3\pi}{2} \leq t \leq 2\pi$ and $\sin t = -\dfrac{3}{4}$, find $\cos t$.

From the way in which the circular sine and cosine functions were developed, you can see that as t increases from 0 to $\dfrac{\pi}{2}$, sin t increases from 0 to 1 and cos t decreases from 1 to 0. Describe the changes in sin t and cos t associated with each of the following changes in t.

19. t increases from $\dfrac{\pi}{2}$ to π.

20. t increases from π to $\dfrac{3\pi}{2}$.

21. t increases from $\dfrac{3\pi}{2}$ to 2π.

22. t increases from 2π to $\dfrac{5\pi}{2}$.

23. t increases from $-\pi$ to $-\dfrac{\pi}{2}$.

24. t increases from $-\dfrac{\pi}{2}$ to 0.

25. In wrapping the real-number line around the unit circle, we have seen that points on the line between 0 and $\dfrac{\pi}{2}$ correspond to points on the unit circle in the first quadrant. Thus, if we choose ϵ so that $0 < \epsilon < \dfrac{\pi}{2}$, then ϵ corresponds to the point P (cos ϵ, sin ϵ) on the unit circle in the first quadrant, as shown below. The line segments $\overline{PQ}$ and $\overline{P'R}$ are perpendicular to the x-axis, and $\overline{P'R}$ lies in the

tangent line to the circle at the point R. It is clear from the diagram that the area of $\triangle OPR$ is less than that of the sector of the circle OPR, which in turn is less than the area of $\triangle OP'R$.

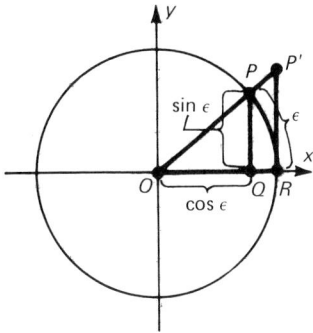

a) Using triangles OPQ and $OP'R$, show that the length of $P'R$ is $\dfrac{\sin \epsilon}{\cos \epsilon}$.

b) Show that the inequality

$$\text{Area of } \triangle OPR < \text{Area of sector } OPR < \text{Area of } \triangle OP'R$$

leads to the inequality

$$\sin \epsilon < \epsilon < \frac{\sin \epsilon}{\cos \epsilon}.$$

c) Show that this last inequality is equivalent to

$$1 < \frac{\epsilon}{\sin \epsilon} < \frac{1}{\cos \epsilon}.$$

and to

$$\cos \epsilon < \frac{\sin \epsilon}{\epsilon} < 1.$$

d) Show that this last inequality may be used to prove that

$$\lim_{\epsilon \to 0} \frac{\sin \epsilon}{\epsilon} = 1.$$

6.3 Properties of the Sine Function

In this section, we will consider the sine function and its properties. Then, in the next section, we will examine the cosine function. Since the origin of the real-number line corresponds to the point $(1, 0)$ on the unit circle, we see that $\sin 0 = 0$, and, since the point 2π on the real-number line also corresponds to the point $(1, 0)$, we see that $\sin 2\pi = 0$.

Furthermore, the point -2π on the real-number line corresponds to the point $(1, 0)$; hence, $\sin(-2\pi) = 0$.

In fact, given any point t on the real line, as we wrap the real line about the unit circle, the point 2π units to the left of t, point $t - 2\pi$, or 2π units to the right of t, point $t + 2\pi$, must correspond to the same point on the unit circle as the point t. This is true, of course, because the circumference of the unit circle is 2π. This fact produces an interesting characteristic of the sine function; namely,

$$\sin(t + n(2\pi)) = \sin t,$$

where t is any real number and n is any integer.

In general, if $a \in R$ and a function h has the property that

$$h(t + a) = h(t)$$

for every real number t, then h is said to be a *periodic function* with period a. It is true that, if a function h has a period a, then it also has a period na, where n is any nonzero integer. This follows from the fact that

$$h(t + na) = h(t + (n - 1)a + a) = h(t + (n - 1)a) =$$
$$h(t + (n - 2)a + a) = h(t + (n - 2)a) = \ldots = h(t + a) = h(t).$$

The *fundamental period* of a function h is the smallest positive value of a such that

$$h(t + a) = h(t)$$

for every real number t. We see that the fundamental period for the sine function is 2π because 2π is the smallest positive real number a such that $\sin(t + a) = \sin t$.

We can use the fact that the fundamental period of the sine function is 2π and some known ordered pairs belonging to the function to plot a few points of its graph, as shown in Figure 6–11. Notice that we have labeled the horizontal axis as t and the vertical axis as y so that the sine function is defined by the equation $y = \sin t$.

Figure 6-11

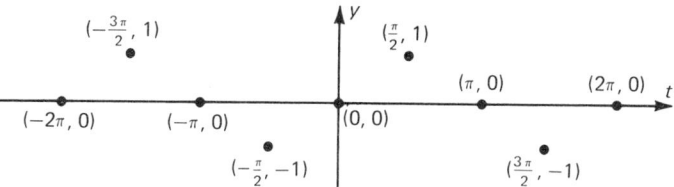

These few points do not give us much of a feel for the shape of the graph, but the locations of these points and the fact that the sine of any real number is the ordinate of a point on the unit circle enable us to make the following observations:

1. Since the greatest and least ordinates of points on the unit circle are 1 and -1, respectively, it follows that the greatest and least

ordinates of the graph of the sine function will be 1 and -1. In other words, the graph of the sine function will never be above the line $y = 1$ nor below the line $y = -1$.

2. From the periodic nature of the sine function, we see that if we know the shape of its graph for any interval of length 2π, then we know what the entire graph looks like.

3. Points on the unit circle that correspond to the real numbers t and $-t$ in the real line have ordinates that are the negatives of one another. (See Figure 6–12.) For example, $\sin \dfrac{\pi}{2} = 1$ and $\sin\left(\dfrac{-\pi}{2}\right) = -1$; $\sin \dfrac{3\pi}{2} = 1$ and $\sin\left(\dfrac{-3\pi}{2}\right) = -1$. In general, $\sin t = -\sin(-t)$ for all real numbers t. This, of course, means that the sine function is an odd function, and its graph is symmetric with respect to the origin. (See Definition 2.15a, page 100 and Theorem 2.15b, page 102.)

Figure 6-12

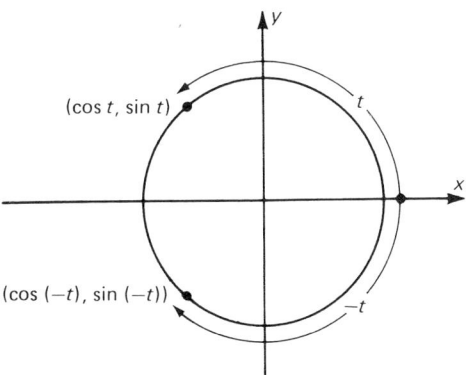

4. The way in which we developed the definition of the sine function, which involved wrapping the real-number line around the unit circle, would lead us to expect that it is a continuous function and that its graph is a continuous curve. The fact that the sine function is continuous and the fact that none of the ordinates of the corresponding points is less than -1 or greater than 1 imply that the range is the set of all real numbers in the interval $-1 \le t \le 1$.

5. It should also be apparent that, since the circumference of the unit circle is 2π, the points on the circle corresponding to the real numbers t and $t + \pi$ are diametrically opposite one another. (See Figure 6–13.) This means that the ordinates of these two points have the same absolute value but opposite signs; that is, they are the negatives of one another. Therefore,

$$\sin(t + \pi) = -\sin t.$$

6.3 PROPERTIES OF THE SINE FUNCTION

Figure 6-13

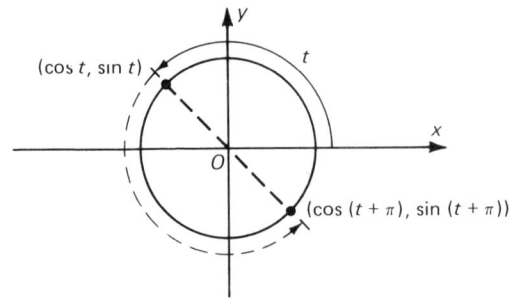

6. From our geometric definition, it seems reasonable to assume that the graph of the sine function should be smooth; that is, it seems reasonable to expect that the graph would have a tangent line at each point.
7. On the basis of our definition, it is also clear that the sine function is increasing for $-\frac{\pi}{2} \leq t \leq \frac{\pi}{2}$ and decreasing for $\frac{\pi}{2} \leq t \leq \frac{3\pi}{2}$. Furthermore, from the periodicity of the sine function, we can deduce that it is an increasing function over every interval of the form

$$-\frac{\pi}{2} + 2n\pi \leq t \leq \frac{\pi}{2} + 2n\pi$$

and is a decreasing function over every interval of the form

$$\frac{\pi}{2} + 2n\pi \leq t \leq \frac{3\pi}{2} + 2n\pi,$$

where n is an integer.

Because the sine function has period 2π and because it is an odd function, it follows that, if we know the nature of its graph over the interval $0 \leq t \leq \pi$, we will know the nature of its entire graph.

More specifically, if we know its graph over the interval $0 \leq t \leq \pi$, we can use the fact that the graph is symmetric with respect to the origin to determine its graph over the interval $-\pi \leq t \leq 0$. But then we have the graph over the interval $-\pi \leq t \leq \pi$. Since this interval has length 2π, we can then use the fact that the period of the sine function is 2π to determine its entire graph. In fact, because the graph of the sine function is symmetric with respect to the line $t = \frac{\pi}{2}$, we really need to know only the nature of its graph over the interval $0 \leq t \leq \frac{\pi}{2}$ in order to know its entire graph.

To see that the graph is symmetric with respect to this line, suppose that t is a real number less than $\frac{\pi}{2}$. (See Figure 6–14.) Then $\frac{\pi}{2} + \frac{\pi}{2} - t =$

Figure 6-14

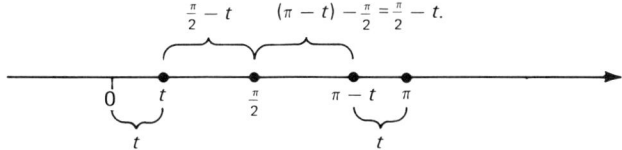

$\pi - t$ is the abscissa of the point that is the same distance to the right of the line $t = \frac{\pi}{2}$ as the point with abscissa t is to the left of this line. But, from paragraph 5 on page 251, we know that $\sin(t + \pi) = -\sin t$. If we replace t by $-t$, we have

$$\sin(\pi - t) = -\sin(-t).$$

Now, since the sine function is odd, we know that

$$\sin t = -\sin(-t).$$

Therefore,

$$\sin(\pi - t) = \sin t.$$

What we have just shown is that points of the graph of the sine function which are on the opposite sides of, but equidistant from the line $t = \frac{\pi}{2}$ have the same ordinates. This means, of course, that the graph is symmetric with respect to the line $t = \frac{\pi}{2}$. With respect to what other lines is the graph of the sine function symmetric? Is this graph symmetric with respect to any points other than the origin?

To summarize this discussion, if we know the shape of the graph of the sine function over the interval $0 \leq t \leq \frac{\pi}{2}$, then we know the shape of the entire graph. The following properties of the sine function and its graph justify this assertion.

1. The shape of the graph for the interval $0 \leq t \leq \frac{\pi}{2}$ indicates the shape of the graph for the interval $\frac{\pi}{2} \leq t \leq \pi$, because the graph is symmetric with respect to the line $t = \frac{\pi}{2}$.

2. The shape of the graph for the interval $0 \leq t \leq \pi$ gives us the shape of the graph for the interval $-\pi \leq t \leq 0$, because the sine

function is odd, so that its graph is symmetric with respect to the origin.

3. The shape of the graph for the interval $-\pi \leq t \leq \pi$ indicates the shape of the entire graph, since the sine function is periodic, with a fundamental period of 2π.

We will now learn all that we can about the graph of the sine function over the interval $0 \leq t \leq \frac{\pi}{2}$ because this will give us information about the entire graph. First of all, the sine function is an increasing function over the interval $0 \leq t \leq \frac{\pi}{2}$, because the ordinates of points on the unit circle increase from 0 to 1 in the first quadrant as we move around the circle in a counterclockwise direction. Second, the graph of the sine function is concave downward over the interval $0 \leq t \leq \pi$. Once this fact is established, it is not difficult to show that the graph is concave downward over any interval of the form

$$2n\pi \leq x \leq (2n+1)\pi, \text{ where } n \text{ is an integer,}$$

and concave upward over any interval of the form

$$(2n-1)\pi \leq x \leq 2n\pi, \text{ where } n \text{ is an integer.}$$

The fact that the graph of the sine function is concave downward over the interval $0 \leq t \leq \pi$ is a consequence of the fact that its derivative, or slope function, is decreasing over that interval. (See Theorem 4.6a.) In Section 6.6 we shall see that the derivative of the sine function is the cosine function. From the definition of the cosine function it is evident that it is a decreasing function on $0 \leq t \leq \pi$, decreasing from a value of 1 when $t = 0$ to a value of -1 when $t = \pi$.

With all the information gained so far about the graph of the sine function, we know that the curve pictured in Figure 6–15 is a fairly good sketch of the graph.

Figure 6-15

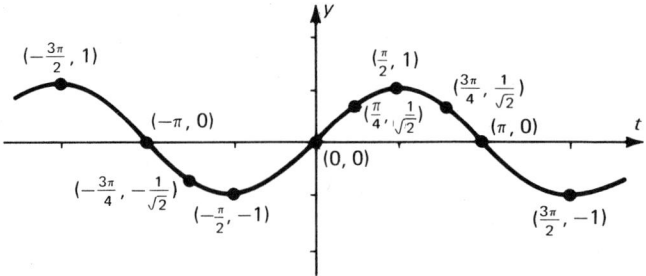

6.3 EXERCISES

For each of Exercises 1–4, first sketch the graph of the sine function for the interval $0 \leq t \leq 2\pi$. Then, on the same set of axes, use the relationship between the graphs of the functions f and af, where a is a real number, to sketch the graph of the given function. Finally, give the fundamental period and the greatest ordinate of the graph.

1. $g:g(t) = 2 \sin t$.
2. $G:G(t) = \tfrac{1}{2} \sin t$.
3. $h:h(t) = -\sin t$.
4. $H:H(t) = -\tfrac{3}{2} \sin t$.

For each of Exercises 5 and 6, first graph the sine function for the interval $0 \leq t \leq 2\pi$. Then, on the same set of axes, use the relationship between the graphs of functions f and $f \circ g$, where $g(t) = t + a$, and a is a real number, to sketch the graph of the given function. Finally, give the fundamental period and greatest ordinate of the graph.

5. $h:h(t) = \sin(t - \pi)$.
6. $H:H(t) = \sin\left(t + \dfrac{\pi}{2}\right).$

Sketch the graphs of the following functions over the interval $-2\pi \leq t \leq 2\pi$.

7. $g:g(t) = 2 \sin(t - \pi)$.
8. $G:G(t) = -\tfrac{1}{2} \sin\left(t + \dfrac{\pi}{2}\right).$

Sketch the graphs of the following functions over the interval $-2\pi \leq t \leq 2\pi$.

9. $h:h(t) = \sin(2t)$.
10. $H:H(t) = \sin(\tfrac{1}{2}t)$.
11. P is the point on the unit circle corresponding to the real number t and R is the point corresponding to $\dfrac{\pi}{2} + t$.

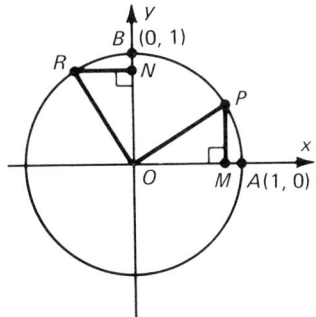

a) Prove that $\triangle OPM \cong \triangle ORN$. ($\cong$ means "is congruent to.")
b) What does this show about the values of $\cos\left(\dfrac{\pi}{2} + t\right)$ and $\sin t$ when $0 < t < \dfrac{\pi}{2}$?

12. Point P corresponds to the real number t, where $0 < t < \frac{\pi}{2}$. Point R corresponds to the number $\frac{\pi}{2} - t$.

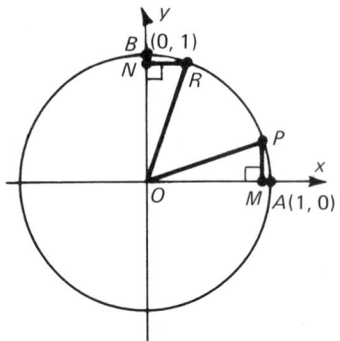

   a) Prove that $\triangle OPM \cong \triangle ORN$.
   b) What does this show about the values of $\cos\left(\frac{\pi}{2} - t\right)$ and $\sin t$?

13. Let P denote the point on the unit circle corresponding to t, where $0 < t < \frac{\pi}{2}$; and let R denote the point corresponding to $\frac{3\pi}{2} + t$.

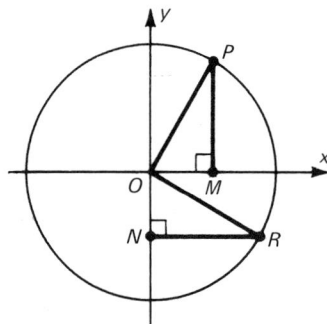

   a) Prove that $\triangle OMP \cong \triangle ONR$.
   b) What does part a suggest about the values of $\cos\left(\frac{3\pi}{2} + t\right)$ and $\sin t$?

14. Use the fact that $\sin(-t) = -\sin t$ to show that if ϵ is negative, then $\lim\limits_{\epsilon \to 0} \frac{\sin \epsilon}{\epsilon} = 1$. (See Exercise 25, page 248.)

6.4 Properties of the Cosine Function

As you recall, the cosine function is the set of ordered pairs whose first components are real numbers and whose second components are the ab-

256 TRIGONOMETRIC FUNCTIONS; SINE AND COSINE

scissas of the corresponding points on the unit circle. The correspondence between the points on the real-number line and points on the unit circle is obtained by wrapping the line around the unit circle in the manner described on pages 243 and 244. See Figure 6–16.

Figure 6-16

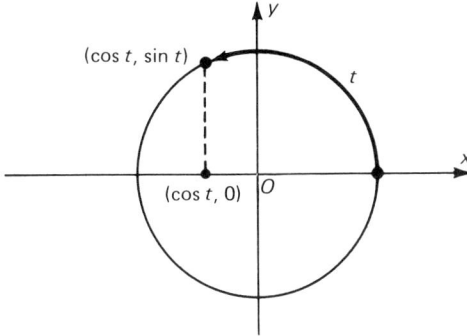

By an analysis of the cosine function similar to the one made for the sine function, we see that the cosine function is periodic, with a fundamental period of 2π. This fact should be clear from the definition of this function in terms of a correspondence between points in the real-number line and points in the unit circle, whose circumference is 2π. We also see that the cosine function has the following properties.

1. Since the greatest and least abscissas of points on the unit circle are 1 and -1, respectively, it follows that these will be the greatest and least *ordinates* of the cosine function. In fact, the range of the cosine function is the same as the range of the sine function; namely, the set of all real numbers in the interval $-1 \leq t \leq 1$.

2. From the fact that the points on the unit circle corresponding to the real numbers t and $-t$ have the same abscissa (see Figure 6–17), we deduce that
$$\cos t = \cos(-t).$$
That is, the cosine function is an even function, and its graph is symmetric with respect to the y-axis.

Figure 6-17

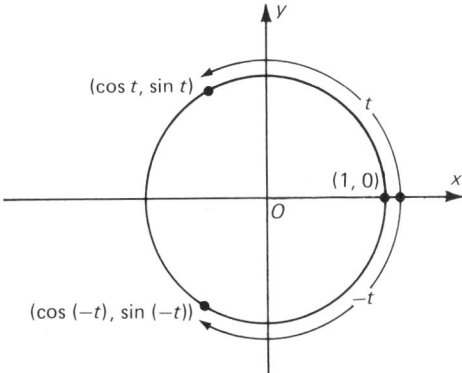

6.4 PROPERTIES OF THE COSINE FUNCTION

3. As was true for the sine function, cos t and cos $(t + \pi)$ are the negatives of one another. That is, cos $(t + \pi) = -\cos t$. (See Figure 6-18.)

Figure 6-18

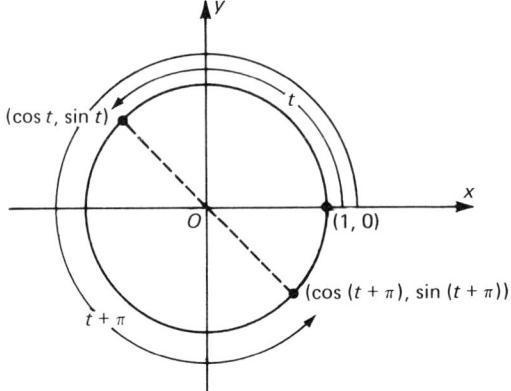

4. The cosine function is an increasing function for $(2n - 1)\pi \leq t \leq 2n\pi$ and is a decreasing function for $2n\pi \leq t \leq (2n + 1)\pi$, where n is an integer.

5. The graph of the cosine function is concave upward for intervals of the form

$$\frac{\pi}{2} + 2n\pi \leq t \leq \frac{3\pi}{2} + 2n\pi, \text{ where } n \text{ is an integer.}$$

The graph is concave downward for intervals of the form

$$-\frac{\pi}{2} + 2n\pi \leq t \leq \frac{\pi}{2} + 2n\pi, \text{ where } n \text{ is an integer.}$$

Using this information and the location of several known points, we can make a fairly good sketch of the graph of the cosine function, as shown in Figure 6-19.

Figure 6-19

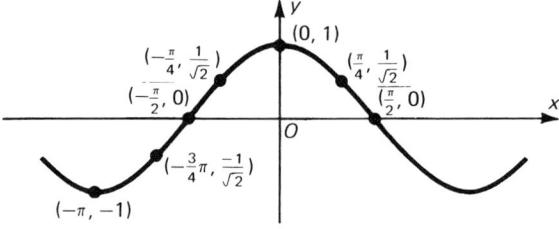

If you compare the sketch of the graph of the cosine function with the sketch of the graph of the sine function given in Figure 6-15, on page 254, you may suspect that

$$\cos t = \sin\left(t + \frac{\pi}{2}\right).$$

This is indeed the case, as we shall presently show. Another way of stating this is to say that the cosine function is obtained from the composition of the sine function and the linear function l defined by

$$l(t) = t + \frac{\pi}{2}, \quad \text{where } t \in R.$$

From our earlier work with translations, we know that this means that the graph of the cosine function is the graph of the sine function translated $\frac{\pi}{2}$ units to the left.

Figure 6-20

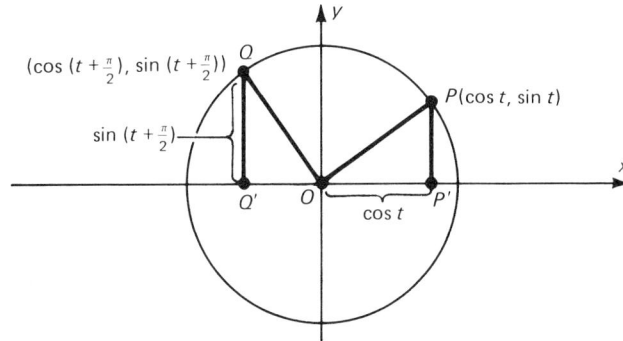

To see that $\cos t = \sin\left(t + \frac{\pi}{2}\right)$, we first choose a point on the unit circle in the first quadrant, as shown in Figure 6–20. The following steps verify that triangles $OQ'Q$ and $PP'O$ are congruent:

1. $\triangle OQ'Q$ and $\triangle PP'O$ are right triangles.
2. $\measuredangle P'OP$ and $\measuredangle Q'OQ$ are complementary, as are $\measuredangle Q'OQ$ and $\measuredangle OQQ'$. Hence, $\measuredangle OQQ' \cong \measuredangle P'OP$.
3. The hypotenuses of these two right triangles, $\overline{OQ}$ and $\overline{OP}$, have length 1, which is the radius of the unit circle. Thus, $\overline{OQ} \cong \overline{OP}$, and $\triangle OQ'Q \cong \triangle PP'O$.

Hence, $\overline{OP'} \cong \overline{QQ'}$, and it follows that $\cos t = \sin\left(t + \frac{\pi}{2}\right)$. You should be able to verify that $\cos t = \sin\left(t + \frac{\pi}{2}\right)$ for the cases when P is in the other quadrants.

Properties 1 through 5 of the cosine function, discussed on pages 257 and 258 can also be verified by using the corresponding properties of the sine function and the fact that $\cos t = \sin\left(t + \frac{\pi}{2}\right)$. You may find it helpful to carry out the steps of these verifications.

6.4 EXERCISES

For each of Exercises 1–10, first sketch the graph of the cosine function for the interval $-2\pi \leq t \leq 2\pi$. Then, on the same set of axes, sketch the graph of the given funciton. Finally, give the fundamental period and greatest ordinate of the graph.

1. $f{:}f(t) = 3\cos t$.
2. $F{:}F(t) = \frac{1}{3}\cos t$.
3. $g{:}g(t) = -\cos t$.
4. $G{:}G(t) = -2\cos t$.
5. $h{:}h(t) = \cos\left(t - \frac{\pi}{2}\right)$.
6. $H{:}H(t) = \cos(t + \pi)$.
7. $S{:}S(t) = \frac{1}{2}\cos(t - \pi)$.
8. $V{:}V(t) = -2\cos\left(t + \frac{\pi}{2}\right)$.
9. $v{:}v(t) = \cos(4t)$.
10. $W{:}W(t) = \cos(\frac{1}{4}t)$.
11. a) Show that if $P(\cos t, \sin t)$ is a point in the second quadrant, then
    $$\cos t = \sin\left(t + \frac{\pi}{2}\right).$$ [Hint: see Exercise 11 on page 255.]

    b) Show that this identity holds when P is a point in some other quadrant.

Show that each of the following identities holds when $P(\cos t, \sin t)$ is a point in some other quadrant than the first quadrant. [See Exercises 12 and 13 on page 256.]

12. $\sin\left(\frac{\pi}{2} - t\right) = \cos t$.

13. $\sin\left(\frac{3\pi}{2} + t\right) = -\cos t$.

14. $\sin\left(\frac{3\pi}{2} - t\right) = -\cos t$.

15. Over which of the intervals, $\left\{t \mid 0 \leq t \leq \frac{\pi}{2}\right\}$, $\left\{t \mid \frac{\pi}{2} \leq t \leq \pi\right\}$, $\left\{t \mid \pi \leq t \leq \frac{3\pi}{2}\right\}$, and $\left\{t \mid \frac{3\pi}{2} \leq t \leq 2\pi\right\}$, is each of the following statements true?

    a) Both $\sin t$ and $\cos t$ increase as t increases.
    b) Both $\sin t$ and $\cos t$ decrease as t increases.
    c) As t increases, $\sin t$ increases but $\cos t$ decreases.
    d) As t increases, $\cos t$ increases but $\sin t$ decreases.

Indicate whether each of the following statements is true or false.

16. The graph of the sine function is symmetric with respect to each of the points at which it intersects the horizontal axis.
17. The graph of the cosine function is symmetric with respect to each of the points at which it intersects the horizontal axis.
18. The graph of the sine function is symmetric with respect to any vertical line that passes through one of its maximum or minimum points.
19. The graph of the cosine function is symmetric with respect to any vertical line that passes through one of its maximum or minimum points.
20. The graph of the sine function intersects the horizontal axis at points whose x-coordinates are integral multiples of π.
21. The graph of the cosine function intersects the horizontal axis at points whose x-coordinates are odd multiples of $\frac{\pi}{2}$.

6.5 Some Identities Involving Cosine and Sine

We have seen that the sine and cosine functions are related by the formula $\cos t = \sin\left(t + \frac{\pi}{2}\right)$. We now turn to some further connections between these two functions.

As you know, corresponding to any real number t, there is a point on the unit circle having coordinates $(x, y) = (\cos t, \sin t)$. Since the coordinates of any point $P(x, y)$ on the unit circle C satisfy the equation $x^2 + y^2 = 1$ it follows that for any real number t

$$(\cos t)^2 + (\sin t)^2 = 1.$$

This identity is often called the Pythagorean Identity since it is the statement of the Pythagorean Theorem for a right triangle whose hypotenuse is a radius of the unit circle C. (See Figure 6–21.)

Figure 6-21

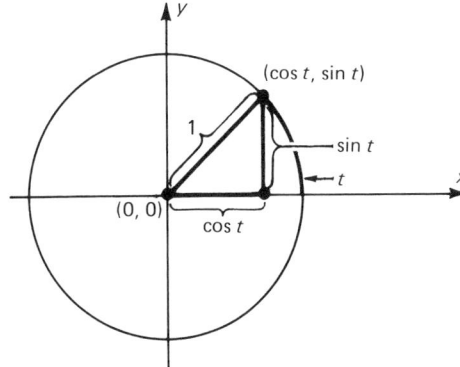

The conventional, although somewhat misleading, way to write the expression $(\cos t)^2$ and $(\sin t)^2$ is $\cos^2 t$ and $\sin^2 t$ respectively. Thus the Pythagorean Identity usually appears in the form

$$\cos^2 t + \sin^2 t = 1.$$

Another important identity is

$$\cos(t - s) = \cos t \cos s + \sin t \sin s.$$

To verify this identity we first suppose that $0 \leq s < t < 2\pi$, and set $S = (\cos s, \sin s)$ and $T = (\cos t, \sin t)$. The length of the arc $\widehat{ST}$ is $t - s$. The length of the arc from $(1, 0)$ to $(\cos(t - s), \sin(t - s))$ is also $t - s$. Since we know from geometry that equal circular arcs intercept equal chords, the distance from S to T must be equal to the distance from $(1,0)$ to $(\cos(t - s), \sin(t - s))$, or

$$\sqrt{(\cos t - \cos s)^2 + (\sin t - \sin s)^2} = \sqrt{(\cos(t - s) - 1)^2 + \sin^2(t - s)}$$

Squaring both sides, we obtain

$$(\cos t - \cos s)^2 + (\sin t - \sin s)^2 = (\cos(t - s) - 1)^2 + \sin^2(t - s)$$

or

$$\cos^2 t - 2\cos t \cos s + \cos^2 s + \sin^2 t - 2\sin t \sin s + \sin^2 s =$$
$$\cos^2(t - s) - 2\cos(t - s) + 1 + \sin^2(t - s).$$

Using the Pythagorean Identity to help in simplifying this last identity, we reduce it to

$$\cos(t - s) = \cos t \cos s + \sin t \sin s.$$

Verification that this identity holds for arbitrary real numbers s and t is left to the exercises. This identity is sometimes called the *difference formula* for the cosine function.

We get a special case of this difference formula by setting $t = \dfrac{\pi}{2}$ and $s = x$.

$$\cos\left(\frac{\pi}{2} - x\right) = \cos\frac{\pi}{2} \cos x + \sin\frac{\pi}{2} \sin x$$

Since $\cos\dfrac{\pi}{2} = 0$ and $\sin\dfrac{\pi}{2} = 1$, we obtain the following useful identity.

$$\cos\left(\frac{\pi}{2} - x\right) = \sin x$$

From this identity we obtain

$$\cos\left(\frac{\pi}{2} - \left(\frac{\pi}{2} - x\right)\right) = \sin\left(\frac{\pi}{2} - x\right)$$

or
$$\sin\left(\frac{\pi}{2} - x\right) = \cos x.$$

Using these last two identities we can derive the difference formula for the sine function from that of the cosine function as follows:

$$\sin(t - s) = \cos\left(\frac{\pi}{2} - (t - s)\right) = \cos\left(\left(\frac{\pi}{2} - t\right) - (-s)\right)$$
$$= \cos\left(\frac{\pi}{2} - t\right)\cos(-s) + \sin\left(\frac{\pi}{2} - t\right)\sin(-s).$$

Using the last two identities and the identities $\cos(-s) = \cos s$ and $\sin(-s) = -\sin s$ we obtain the standard difference formula for the sine function.

$$\sin(t - s) = \sin t \cos s - \cos t \sin s.$$

In the exercises you will use the difference formulas for the cosine function and sine function to obtain the following *addition formulas* for these functions.

$$\cos(t + s) = \cos t \cos s - \sin t \sin s$$

and

$$\sin(t + s) = \sin t \cos s + \cos t \sin s.$$

By setting $s = t$ in the addition formulas for the cosine and sine functions we obtain the following identities:

$$\cos 2t = \cos^2 t - \sin^2 t$$

$$\sin 2t = 2 \sin t \cos t.$$

From their use in trigonometry these identities are often called the *double angle formulas*.

6.5a EXERCISES

Express each of the following in terms of sin t or cos t.

1. $\sin\left(\frac{\pi}{2} + t\right)$
2. $\cos\left(\frac{\pi}{2} + t\right)$
3. $\sin(\pi - t)$
4. $\cos(t + \pi)$
5. $\cos\left(\frac{3\pi}{2} + t\right)$
6. $\sin\left(\frac{3\pi}{2} - t\right)$

If s and t are positive real numbers less than $\frac{\pi}{2}$, find $\sin(t+s)$, $\cos(t-s)$, $\sin 2t$, and $\cos 2s$ in each of the following exercises.

7. $\sin s = \frac{3}{5}$ and $\cos t = \frac{5}{13}$.
8. $\cos s = \frac{8}{17}$ and $\sin t = \frac{4}{5}$.
9. $\sin s = \frac{\sqrt{2}}{2}$ and $\sin t = \frac{1}{2}$,
10. $\cos s = \frac{\sqrt{3}}{2}$ and $\cos t = \frac{7}{25}$.
11. Beginning with the formulas for $\cos(t-s)$ and $\sin(t-s)$, replace s by $-s$ and obtain the formulas for $\cos(t+s)$ and $\sin(t+s)$.
12. Prove that $\cos 2t = 2\cos^2 t - 1 = 1 - 2\sin^2 t$.
13. Let $t = \frac{1}{2}\theta$ in Exercise 12 and derive the following identities (called the *half-angle formulas*).

$$|\sin \tfrac{1}{2}\theta| = \sqrt{\frac{1-\cos\theta}{2}} \qquad |\cos \tfrac{1}{2}\theta| = \sqrt{\frac{1+\cos\theta}{2}}$$

14. Prove that $\sin(t+s) - \sin(t-s) = 2\cos t \sin t$.
15. In the identity in Exercise 14, let $t+s = A$ and $t-s = B$ and show that

$$\sin A - \sin B = 2\cos\frac{A+B}{2} \sin\frac{A-B}{2}.$$

Identities will be most useful to you in more advanced courses in mathematics. One of the uses of identities is in the simplifying of expressions so that the relationships they express can be analyzed more easily.

Example 1 Consider the expression $\dfrac{\sin(t_1+t_2) + \sin(t_1-t_2)}{\cos(t_1+t_2) + \cos(t_1-t_2)}$. By the addition and difference formulas, this expression may be written

$$\frac{(\sin t_1 \cos t_2 + \cos t_1 \sin t_2) + (\sin t_1 \cos t_2 - \cos t_1 \sin t_2)}{(\cos t_1 \cos t_2 - \sin t_1 \sin t_2) + (\cos t_1 \cos t_2 + \sin t_1 \sin t_2)},$$

which simplifies to

$$\frac{2\sin t_1 \cos t_2}{2\cos t_1 \cos t_2}.$$

If $\cos t_2 \neq 0$, this becomes

$$\frac{\sin t_1}{\cos t_1}.$$

Thus, the quotient expressed above, which appears to be so complicated, is greatly simplified by applying the various identities.

Example 2 Consider the expression $\dfrac{(\cos^2 t)(1 - \cos 2t)}{1 - \sin^2 t}$. Since, by the Pythagorean Identity, $\sin^2 t + \cos^2 t = 1$, we have

$$\frac{(\cos^2 t)(1 - \cos 2t)}{1 - \sin^2 t} = \frac{(\cos^2 t)(1 - \cos 2t)}{\cos^2 t}.$$

If $\cos^2 t \neq 0$, this becomes $1 - \cos 2t$. By a double angle formula, we have

$$\begin{aligned}
1 - \cos 2t &= 1 - (\cos^2 t - \sin^2 t) \\
&= (1 - \cos^2 t) + \sin^2 t \\
&= \sin^2 t + \sin^2 t \\
&= 2 \sin^2 t.
\end{aligned}$$

A second use for identities is in connection with the solution of equations involving the circular or trigonometric functions.

Example 3 Find the set of solutions of $3 - 3 \sin t - \cos^2 t = 0$.

$$\begin{aligned}
3 - 3 \sin t - \cos^2 t &= 3 - 3 \sin t - (1 - \sin^2 t) \\
&= 2 - 3 \sin t + \sin^2 t \\
&= (2 - \sin t)(1 - \sin t).
\end{aligned}$$

Hence, $3 - 3 \sin t - \cos^2 t = 0$ if $\sin t = 2$ or if $\sin t = 1$. Since $\sin$ cannot exceed 1, the only value of t in the interval $0 \leq t < 2\pi$ which is a solution of the equation is $\dfrac{\pi}{2}$. Hence, the set of solutions is

$$\left\{ t \mid t = \frac{\pi}{2} + 2n\pi,\ n \in I \right\}.$$

Example 4 Find the set of solutions of $\sin 2t - \cos t = 0$.

$$\sin 2t - \cos t = 2 \sin t \cos t - \cos t = (\cos t)(2 \sin t - 1).$$

Hence, $\sin 2t - \cos t = 0$ if $\cos t = 0$ or if $\sin t = \tfrac{1}{2}$. The solutions of $\cos t = 0$ or $\sin t = \tfrac{1}{2}$ in the interval $0 \leq t < 2\pi$ are $\dfrac{\pi}{6}, \dfrac{\pi}{2}, \dfrac{5\pi}{6},$ and $\dfrac{3\pi}{2}$. Hence, the set of solutions of $\sin 2t - \cos t = 0$ is

$$\left\{ t \mid t = \frac{\pi}{6} + 2n\pi \text{ or } t = \frac{\pi}{2} + 2n\pi \text{ or } t = \frac{5\pi}{6} + 2n\pi \text{ or } t = \frac{3\pi}{2} + 2n\pi,\ n \in I \right\}.$$

6.5b EXERCISES

Prove that each of the following is an identity (that is, true for all values of t for which the expression is defined) by transforming the left member until it is identical with the right member.

1. $(\sin t + \cos t)^2 = 1 + \sin 2t.$
2. $\dfrac{(1 + \sin t)^2}{\cos^2 t} = \dfrac{1 + \sin t}{1 - \sin t}.$
3. $\dfrac{\sin 2t + \sin t}{\cos 2t + \cos t + 1} = \dfrac{\sin t}{\cos t}.$
4. $\dfrac{\sin t}{1 + \cos t} + \dfrac{\cos t}{\sin t} = \dfrac{1}{\sin t}.$
5. $\cos\left(\dfrac{\pi}{6} - t\right) - \cos\left(\dfrac{\pi}{6} + t\right) = \sin t.$
6. $\dfrac{\sin 2t + 1}{\cos 2t} = \dfrac{\cos t + \sin t}{\cos t - \sin t}.$
7. $\dfrac{\cos(t_1 + t_2) + \cos(t_1 - t_2)}{2 \cos t_2} = \cos t_1.$
8. $\dfrac{\cos 2t}{\sin t} + \dfrac{\sin 2t}{\cos t} = \dfrac{1}{\sin t}.$
9. $1 - 2 \sin^2 \dfrac{t}{2} = \cos t.$
10. $\dfrac{\cos^2 \dfrac{t}{2} - \cos t}{\sin^2 \dfrac{t}{2}} = 1.$

In each of the following exercises, determine the values of t for which the statement is true.

11. $4 \sin t + 3 = 1.$
12. $2 (\cos t)^2 = 1.$
13. $4 (\sin t)^2 = 3.$
14. $2 (\cos t)^2 + \cos t = 0.$
15. $2 (\sin t)^2 - \sqrt{3} \sin t = 0.$
16. $2 (\cos t)^2 = \sqrt{2} \cos t.$
17. $\sin t \cos t + (\cos t)^2 = 0.$
18. $\sin t = \dfrac{1}{\cos t}.$
19. $2 \sin^2 t - \cos t - 1 = 0.$
20. $4 \cos^2 t + 4 \sin t = 5.$

21. $\sin 2t + 2 \sin t \cos t = 0$.
22. $\sin 2t + \sin t = 0$.
23. $\cos 2t = \cos^2 t$.
24. $\cos 2t + \sin t - 1 = 0$.

6.6 The Slope Functions of the Sine and Cosine Functions

In Section 6.3 we asserted that the slope function of the sine functions is the cosine function. To prove this fact, we will use Theorem 3.10a, which states that if $\lim\limits_{x \to x_0} \dfrac{f(x) - f(x_0)}{x - x_0}$ exists, then $f'(x_0)$ exists and has this limit as it value.

Let us consider any point $(t_0, \sin t_0)$ on the graph of the sine function. If t is close to t_0, but $t \neq t_0$, then $t = t_0 + \epsilon$, where ϵ is a small number, positive or negative. We can therefore write $\dfrac{\sin t - \sin t_0}{t - t_0}$ as $\dfrac{\sin (t_0 + \epsilon) - \sin t_0}{(t_0 + \epsilon) - t_0}$. Now, $\sin A - \sin B = 2 \cos \dfrac{A + B}{2} \sin \dfrac{A - B}{2}$ (see Exercise 15 on page 264). Hence,

$$\frac{\sin (t_0 + \epsilon) - \sin t_0}{(t_0 + \epsilon) - t_0} = \frac{2 \cos (t_0 + \epsilon/2) \sin (\epsilon/2)}{\epsilon} = \cos (t_0 + \epsilon/2) \frac{\sin (\epsilon/2)}{\epsilon/2}$$

As $t \to t_0$, $\epsilon \to 0$ and $\epsilon/2 \to 0$. Hence,

$$\lim_{t \to t_0} \frac{\sin t - \sin t_0}{t - t_0} = \lim_{\epsilon/2 \to 0} \cos (t_0 + \epsilon/2) \frac{\sin (\epsilon/2)}{\epsilon/2}$$

$$= \lim_{\epsilon/2 \to 0} \cos (t_0 + \epsilon/2) \cdot \lim_{\epsilon/2 \to 0} \frac{\sin (\epsilon/2)}{\epsilon/2}$$

Now, $\lim\limits_{\epsilon \to 0} \dfrac{\sin (\epsilon)}{\epsilon} = 1$ (see Exercise 25 on page 248 and Exercise 14 on page 256). Thus,

$$\lim_{t \to t_0} \frac{\sin t - \sin t_0}{t - t_0} = (\cos t_0)(1) = \cos t_0.$$

Hence $\sin'(t_0)$ is defined and has the value $\cos t_0$. Since $(t_0, \sin t_0)$ is an arbitrary point on the graph of the sine function, it follows that $\sin' t = \cos t$.

Just as we used the slope functions of quadratic functions and other polynomial functions to gain information about their graphs, we can use the slope function of the sine function to gain information about its graph. For example, the slope of the tangent line to the graph of the sine function at the point $(0, 0)$ is $\cos (0) = 1$. That is, the graph has the line defined by $f(t) = t$ as its best linear approximation at the origin. This bit of information tells us that, near the origin, the graph of the sine function must look something like Figure 6–22a and not, as it is sometimes drawn, like Figure 6–22b.

Figure 6-22

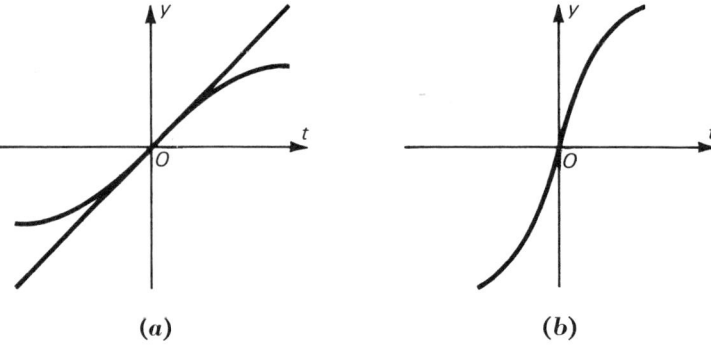

(a)   (b)

Since we know that $\cos t$ is positive for intervals of the form

$$-\frac{\pi}{2} + 2n\pi < t < \frac{\pi}{2} + 2n\pi, \quad \text{where } n \in I,$$

we know that the sine function is increasing over these intervals. Furthermore, since $\cos t$ is negative for intervals of the form

$$\frac{\pi}{2} + 2n\pi < t < \frac{3\pi}{2} + 2n\pi,$$

we know that the sine function is decreasing over these intervals. Of course, we already had obtained this information in other ways, but it is important to know that the slope function does provide this information. because there are often occasions when using the slope function is the easiest way of determining the intervals over which a function is increasing or decreasing.

In earlier chapters, we saw that the slope function of the slope function enables us to determine the concavity of the graph of a function. The question now arises as to whether we can learn about the concavity of the graph of the sine function from the slope function of its slope function. More importantly, does the slope function of the sine function have a slope function?

The answer to this question is obvious once we recall that the graph of the cosine function is the graph of the sine function translated $\frac{\pi}{2}$ units to the left. This means that at the point $(t_0, \cos t_0) = \left(t_0, \sin \left(t_0 + \frac{\pi}{2}\right)\right)$, the slope of the tangent line to the graph of the cosine function is equal to the slope of the tangent line to the graph of the sine function at the point $\frac{\pi}{2}$ units to the right of $(t_0, \cos t_0)$. (See Figure 6-23.) This is, of course, the point $\left(t_0 + \frac{\pi}{2}, \sin\left(t_0 + \frac{\pi}{2}\right)\right)$. But we know that the slope of the tangent line to the graph of the sine function at $\left(t_0 + \frac{\pi}{2}, \sin\left(t_0 + \frac{\pi}{2}\right)\right)$ is $\cos\left(t_0 + \frac{\pi}{2}\right)$. Because

$$\cos\left(t_0 + \frac{\pi}{2}\right) = \cos t_0 \cos \frac{\pi}{2} - \sin t_0 \sin \frac{\pi}{2} = -\sin t_0,$$

and because the point $(t_0, \cos t_0)$ is arbitrarily chosen, this means that, for all t,

$$\cos'(t) = -\sin t.$$

Figure 6-23

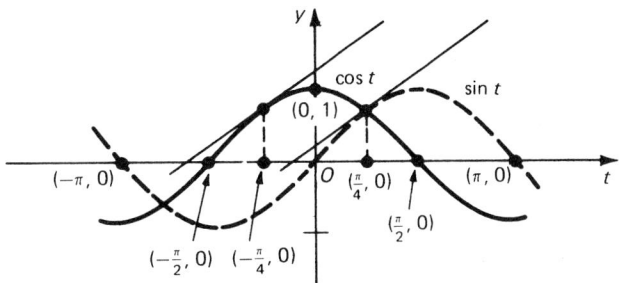

To summarize: (1) the slope function of the sine function is the cosine function; (2) the slope function of the cosine function is the negative of the sine function; therefore, (3) the slope function of the slope function of the sine function is the negative of the sine function. In symbols, $\sin''(t) = (\sin')'(t) = \cos'(t) = -\sin t$. Now, if $\sin t < 0$, $-\sin t > 0$; and if $\sin t > 0$, then $-\sin t < 0$. Hence, by Theorem 4.6a (page 194), the graph of the sine function will be concave downward over intervals where $\sin t$ is positive and will be concave upward over intervals where $\sin t$ is negative.

7

Other Trigonometric Functions

Introduction

In this chapter, we will consider some trigonometric functions, called the tangent, cotangent, secant, and cosecant, that are all derived, or "generated," from operations on the sine, cosine, and unit functions. Of these several functions, the tangent function is the most used, and we will accordingly devote more space to considering its properties.

7.1 The Tangent Function

The first of the new trigonometric functions that we shall consider is the tangent function, which is defined as the quotient of the sine function and the cosine function.

Definition 7.1 The **tangent function** is defined by

$$\tan x = \frac{\sin x}{\cos x}, \quad \text{with } x \in R \text{ and } \cos x \neq 0.$$

Note that the tangent function is not defined when $\cos x = 0$, which is when $x = (2n + 1)\frac{\pi}{2}$, where n is an integer $\left(\text{that is, when } x \text{ is an odd multiple of } \frac{\pi}{2}\right)$.

Because the sine and cosine functions are periodic, it seems logical to assume that the tangent function should also be periodic. The tangent function is periodic; however, it has a fundamental period of π rather than 2π, as the sine and cosine functions do. This fact can be verified by using the addition formulas (see page 263) for the sine and cosine functions. By the definition of the tangent function and the addition formulas, we have

$$\tan(x + \pi) = \frac{\sin(x + \pi)}{\cos(x + \pi)} = \frac{\sin x \cos \pi + \sin \pi \cos x}{\cos x \cos \pi - \sin x \sin \pi}.$$

Since $\sin \pi = 0$ and $\cos \pi = -1$, this becomes

$$\tan(x + \pi) = \frac{-\sin x}{-\cos x} = \frac{\sin x}{\cos x} = \tan x.$$

From the equality $\tan(x + \pi) = \tan x$, it follows that the tangent function has a period of π. Furthermore, because π is the smallest positive number a such that $\tan(x + a) = \tan x$, the fundamental period of the tangent function is π.

The tangent function is defined for all real numbers x in the interval $-\frac{\pi}{2} < x < \frac{\pi}{2}$, and this interval is of length π; therefore, we can learn about the properties of the tangent function in general by studying its properties over this interval. This is so because the tangent function is periodic with a fundamental period of π, and its graph over any interval $(2n - 1)\frac{\pi}{2} < x < (2n + 1)\frac{\pi}{2}$ will be a translate ($n\pi$ units to the right if n is a positive integer and $|n|\pi$ units to the left if n is negative) of the graph of this function on the interval $-\frac{\pi}{2} < x < \frac{\pi}{2}$.

What are the zeros of the tangent function over the interval

$$-\frac{\pi}{2} < x < \frac{\pi}{2}?$$

From the definition of the tangent function, $\tan x = 0$ if and only if $\sin x = 0$ and $\cos x \neq 0$. Because the cosine function has no zeros on this interval and the only zero of the sine function is $x = 0$, we know that 0 is the only zero of the tangent function on the interval

$$-\frac{\pi}{2} < x < \frac{\pi}{2}.$$

From the fact that the sine function is an odd function [that is, $\sin(-x) = -\sin x$ for all x] and the fact that the cosine function is an even function [that is, $\cos(-x) = \cos x$ for all x], we deduce that the tangent function is an odd function. In symbols,

$$\tan(-x) = \frac{\sin(-x)}{\cos(-x)} = \frac{-\sin x}{\cos x} = -\tan x.$$

Hence, the graph of the tangent function is symmetric with respect to the origin. We could, therefore, determine the nature of the entire graph of the tangent function by determining its nature on the interval

$0 \leq x < \frac{\pi}{2}$. However, it will be convenient in the discussion that follows to consider the tangent function sometimes on the interval $0 \leq x < \frac{\pi}{2}$ and sometimes on the interval $-\frac{\pi}{2} < x \leq 0$.

From our studies of the sine and cosine functions, we know that $\cos x > 0$ for $0 < x < \frac{\pi}{2}$ and that $\sin x > 0$ for $0 < x < \frac{\pi}{2}$. It follows that $\tan x > 0$ for $0 < x < \frac{\pi}{2}$. Furthermore, as the following table indicates, as x increases from 0 to $\frac{\pi}{2}$ but remains less than $\frac{\pi}{2}$, the value of $\tan x$ increases.

x	$\sin x$	$\cos x$	$\tan x$
0	0	1	0
$\frac{\pi}{6}$	$\frac{1}{2}$	$\frac{\sqrt{3}}{2}$	$\frac{\sqrt{3}}{3}$
$\frac{\pi}{4}$	$\frac{\sqrt{2}}{2}$	$\frac{\sqrt{2}}{2}$	1
$\frac{\pi}{3}$	$\frac{\sqrt{3}}{2}$	$\frac{1}{2}$	$\sqrt{3}$
.	.	.	.
.	.	.	.
.	.	.	.

Since $\tan x$ is not defined at $x = \frac{\pi}{2}$, the question is: Do the values of $\tan x$, $0 \leq x < \frac{\pi}{2}$, have an upper bound? We see that, as x approaches $\frac{\pi}{2}$, $\sin x$ gets close to 1 and $\cos x$ gets close to 0. This means, as the following table indicates, that the values of $\frac{\sin x}{\cos x} = \tan x$ get extremely large.

x	$\sin x$	$\cos x$	$\tan x$
$\frac{4\pi}{9}$	.9848	.1736	5.671
$\frac{17\pi}{36}$	.9962	.0872	11.43
$\frac{89\pi}{180}$	.9998	.0175	57.29
.	.	.	.
.	.	.	.
.	.	.	.

Indeed, given any positive number M, no matter how large, if we require that x be sufficiently close to $\frac{\pi}{2}$, but less than $\frac{\pi}{2}$, then we find that $\tan x > M$. Hence, the values of $\tan x$ as x approaches $\frac{\pi}{2}$ from the left have no upper bound. For this reason, we say that, as x approaches $\frac{\pi}{2}$ from the left, the value of $\tan x$ increases without bound. What happens to $\tan x$ as x approaches $-\frac{\pi}{2}$ from the right?

7.1a EXERCISES

Use the values of $\sin x$ and $\cos x$ to find the value of $\tan x$ for each of the following values of x.

1. $\dfrac{\pi}{6}$
2. $\dfrac{2\pi}{3}$
3. $\dfrac{5\pi}{4}$
4. $-\dfrac{\pi}{3}$
5. $\dfrac{11\pi}{6}$
6. $\dfrac{\pi}{3}$

7. $\dfrac{\pi}{4}$
8. $-\dfrac{7\pi}{6}$
9. $-\dfrac{2\pi}{3}$
10. $\dfrac{3\pi}{4}$
11. $-\dfrac{9\pi}{4}$
12. $-\dfrac{5\pi}{6}$

Use the definition of the tangent function to prove that each of the following statements is true for all values of x for which the expressions are defined.

13. $\tan(\pi - x) = -\tan x.$
14. $\tan\left(\dfrac{\pi}{2} + x\right) = -\dfrac{1}{\tan x}.$
15. $\tan\left(\dfrac{\pi}{2} - x\right) = \dfrac{1}{\tan x}.$
16. $\tan\left(\dfrac{3\pi}{2} + x\right) = -\dfrac{1}{\tan x}.$

17. Let P be the point, on the unit circle about the origin, that corresponds to the real number t and let Q be the projection of P on the x-axis. Further, let the line through P and the origin intersect the vertical line through the point $(1, 0)$ at the point T, as shown in the figures below.

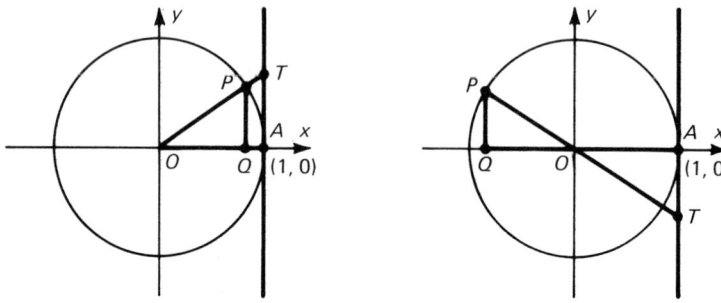

a) Draw two figures corresponding to the preceding ones showing P in the third and fourth quadrants, respectively.

b) Prove that in all four cases $\triangle OQP$ is similar to $\triangle OAT$.

c) Use part b to prove that the ordinate of point T is tan t.

d) Prove that if $t = n\pi$, where $n \in I$, then the ordinate of point T is tan t.

e) How does this geometric interpretation of tan t show that tan t is not defined when $t = (2n + 1)\dfrac{\pi}{2}$, where $n \in I$?

f) Describe the behavior of point T as t increases over each of the following intervals. Then describe the corresponding behavior of tan t.

[1] $0 \le t < \dfrac{\pi}{2}$. [3] $\pi \le t < \dfrac{3\pi}{2}$. [5] $-\pi \le t < -\dfrac{\pi}{2}$.

[2] $\dfrac{\pi}{2} < t \le \pi$. [4] $\dfrac{3\pi}{2} < t \le 2\pi$. [6] $-\dfrac{\pi}{2} < t \le 0$.

To determine more characteristics of the graph of the tangent function, it would be helpful to know if it has a slope function, and if so, what that slope function is. The tangent function does have a slope function, as we shall show in a later section, that is defined by

$$\tan' x = \frac{1}{\cos^2 x}, \quad \text{where } x \in R \text{ and } \cos x \ne 0.$$

For the present we shall accept this definition and make use of it.

First, observe that $\tan' x > 0$ for all x in the interval $-\frac{\pi}{2} < x < \frac{\pi}{2}$ and that, therefore, the tangent function is strictly increasing on this interval. Second, note that when $x = 0$,

$$\tan' 0 = \frac{1}{\cos^2 0} = 1.$$

Therefore, the slope of the graph of the tangent function is 1 at the origin.

Third, notice that, although $\tan' x$ is not defined at $x = -\frac{\pi}{2}$, it is defined for any value of x close to $-\frac{\pi}{2}$ and slightly greater than $-\frac{\pi}{2}$. For such values of x, $\tan' x = \frac{1}{\cos^2 x}$ is quite large. This is so because, for values of x just slightly greater than $-\frac{\pi}{2}$, $\cos x$ is very small and therefore $\cos^2 x$ is very small. (Why?) In fact, it should be clear that, if $-\frac{\pi}{2} < x \leq 0$, then, the closer x is to $-\frac{\pi}{2}$, the larger $\tan' x$ is. Thus, we see that, for values of x slightly greater than $-\frac{\pi}{2}$, $\tan' x$ is large and that, at $x = 0$, $\tan' x = 1$.

To discover what happens to $\tan' x$ as x increases from $-\frac{\pi}{2}$ to 0, consider the behavior of $\cos^2 x$ as x increases from $-\frac{\pi}{2}$ to 0. We know that the cosine function is increasing on this interval and $\cos x$ increases from 0 to 1. Hence, if x_1 and x_2 are values in the interval $-\frac{\pi}{2} < x \leq 0$ such that $x_1 > x_2$, then $\cos x_1 > \cos x_2$ and $\cos^2 x_1 > \cos^2 x_2$. From this it follows that

$$\frac{1}{\cos^2 x_1} < \frac{1}{\cos^2 x_2}.$$

Since $\tan' x = \frac{1}{\cos^2 x}$, this means that the function $\tan' x$ is decreasing on the interval $-\frac{\pi}{2} < x \leq 0$.

If we now assume (correctly) that $\tan'$, the slope function of the tangent function, has a slope function, $\tan''$, on this interval, we see that $\tan'' x$ must be negative on this interval. (Why?) We can then use Theorem 4.6a, page 194, to infer that the graph of the tangent function is concave downward on the interval $-\frac{\pi}{2} < x \leq 0$.

Let us now review what we know about the graph of the tangent function on the interval $-\frac{\pi}{2} < x \leq 0$:

1. The graph lies below the x-axis for $-\frac{\pi}{2} < x < 0$, since $\tan x < 0$ if $-\frac{\pi}{2} < x < 0$.

2. The graph intersects the x-axis at $x = 0$, since $\tan 0 = 0$. The graph has slope 1 at this intercept, since $\tan' 0 = \frac{1}{\cos^2 0} = 1$.

3. The graph is rising as x increases on the interval $-\frac{\pi}{2} < x < 0$, since $\tan' x > 0$ on this interval.

4. The graph has a very steep slope and a negative ordinate which is large in absolute value for points near the line $x = -\frac{\pi}{2}$. This is true since $\tan' x$ is very large and positive and $\tan x$ is negative and large in absolute value for values of x slightly greater than $-\frac{\pi}{2}$.

5. The graph is concave downward for $-\frac{\pi}{2} < x < 0$, since $\tan'' x < 0$ on this interval.

The preceding properties, which we discovered from an informal analysis, can be combined with the earlier established fact that the graph of the tangent function is symmetric with respect to the origin to produce a reasonable approximation of the graph of the tangent function.

Since the fundamental period of the tangent function is π, its graph on any interval of the form $(2n - 1)\frac{\pi}{2} < x < (2n + 1)\frac{\pi}{2}$ is just a replica of the graph on the interval $-\frac{\pi}{2} < x < \frac{\pi}{2}$, as Figure 7-1 illustrates.

Figure 7-1

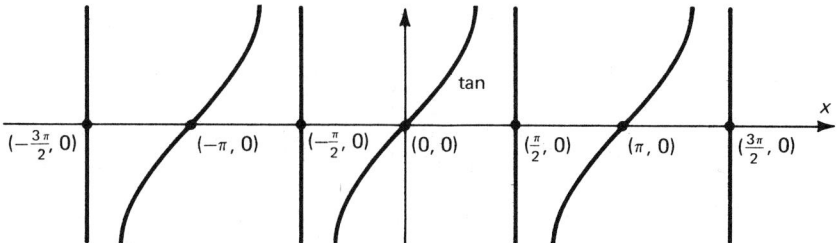

The word *asymptotic* is used to describe the behavior of the graph of the tangent function near the vertical lines $x = -\frac{\pi}{2}$, $x = \frac{\pi}{2}$, ..., $x = (2n+1)\frac{\pi}{2}$, where n is an integer. We say that the graph of the tangent function is asymptotic to each of the lines $x = (2n+1)\frac{\pi}{2}$ and that each of these lines is a vertical *asymptote* of the graph.

We shall now give informal definitions of these terms. Suppose that a curve is traced out by a point in such a way that the point gets arbitrarily far from the origin. Also, suppose that there is a line l such that this point becomes and remains arbitrarily close to l as it gets sufficiently far from the origin. Then we say that the curve is *asymptotic* to the line l and that the line l is an *asymptote* of the curve.

7.1b EXERCISES

Sketch the graphs of the following functions over the interval $-\pi \leq x \leq 2\pi$.

1. $f:f(x) = -\tan x$.
2. $g:g(x) = \tan\left(x + \frac{\pi}{4}\right)$.
3. $h:h(x) = \tan x - 2$.
4. $F:F(x) = \tan(\frac{1}{2}x)$.
5. $G:G(x) = \frac{1}{2}\tan 2x$.
6. $H:H(x) = \tan |x|$.

For each of the following exercises, write an equation for the tangent line to the graph of the tangent function at the indicated point. Then sketch the graph of the tangent function and draw the three tangent lines on the graph.

7. $\left(\frac{\pi}{4}, 1\right)$
8. $\left(-\frac{\pi}{3}, -\sqrt{3}\right)$
9. $\left(\frac{\pi}{6}, \frac{\sqrt{3}}{3}\right)$

In each of the following exercises, use the given information to find the value of $\tan x$.

10. $\sin x = \frac{5}{13}$ and $0 < x < \frac{\pi}{2}$.
11. $\cos x = -.6000$ and $\pi < x < \frac{3\pi}{2}$.

12. $\sin x = -\dfrac{15}{17}$ and $-\dfrac{\pi}{2} < x < 0$.

13. $\cos x = .9600$ and $\dfrac{3\pi}{2} < x < 2\pi$.

14. By an argument similar to the one presented on page 275, show that, if $f(x)$ is positive on an interval and if f is a strictly increasing function on that interval, then $\dfrac{1}{f}$ is a strictly decreasing function on that interval.

15. a) Sketch the graph of the function f which is the restriction of the tangent function to the interval $-\dfrac{\pi}{2} < x < \dfrac{\pi}{2}$ that is, the function defined by $f(x) = \tan x$, $-\dfrac{\pi}{2} < x < \dfrac{\pi}{2}$.

   b) Does f have an inverse function, f^{-1}? Explain your answer.
   c) Sketch the graph of f^{-1} and state its domain and range.
   d) Write an equation for the tangent line to the graph of f^{-1} at the point where $x = 0$. At the point where $x = 1$. At the point where $x = -\sqrt{3}$.

7.2 Derivation of the Slope Function of the Tangent Function (*Optional*)

In our analysis of the tangent function in Section 7.1, we used the fact that $\tan' x = \dfrac{1}{\cos^2 x}$. Here, we will show how the formula for the slope function of the tangent function is derived.

Let $(x_0, \tan x_0)$ be a point on the graph of the tangent function. Now consider the expression

$$\dfrac{\tan x - \tan x_0}{x - x_0}.$$

From the definition $\tan x = \dfrac{\sin x}{\cos x}$, we see that this expression can be written as

$$\dfrac{\dfrac{\sin x}{\cos x} - \dfrac{\sin x_0}{\cos x_0}}{x - x_0}.$$

This expression can be shown to be equal to

$$\dfrac{\sin x \cos x_0 - \sin x_0 \cos x}{x - x_0} \cdot \dfrac{1}{\cos x \cos x_0}.$$

From the addition formula for the sine function, we know that $\sin x \cos x_0 - \sin x_0 \cos x = \sin(x - x_0)$; therefore,

$$\frac{\tan x - \tan x_0}{x - x_0} = \frac{\sin(x - x_0)}{x - x_0} \cdot \frac{1}{\cos x \cos x_0}.$$

As x gets close to x_0, $x - x_0$ approaches 0, and, therefore $\frac{\sin(x - x_0)}{x - x_0}$ approaches 1 (see page 267). Also, as x gets close to x_0, $\cos x$ approaches $\cos x_0$, and therefore, $\frac{1}{\cos x \cos x_0}$ approaches $\frac{1}{\cos^2 x_0}$. Thus, we see that $\frac{\tan x - \tan x_0}{x - x_0}$ gets arbitrarily close to $\frac{1}{\cos^2 x_0}$ as x gets sufficiently close to x_0.

From this we see that the slope of the tangent line to the graph of the tangent function at the point $(x_0, \tan x_0)$ is $\frac{1}{\cos^2 x_0}$. Since x_0 was any value such that $\tan x$ is defined, we therefore have

$$\tan' x = \frac{1}{\cos^2 x}, \quad \text{where } \cos x \neq 0.$$

7.2 EXERCISES

Find the values of x in the interval $-\frac{\pi}{2} < x < \frac{\pi}{2}$ that satisfy the condition given in each of the following exercises.

1. $3 \tan x - 1 = 2$.
2. $3 \tan x - \sqrt{3} \leq 0$.
3. $2 \tan x = 2\sqrt{3}$.
4. $\sqrt{3} \tan x + 1 > 0$.
5. $\dfrac{1}{\tan x} = -1$.
6. $\dfrac{4 - \tan x}{\tan x} < 3$.
7. $3 \tan^2 x = 1$.
8. $\tan^2 x > 3$.
9. $3 \tan^2 x = \sqrt{3} \tan x$.
10. $\tan^2 x - \tan x \leq 0$.

The solution of certain conditions involving the tangent function requires the use of the table of values for circular functions given on page 386. Find the set of solutions for each of the following conditions. When necessary, use the table of values for circular functions.

11. $\tan^2 x - 4\tan x + 3 = 0$.
12. $\tan^2 x - 2\tan x > 3$.
13. $\tan^2 x = 14 + 5\tan x$.
14. $2\tan^2 x + \tan x < 3$.
15. $3\tan^2 x - 16\tan x + 5 = 0$.
16. $4\tan^2 x - \tan x \leq 3$.
17. Find the equations of the tangent lines to the graph of the tangent function in the interval $-\dfrac{\pi}{2} < x < \dfrac{\pi}{2}$ that are parallel to the line $y = 2x$. Sketch the graph of the tangent function and draw the tangent lines.
18. Use the areas of $\triangle OAP$, sector OAP, and $\triangle OAT$ in the accompanying figure to prove that, if $0 < x < \dfrac{\pi}{2}$, then $\sin x < x < \tan x$.

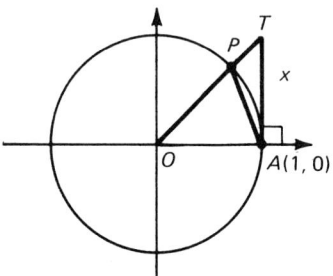

19. Use the definition of the tangent function to prove that, for all values of x for which $|\tan x|$ and $|\sin x|$ are defined, $|\tan x| \geq |\sin x|$.

7.3 The Cotangent Function

The cotangent function bears the same relationship to the tangent function as the cosine function does to the sine function. By this we mean that the cotangent function is the composition of the tangent function with the linear function l defined by $l(x) = \dfrac{\pi}{2} - x$.

Definition 7.3 The **cotangent function** is defined by

$$\cot x = \tan\left(\dfrac{\pi}{2} - x\right), \qquad x \neq n\pi,\ n \in I.$$

Since $\tan\left(\dfrac{\pi}{2} - x\right) = \dfrac{\sin\left(\dfrac{\pi}{2} - x\right)}{\cos\left(\dfrac{\pi}{2} - x\right)} = \dfrac{\cos x}{\sin x}$, we see that we could have defined the cotangent function by $\cot x = \dfrac{\cos x}{\sin x}$, when $\sin x \neq 0$.

We also see that if $\cos x \neq 0$, that is, if $x \neq (2n + 1)\dfrac{\pi}{2}$, where n is an integer, and $\cot x$ is defined, then

$$\cot x = \dfrac{\cos x}{\sin x} = \dfrac{1}{\dfrac{\sin x}{\cos x}} = \dfrac{1}{\tan x}.$$

To simplify the determination of the graph of the cotangent function, we can use a small trick that enables us to use what we know about translations and reflections of graphs. Since we have seen that $\tan x = -\tan(-x)$, it follows that

$$\cot x = \tan\left(\dfrac{\pi}{2} - x\right) = -\tan\left(x - \dfrac{\pi}{2}\right).$$

From our earlier work with translations, we know that the graph of the function defined by $f(x - a)$ is the graph of f translated $|a|$ units, to the right if $a > 0$ and to the left if $a < 0$. We also know that the graph of $-f$ is the graph of f reflected in the x-axis. From the fact that $\cot x = -\tan\left(x - \dfrac{\pi}{2}\right)$, we can therefore conclude that the graph of the cotangent function is the graph of the tangent function translated $\dfrac{\pi}{2}$ units to the right and then reflected in the x-axis. The graph of the cotangent function is shown in Figure 7-2.

Figure 7-2

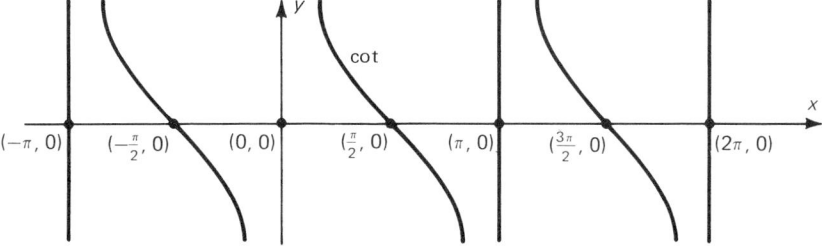

From the graph of the cotangent function that we just constructed, we can read off many properties of this function. For example, it is clear that the function is periodic with a fundamental period of π. Further, the graph of the cotangent function over any interval $n\pi < x < (n+1)\pi$ is a translate to the right or left of the portion of the graph of the function over the interval $0 < x < \pi$. Thus, we confine our attention to the graph of the cotangent function on this interval.

On the interval $0 < x < \pi$, we see that the cotangent function is strictly decreasing and has a single x-intercept at $\left(\frac{\pi}{2}, 0\right)$. Further, the graph of this function is concave upward for $0 < x < \frac{\pi}{2}$ and concave downward for $\frac{\pi}{2} < x < \pi$. Figure 7–2 also shows that the graph of the cotangent function is asymptotic to each of the lines $x = n\pi$, where n is an integer.

We could have derived all of these facts about the graph of the cotangent function by analyzing this function in the same way that we analyzed the tangent function. It is much easier, however, to obtain information about the graph of the cotangent function by using the fact that its graph is the graph of the tangent function translated $\frac{\pi}{2}$ units to the right and then reflected in the x-axis.

7.3 EXERCISES

Find the value of cot x for each of the following values of x.

1. $\frac{\pi}{3}$

2. $\frac{5\pi}{3}$

3. $\frac{5\pi}{6}$

4. $\frac{\pi}{2}$

5. $\frac{2\pi}{3}$

6. $\frac{11\pi}{6}$

7. $\frac{5\pi}{4}$

8. $-\frac{\pi}{4}$

9. $-\dfrac{2\pi}{3}$

10. $\dfrac{9\pi}{4}$

11. $-\dfrac{5\pi}{6}$

12. $-\dfrac{5\pi}{4}$

Sketch the graph of each of the following functions over the interval $-\pi < x < 2\pi$.

13. $f : f(x) = \cot x + 1$.
14. $g : g(x) = \cot\left(x - \dfrac{\pi}{4}\right)$.
15. $F : F(x) = \cot 2x$.
16. $G : G(x) = \tan x + \cot x$.

Prove that each of the following statements is true for all values of x for which the expressions are defined.

17. $\cot\left(\dfrac{\pi}{2} + x\right) = -\tan x$.
18. $\cot\left(\dfrac{3\pi}{2} - x\right) = \tan x$.
19. $\cot\left(\dfrac{3\pi}{2} + x\right) = -\tan x$.
20. $\cot x + \tan x = \dfrac{1}{\sin x \cos x}$.

Find the values of x in the interval $0 < x < \pi$ *that satisfy each of the following conditions.*

21. $2(\cot x - 2) + 2 = 0$.
22. $6 \cot x - \sqrt{3} = \sqrt{3}$.
23. $\cot x = -\sqrt{3}$.
24. $-3 \cot x = \sqrt{3}$.
25. $\cot^2 x = \cot x$.
26. $\cot^2 x + 1 = \dfrac{2}{\tan x}$.

Use the definition of the cotangent function to prove each of the following theorems.

27. The cotangent function is an odd function.
28. The cotangent function is periodic with fundamental period π.
29. In the interval $0 < x < \pi$, the cotangent function is strictly decreasing.

7.4 The Secant and Cosecant Functions

The reciprocal of the cosine function occurs quite often in working with the circular functions, sine and cosine. Thus, it becomes convenient to give a name to the function $f:f(x) = \dfrac{1}{\cos x}$, where $\cos x \neq 0$. This function, called the *secant function*, is defined as follows:

Definition 7.4 The **secant function** is defined by

$$\sec x = \frac{1}{\cos x}, \quad \text{with } x \in R \text{ and } \cos x \neq 0.$$

The secant function is not defined when $\cos x = 0$; that is, when $x = (2n + 1)\dfrac{\pi}{2}$, where $n \in I$.

We know that the cosine function has a fundamental period of 2π; that is, we know that $\cos x = \cos(x + 2n\pi)$, where $n \in I$. It follows that $\sec x = \dfrac{1}{\cos x} = \dfrac{1}{\cos(x + 2n\pi)} = \sec(x + 2n\pi)$, where $n \in I$ and $\cos x \neq 0$. Thus, because $\sec x = \sec(x + 2n\pi)$ and because 2π is the smallest positive number a such that $\sec x = \sec(x + a)$, the secant function also has a fundamental period of 2π. This being the case, we will confine our attention to the behavior of the secant function on the interval $-\pi \leq x \leq \pi$. Whatever we learn about the behavior of the secant function and the graph of the secant function on this interval can be used to determine the behavior of the secant function and its graph over any interval of the form $(2n - 1)\pi \leq x \leq (2n + 1)\pi$.

We can learn much about the secant function over the interval $-\pi \leq x \leq \pi$ just from the fact that it is the reciprocal of the cosine function. For example, $\cos x < 0$ if $-\pi \leq x < -\dfrac{\pi}{2}$ or if $\dfrac{\pi}{2} < x \leq \pi$. Therefore, $\sec x < 0$ over these same intervals. For $-\dfrac{\pi}{2} < x < \dfrac{\pi}{2}$, $\cos x > 0$; and, therefore, $\sec x > 0$ on this interval. We also see that, since $0 < \cos x \leq 1$ for $-\dfrac{\pi}{2} < x < \dfrac{\pi}{2}$, $\sec x \geq 1$ on this interval.

Similarly, since $-1 \leq \cos x < 0$ on the intervals $-\pi \leq x < -\frac{\pi}{2}$ and $\frac{\pi}{2} < x \leq \pi$, we see that $\sec x \leq -1$ on these intervals.

From these observations, we can conclude that, for $-\pi \leq x \leq \pi$, the graph of the secant function must lie in those sections of the plane shown in Figure 7-3 that are shaded. One point is clear even with this minimal amount of information; namely, $\sec x$ is never equal to zero.

Figure 7-3

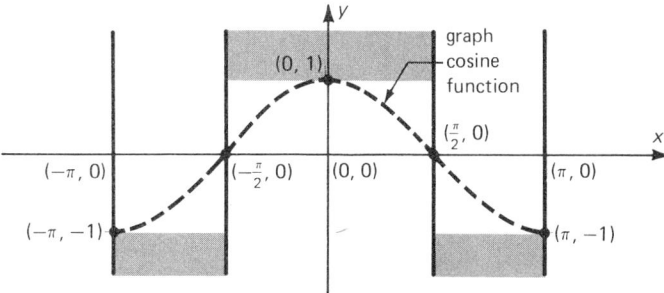

Since $\sec x = \dfrac{1}{\cos x} = \dfrac{1}{\cos(-x)} = \sec(-x)$, the secant function is an even function, which means that its graph is symmetric with respect to the origin. There are some other observations that we can make about the secant function and its graph based on the fact that this function is the reciprocal of the cosine function. For example, since the graph of cosine is concave downward on the interval $-\dfrac{\pi}{2} < x < \dfrac{\pi}{2}$, it would seem apparent that the graph of secant would be concave upward over this interval. Also, since $\cos x$ increases as x increases on the interval $-\dfrac{\pi}{2} < x < 0$, $\sec x$ must decrease on this interval. A similar analysis indicates that the secant function is increasing on the interval $0 < x < \dfrac{\pi}{2}$. If we investigate the behavior of $\sec x$ for values of x near odd multiples of $\dfrac{\pi}{2}$, we see that lines of the form $x = (2n+1)\dfrac{\pi}{2}$ are vertical asymptotes of the graph of the secant function.

On the basis of the information given above, we can, after plotting a few critical points, construct a fairly good characterization of the graph of the secant function. A dotted-line graph of the cosine function is included in Figure 7-3 for comparison.

On page 286 we list various properties of the secant function. After reading each statement of a property, check the graph in Figure 7-4 for the corresponding geometrical property.

Figure 7-4

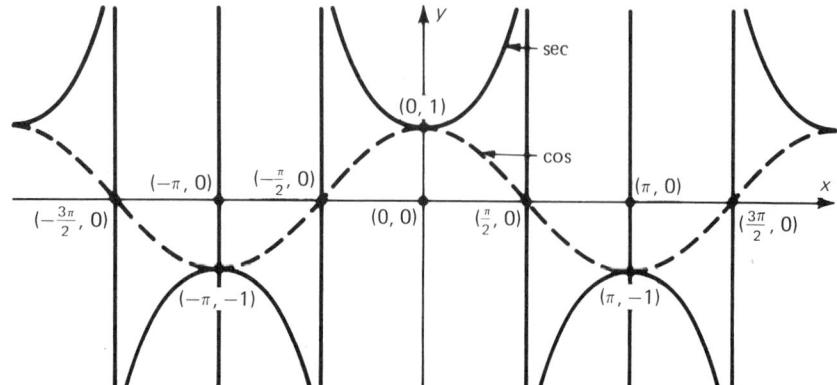

1. The secant function is periodic with a fundamental period of 2π.
2. The secant function is an increasing function over the intervals $2n\pi \leq x < (2n + \frac{1}{2})\pi$ and $(2n + \frac{1}{2})\pi < x \leq (2n + 1)\pi$, where $n \in I$; that is, where $n = 0, \pm 1, \pm 2, \ldots$.
3. The secant function is a decreasing function over the intervals $(2n - 1)\pi \leq x < (2n - \frac{1}{2})\pi$ and $(2n - \frac{1}{2})\pi < x \leq 2n\pi$, where $n \in I$.
4. On the interval $(2n - \frac{1}{2})\pi < x < (2n + \frac{1}{2})\pi$, sec x has a minimum value of 1 when $x = 2n\pi$, where $n \in I$.
5. On the interval $(2n + \frac{1}{2})\pi < x < (2n + \frac{3}{2})\pi$, sec x has a maximum value of -1 when $x = (2n + 1)\pi$, where $n \in I$.
6. The secant function is an even function.
7. The graph of the secant function is
   concave upward for $(2n - \frac{1}{2})\pi < x < (2n + \frac{1}{2})\pi$ and is
   concave downward for $(2n + \frac{1}{2})\pi < x < (2n + \frac{3}{2})\pi$, where $n \in I$.
8. The graph of the secant function is asymptotic on both the right and the left to each of the vertical lines $x = (2n + 1)\frac{\pi}{2}$, where $n \in I$. When $x = (2n + 1)\frac{\pi}{2}$, sec x is not defined.

7.4a EXERCISES

Determine sec x for each of the following values of x.

1. $\dfrac{\pi}{6}$
2. $\dfrac{5\pi}{4}$
3. $\dfrac{2\pi}{3}$
4. $-\pi$
5. $\dfrac{7\pi}{4}$
6. $-\dfrac{5\pi}{6}$

Sketch the graph of each of the following functions over the interval $-\pi \leq x \leq 2\pi$.

7. $f: f(x) = \sec(\pi + x)$.
8. $g: g(x) = \sec \dfrac{x}{\pi}$.
9. $F: F(x) = |\sec x|$.
10. $G: G(x) = \tfrac{1}{2} \sec\left(x - \dfrac{\pi}{2}\right)$.

Prove that each of the following statements is true for all values of x for which the expression is defined.

11. $\sec(\pi - x) = -\sec x$.
12. $\sec(\pi + x) = -\sec x$.
13. $\sec(2\pi - x) = \sec x$.
14. $\sin x \sec x = \tan x$.

Find the values of x in the interval $0 < x < 2\pi$ for which each of the following statements is true.

15. $\sec x < 2$.
16. $\sec^2 x = 2$.
17. $3 \sec^2 x \geq 4$.
18. $5 \sec^2 x < 5$.
19. $\sec^2 x = 2 \sec x$.
20. $\sec^2 x - \sqrt{2} \sec x > 0$.

For each of the following exercises, write an expression involving only $\sin x$ or $\cos x$ that is equivalent to the given expression for all values of x for which the expressions are defined.

21. $\dfrac{\tan x}{\sec x}$
22. $\dfrac{\sec x}{\cot x}$
23. $\sec^2 x - 1$
24. $(\cot x)(\sin x - \sec x)$
25. Use the definition of the secant function to prove that it is strictly increasing on the interval $0 \leq x < \dfrac{\pi}{2}$.
26. Prove that, for all values of x for which the expressions are defined, $|\sec x| \geq |\tan x|$.

Just as the cotangent and cosine functions can be defined as the composition of the tangent and sine functions, respectively, with the linear function $l : l(x) = \frac{\pi}{2} - x$, so the *cosecant function*, denoted by csc, can be defined as the composition of the secant function with l.

Definition 7.4b The **cosecant function** is defined by
$$\csc x = \sec\left(\frac{\pi}{2} - x\right), \qquad \text{with } x \in R \text{ and } x \neq n\pi, \text{ where } n \in I.$$

Since $\sec\left(\frac{\pi}{2} - x\right) = \dfrac{1}{\cos\left(\frac{\pi}{2} - x\right)} = \dfrac{1}{\sin x}$, where $x \neq n\pi$ and $n \in I$,

we see that we could have defined the cosecant function by
$$\csc x = \frac{1}{\sin x}, \qquad \text{where } x \neq n\pi \text{ and } n \in I.$$

The graph of the cosecant function is just the graph of the secant function translated $\frac{\pi}{2}$ units to the right, as we will show in the following discussion. We will subsequently be able to use this fact to deduce the important properties of the cosecant function directly from those of the secant function.

We mentioned earlier in this section that the secant function is an even function (see page 285); that is, $\sec x = \sec(-x)$ for all values of x such that $\sec x$ is defined. From this fact, it follows that

$$\sec\left(\frac{\pi}{2} - x\right) = \sec\left(-\left(\frac{\pi}{2} - x\right)\right) = \sec\left(x - \frac{\pi}{2}\right).$$

By the definition of the cosecant function, we have

$$\csc x = \sec\left(\frac{\pi}{2} - x\right) = \sec\left(x - \frac{\pi}{2}\right).$$

We know from our work with translations that the graph of $\sec\left(x - \frac{\pi}{2}\right)$ is just the graph of the secant function translated $\frac{\pi}{2}$ units to the right (see page 133). Thus, the graph of the cosecant function is the graph of the secant function translated $\frac{\pi}{2}$ units to the right.

A portion of the graph of the secant function is given in Figure 7-5. Since the graph of the sine function is the graph of the cosine function translated $\frac{\pi}{2}$ units to the right, it follows that the graph of the cosecant function is related to the graph of the sine function in much the same way as the graph of the secant function is related to the graph of the cosine function (see page 286). A portion of the graph of the sine function is, therefore, included in Figure 7–5 to exhibit this relationship.

Figure 7-5

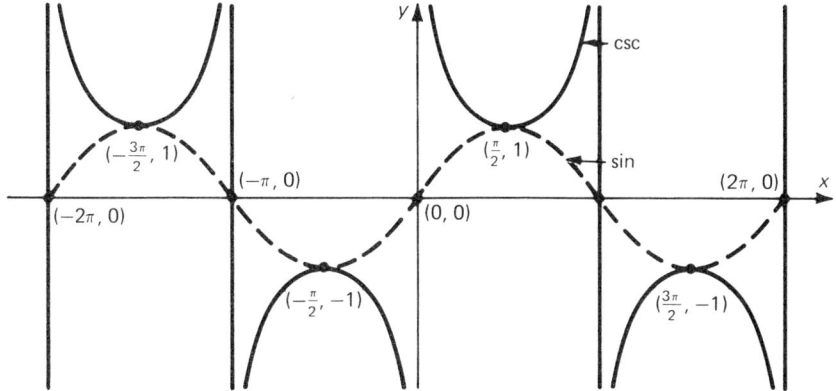

From the list of properties of the secant function given on page 286 and from the fact that $\csc x = \sec\left(x - \frac{\pi}{2}\right)$, you should be able to list a set of properties of the cosecant function which are analogous to those of the secant function. Any property that the secant function has on an interval $a < x < b$ will be a property of the cosecant function on the translation of this interval $\frac{\pi}{2}$ units to the right, namely, on the interval $a + \frac{\pi}{2} < x < b + \frac{\pi}{2}$. This is so because the graph of the cosecant function is the graph of the secant function translated $\frac{\pi}{2}$ units to the right.

7.4b EXERCISES

Find the value of csc x for each of the following values of x.

1. $\dfrac{\pi}{3}$

2. $\dfrac{3\pi}{4}$

3. $\dfrac{11\pi}{6}$

4. $-\dfrac{\pi}{2}$

5. $\dfrac{5\pi}{2}$

6. $-\dfrac{2\pi}{3}$

Sketch the graph of each of the following functions.

7. $f: f(x) = \csc(x - \pi)$.
8. $g: g(x) = \tfrac{1}{2}\csc x$.
9. $F: F(x) = \csc(2x)$.
10. $G: G(x) = 2\csc(\tfrac{1}{2}x)$.

Prove that each of the following statements is true for all values of x for which the expressions are defined.

11. $\csc(\pi - x) = \csc x$.
12. $\csc(\pi + x) = -\csc x$.
13. $\csc\left(\dfrac{\pi}{2} + x\right) = \sec x$.
14. $\csc\left(\dfrac{3\pi}{2} - x\right) = -\sec x$.
15. $\csc\left(\dfrac{3\pi}{2} + x\right) = -\sec x$.

Find the values of x in the interval $0 < x < 2\pi$ that satisfy each of the following conditions.

16. $\csc x = \sqrt{2}$.
17. $\csc^2 x < 4$.
18. $3\csc^2 x = 4$.
19. $10\csc^2 x \geq 10$.
20. $\csc^2 x + \sqrt{2}\csc x = 0$.
21. $\csc^2 x + 2\csc x = 0$.

Use the definition of the cosecant function to prove each of the following theorems.

22. The cosecant function is an odd function.
23. The cosecant function is periodic with a fundamental period of 2π.

For each of the following exercises, write an expression involving only $\sin x$ or $\cos x$ that is equivalent to the given expression for all values of x for which the expression is defined.

24. $\dfrac{\sec x}{\csc x}$

25. $\dfrac{\cot^2 x}{\csc^2 x}$

26. $\tan x \,(\cos x + \csc x)$

27. $\sec x + \csc x$

28. List a set of properties for the cosecant function that are analogous to those of the secant function given on page 286.

29. Prove that, for all values of x for which the expressions are defined, $|\csc x| \geq |\cot x|$.

7.5 More Identities

On pages 259 and 261–263 of Chapter 6, certain identities for the sine and cosine functions were given. Since we have now defined four more circular functions, namely, the tangent, cotangent, secant, and cosecant functions, we can easily obtain some identities for these new functions from the identities already given. In this section, we will derive a few of these identities.

From the Pythagorean Identity, $\cos^2 t + \sin^2 t = 1$, given on page 262, we can conclude that, if $\cos t \neq 0$, then

$$\frac{\cos^2 t + \sin^2 t}{\cos^2 t} = \frac{1}{\cos^2 t},$$

or

$$1 + \left(\frac{\sin t}{\cos t}\right)^2 = \left(\frac{1}{\cos t}\right)^2, \qquad \text{where } t \neq (2n+1)\frac{\pi}{2} \text{ and } n \in I.$$

Since $\dfrac{\sin t}{\cos t} = \tan t$ and $\dfrac{1}{\cos t} = \sec t$, we therefore have

$$\mathbf{1 + \tan^2 t = \sec^2 t.}$$

This identity may also be written as

$$\mathbf{\sec^2 t - \tan^2 t = 1,} \qquad \text{where } t \neq (2n+1)\frac{\pi}{2} \text{ and } n \in I.$$

Similarly, we can divide both sides of the Pythagorean Identity by $\sin^2 t$, provided that $\sin t \neq 0$, to obtain

$$\frac{\cos^2 t + \sin^2 t}{\sin^2 t} = \frac{1}{\sin^2 t},$$

or

$$\left(\frac{\cos t}{\sin t}\right)^2 + 1 = \left(\frac{1}{\sin t}\right)^2, \qquad \text{where } t \neq n\pi \text{ and } n \in I.$$

Because $\cot t = \dfrac{\cos t}{\sin t}$ and $\csc t = \dfrac{1}{\sin t}$, this equation becomes

$$\cot^2 t + 1 = \csc^2 t.$$

This identity is usually written as

$$\csc^2 t - \cot^2 t = 1, \quad \text{where } t \neq n\pi \text{ and } n \in I.$$

From the addition formulas for the sine and cosine functions,

$$\cos(t_1 + t_2) = \cos t_1 \cdot \cos t_2 - \sin t_1 \cdot \sin t_2, \text{ and}$$
$$\sin(t_1 + t_2) = \sin t_1 \cdot \cos t_2 + \sin t_2 \cdot \cos t_1,$$

we can obtain addition formulas for the tangent and cotangent functions. By the definition of the tangent function and the addition formulas for sine and cosine, we have

$$\tan(t_1 + t_2) = \frac{\sin(t_1 + t_2)}{\cos(t_1 + t_2)} = \frac{\sin t_1 \cdot \cos t_2 + \sin t_2 \cdot \cos t_1}{\cos t_1 \cdot \cos t_2 - \sin t_1 \cdot \sin t_2}.$$

We can simplify the equation for $\tan(t_1 + t_2)$ greatly by dividing the numerator and the denominator of the right side of the equation by $\cos t_1 \cos t_2$. Thus, we have

$$\tan(t_1 + t_2) = \frac{\dfrac{\sin t_1 \cdot \cos t_2}{\cos t_1 \cdot \cos t_2} + \dfrac{\sin t_2 \cdot \cos t_1}{\cos t_1 \cdot \cos t_2}}{\dfrac{\cos t_1 \cdot \cos t_2}{\cos t_1 \cdot \cos t_2} - \dfrac{\sin t_1 \cdot \sin t_2}{\cos t_1 \cdot \cos t_2}} = \frac{\dfrac{\sin t_1}{\cos t_1} + \dfrac{\sin t_2}{\cos t_2}}{1 - \dfrac{\sin t_1}{\cos t_1} \cdot \dfrac{\sin t_2}{\cos t_2}},$$

which, by the definition of the tangent function, becomes

$$\tan(t_1 + t_2) = \frac{\tan t_1 + \tan t_2}{1 - \tan t_1 \tan t_2}.$$

Keep in mind that this identity holds if and only if, for given values of t_1 and t_2, *both* sides of the equation are defined.

From the above formula, it is an easy matter to derive the double angle formula for the tangent function simply by letting $t_1 = t_2 = t$ to obtain

$$\tan(t + t) = \frac{\tan t + \tan t}{1 - \tan t \cdot \tan t}, \text{ or}$$

$$\tan 2t = \frac{2 \tan t}{1 - \tan^2 t}.$$

We could go on and derive many more identities. (You are asked to derive a few more in the exercises.) In general, however, it is probably not wise to try to memorize too many such identities, unless you have an infallible memory. For most people, it is too easy to get a wrong sign or forget some part of the identity. A safer course is to remember the basic identities which we have given and to use your algebraic skills to simplify expressions or to derive other identities. The identities for circular functions are extremely valuable in that, by using them, many formulas in calculus and in other branches of mathematics can be greatly simplified. For this reason, it is important to know the basic identities and to acquire some skill in manipulating them algebraically.

7.5 EXERCISES

1. a) Beginning with the formula for $\tan(x_1 + x_2)$, use the fact that $\tan(x_1 - x_2) = \tan(x_1 + (-x_2))$ to develop a formula for $\tan(x_1 - x_2)$.
    b) Beginning with the formulas for $\sin(x_1 - x_2)$ and $\cos(x_1 - x_2)$, develop a formula for $\tan(x_1 - x_2)$.

Develop a formula for each of the following.

2. $\cot(x_1 + x_2)$
3. $\cot(2x)$
4. $\cot(x_1 - x_2)$
5. Beginning with the formulas for $|\sin \frac{1}{2}x|$ and $|\cos \frac{1}{2}x|$, develop formulas for $|\tan \frac{1}{2}x|$ and $|\cot \frac{1}{2}x|$.

Prove each of the following identities for all values of x for which it is defined.

6. $\sin^2 x + (\sin^2 x)\tan^2 x = \tan^2 x$.
7. $(1 - \sin^2 x)(\sec^2 x - 1) = \sin^2 x$.
8. $\dfrac{1 + \sec x}{\tan x} = \cot x + \csc x$.
9. $\dfrac{(\sec x - 1)^2}{\tan^2 x} = \dfrac{\sec x - 1}{\sec x + 1}$.
10. $\dfrac{\sec x + \csc x}{\tan x + \cot x} = \sin x + \cos x$.
11. $\dfrac{2 \tan x}{\sec^2 x - 2 \tan^2 x} = \tan 2x$.
12. $\dfrac{1 - \tan^2 x}{\sec^2 x} = \cos 2x$.
13. $\dfrac{\tan x_1 + \tan x_2}{\tan x_1 - \tan x_2} = \dfrac{\sin(x_1 + x_2)}{\sin(x_1 - x_2)}$.
14. $\dfrac{1 + \sec^2 x}{\tan x} = \dfrac{1 + \cos^2 x}{\sin x \cdot \cos x}$.

8

Exponential and Logarithmic Functions

Introduction

There are four elementary functions from which a great many functions, including nearly all of the functions studied in calculus, can be derived. These four elementary functions are the unit, the identity, the sine and the exponential functions. We have already studied the unit and identity functions in our discussion of constant and linear functions, and the sine function in our discussion of trigonometric functions.

The present chapter will be concerned with the fourth of these elementary functions, the exponential function. We begin with an example from science of a type of natural process that can be described mathematically by an exponential function.

8.1 The Fundamental Law of Growth and Decay

Radioactive substances have the characteristic that they decay into other substances. In this process, electrons or the helium nucleus are emitted from the disintegrating atoms, which change to atoms of other substances. This decay has been found to be predictable, because the rate of decay is proportional to the amount of the substance present. The time that it takes an amount of radioactive substance to decay to one half the original amount depends upon the nature of the substance, but this time is independent of the amount of the substance present. This means that it takes just as long for 2 grams of radium to decay to 1 gram as it takes for 200 grams to decay to 100 grams. This length of time is called the *half-life* of the particular radioactive substance and can vary from microseconds (millionths of a second) in the case of some elements to billions of years in others. For example, the half-life of Thorium C' is about $\frac{1}{3}$ of a microsecond and the half-life of Uranium I is about $4\frac{1}{2}$ billion years. The half-life of radium is about 1620 years.

If a certain radioactive substance had a half-life of one day (Uranium Y, for example, has a half-life of 24.6 hours), then, from time 0 with a weight W_0, the amount present at the end of one day would be $\frac{W_0}{2}$; at the end of two days, $\frac{1}{2}\left(\frac{W_0}{2}\right) = W_0(\frac{1}{2})^2$; . . . ; and at the end of n days, $W_0(\frac{1}{2})^n$. If we denote the weight present at the end of t days by $W(t)$, we have the equation

$$W(t) = W_0(\tfrac{1}{2})^t,$$

where $(\frac{1}{2})^0 = 1$. This equation defines a function W that has the set of nonnegative integers for its domain and a range that is a subset of R. The function W is the set of ordered pairs of the form $(t, W(t))$, where t denotes time in days and $W(t)$ denotes corresponding weights.

It should be clear that, given the same rate of decay, the amount of radioactive material present one day before we started counting time (that is, when $t = -1$) would be twice as great as when $t = 0$. Thus, the amount of material present when $t = -1$ would be $2W_0 = (\frac{1}{2})^{-1}W_0$. Similarly, two days before $t = 0$ (that is, when $t = -2$), the amount present would be $2(2W_0) = 2^2 W_0 = (\frac{1}{2})^{-2} W_0$; and n days before $t = 0$, when $t = -n$, the amount present would be $2^n W_0 = (\frac{1}{2})^{-n} W_0$. That is, for values of t which are negative integers,

$$W(t) = 2^{-t} W_0 = W_0(\tfrac{1}{2})^t.$$

Thus, we can see that the function W previously discussed for this example can be extended to have as its domain the set of all integers, rather than the set of nonnegative integers. We now have the function W defined below for this example:

$$\text{W}:\text{W}(t) = W_0(\tfrac{1}{2})^t, \quad \text{where } t \in I.$$

Example 1 If, for the purposes of illustration, we assume that W_0 is equal to 3, we can give a portion of the graph of this function, which is a set of points in the plane whose first coordinates are integers. This graph is shown in Figure 8–1 on page 296.

Figure 8-1

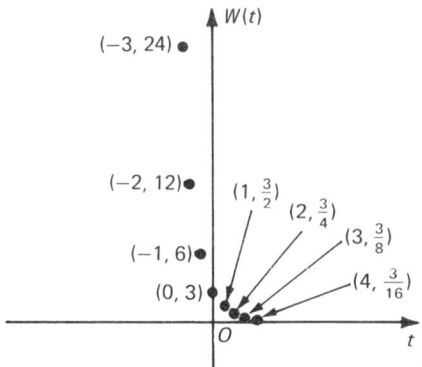

There are other physical processes, like the decaying of radioactive substances just described, which have the property that some measurable quantity associated with the process increases or decreases over a certain period of time at a rate proportional to the amount present. Any process satisfying this property is said to obey the *Fundamental Law of Growth and Decay*. Certain biological cultures, for example, have the property that, under appropriate conditions, any given amount of the culture will double in a fixed interval of time, independently of the amount present. If the weight of the culture at time $t = 0$ is w_0 and if it is assumed that the amount of culture doubles each day, then at the end of one day there will be $2w_0$ of the culture; at the end of two days, $2(2w_0) = 2^2 w_0$ of the culture; . . . ; and, at the end of n days, $2^n w_0$ of the culture.

If we further assume that the culture in this case was subject to the same conditions before we started counting time, then, one day before we started counting time (that is, when $t = -1$), the amount of biological culture present would be $\frac{1}{2}w_0 = 2^{-1}w_0$, and n days before $t = 0$, when $t = -n$, the amount present would be $(\frac{1}{2})^n w_0 = 2^{-n} w_0$. Thus, if we let $w(t)$ denote the amount of culture present at a time t, we have a function w defined on the integers by

$$w(t) = 2^t w_0, \quad \text{with } t \in I.$$

The domain of w, as indicated, is the set of integers, and its range is a subset of R.

Example 2 If we also assume for this example that the parameter w_0 is equal to 3, then a portion of the graph of w would look like the illustration in Figure 8–2.

Of course, the recorded weights for neither the biological culture nor the radioactive substance considered in the first example would conform exactly to the second components of the ordered pairs in the functions w and W, but they would very closely approximate these numbers over some interval of time. The functions w and W and their graphs serve as "mathematical models" of these physical processes and give us an oppor-

Figure 8-2

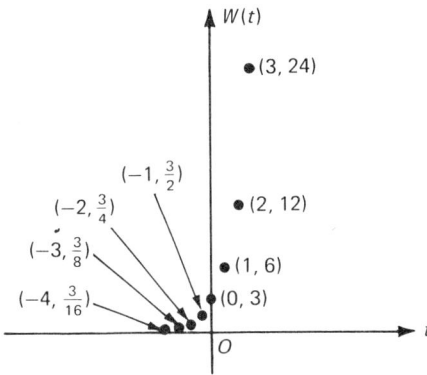

tunity to predict physical events on the basis of the mathematical character of these models.

In conducting experiments or observing physical phenomena, scientists record various measurable quantities such as weight or volume and then plot them against time or some other variable. The resulting graph is a set of points on graph paper, like the one shown in Figure 8-3.

Figure 8-3

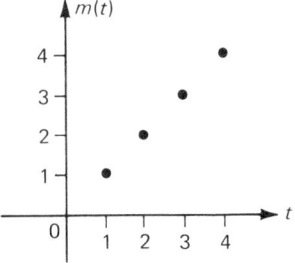

In analyzing this graph of his results for this particular example, the scientist might observe that the points are nearly collinear—that all of the points seem to be on or close to the line defined by the equation $m(t) = t$. In this case, he might express this by saying that the line $m(t) = t$ "fits" the data closely. If he conducts the same experiment a number of times and finds that the line $m(t) = t$ always "fits" the data reasonably well, he will then say that the equation $m(t) = t$ "describes" the behavior with respect to the variable t of the quantity he is measuring. By using the formula $m(t) = t$, he will then predict values of $m(t)$ for values of t other than those he has used in his experiment. Under these circumstances, he is using the function m and its graph as a mathematical model of a physical phenomenon, and he will continue to use this model to describe the particular phenomenon as long as the physical results continue to approximate his predictions from the mathematical model.

To return to our first example, our intuition tells us that the amount of radioactive material decreases steadily rather than instantaneously decreasing to one-half the original amount at the end of a 24-hour period.

Similarly, for the second example, we sense that the biological culture doesn't suddenly double at the end of each day, but that it increases steadily throughout the day.

For the first example, we might, on the basis of our equation $W(t) = W_0(\frac{1}{2})^t$, predict that the amount of radioactive material left at the end of $\frac{1}{2}$ day would be $W_0(\frac{1}{2})^{\frac{1}{2}}$ and that the amount left at the end of $1\frac{1}{2}$ days would be $W_0(\frac{1}{2})^{\frac{3}{2}}$.

The foregoing discussion leads us to expect that the function W, first defined on page 295, would serve as a good model for radioactive decay, if we extend the function by letting its domain be the set of rational numbers. That is, we assume that the function W defined by

$$W(t) = W_0(\tfrac{1}{2})^t, \quad \text{where } t \text{ is a rational number,}$$

would describe the process of radioactive decay in this example.

Experimental results in an actual situation would verify that, for a radioactive substance, the above formula is a good mathematical model if the unit of time t is the half-life of the substance. [Notice that if $t = 1$—that is, in one half-life—the equation yields $W(1) = W_0(\frac{1}{2})^1 = \dfrac{W_0}{2}$. This result is correct, of course, because the substance loses $\frac{1}{2}$ its original weight W_0 in one half-life.]

Similarly, the following formula for the biological culture (see page 296),

$$w(t) = w_0 2^t, \quad \text{where } t \text{ is a rational number,}$$

is a good mathematical model, over certain time intervals, for the growth of the culture.

We might now think of generalizing the results for these examples in two ways. First, when we graph a great many points $(t, W(t))$ of the function $W(t) = W_0(\frac{1}{2})^t$, where t is rational and where W_0 is some given parameter, we obtain a graph like the one shown in Figure 8-4. The shape of this graph suggests that, if the domain of W were extended to the set of all *real* numbers, then the graph of this extended function would be the graph of a continuous function. This is the case, as we shall presently see.

Figure 8-4

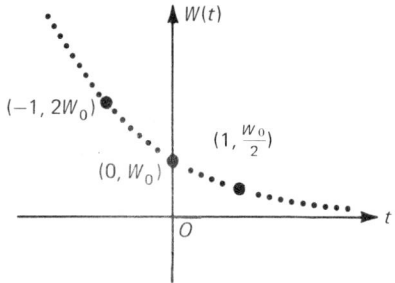

Even though the radioactive decay of a substance is a discrete process, the continuous function defined by

$$W(t) = W_0(\tfrac{1}{2})^t, \quad \text{with } t \in R,$$

serves as a useful model for the prediction of radioactive decay. The disintegration of radioactive substances is discrete rather than continuous because it results from the disintegration of individual atoms as they emit electrons or their helium nucleus. Similarly, the growth of a culture of bacteria (one-celled microorganisms), as described in the second example, is a discrete process that is accomplished by the splitting of one bacterium into two bacteria.

Of course, if we want function W to be defined for all real numbers, we are faced with the problem of giving meaning to an expression like $(\tfrac{1}{2})^{t_0}$, where t_0 is an arbitrary real number. For example, what should be the meaning of the expression $(\tfrac{1}{2})^{\sqrt{2}}$? The discussion in later sections will provide us with a solution to this problem. First, however, we will review and extend your knowledge of integral and rational exponents.

There is a second way in which the functions defined by

$$W(t) = W_0(\tfrac{1}{2})^t \quad \text{and} \quad w(t) = w_0(2)^t$$

might be generalized. For these two examples, the amount of substance present at the end of a time interval of one unit was determined by multiplying the amount present at the beginning of the unit of time by a multiplying factor of $\tfrac{1}{2}$ in one case and 2 in the other. It seems reasonable to expect that, for other examples of natural processes which obey the law of growth and decay, the multiplying factor would be different from $\tfrac{1}{2}$ or 2.

In fact, this multiplying factor might be any positive real number a. This leads us to consider functions defined by equations of the form

$$F(t) = Wa^t, \quad \text{where } t \in R,$$

and where a is any positive real number. Such functions are called *exponential functions*, and we will consider the properties of such functions in the sections of this chapter that follow the material on integral, rational, and real exponents.

8.1 EXERCISES

Suppose that, at the beginning of an experiment, there are 800 milligrams of a radioactive substance whose half-life is 1 day. Solve each of the following problems.

1. What amount of this material will be left 3 days later? 5 days later? 1 week later?
2. What is the ratio of the amount that will be left after 8 days to the amount left after 4 days?

3. What amount of the material was present 3 days before the experiment began?
4. How many days after the beginning of the experiment will the amount remaining be less than 1% of the amount present at the beginning of the experiment?

Consider an experiment involving a weight W_0 of a radioactive substance whose half-life is 2 days. Solve each of the following problems.

5. What will be the weight of the material left 4 days later? 8 days later? 12 days later?
6. Write an equation which expresses the weight $W(t)$ of the material that will be left after t days.
7. The amount left at the end of $2n$ days is how many times as great as the amount left at the end of $2n + 6$ days?
8. If 200 milligrams of the material are left at the end of 2 weeks, after how many days were 1600 milligrams present?

Consider an experiment in which there were 1.6 milligrams of a radioactive substance present at the end of 3 hours and 0.2 milligram left at the end of 4 hours and 30 minutes.

9. What is the half-life of the substance?
10. How many milligrams were present at the end of 4 hours? At the end of 2 hours? At the beginning of the experiment?
11. After how many hours of the experiment were there 25.6 milligrams present?

Suppose that the number of bacteria in a certain culture doubles every 12 hours. Solve each of the following problems.

12. If the number present at the beginning of an experiment is N_0, what is the number present after 24 hours? After 36 hours? After 3 days?
13. Write an equation which expresses the number of bacteria, $N(t)$, present after t hours.
14. What is the ratio of the number of bacteria present 2 days after the experiment began to the number present 1 day before it began?

Suppose that the number of motor vehicles registered in a particular state is roughly doubled every n years. Solve Problems 15 and 16.

15. If there were 200,000 vehicles registered in 1930 and 1,600,000 registered in 1960, find the value of n.
16. If this rate of growth continues, how many motor vehicles will be registered in 1980? In 1990? In 2000?
17. If an amount of P dollars is deposited in a bank and draws interest at the nominal rate of r per cent per year, compounded n times per

year, the amount (including interest) on deposit at the end of t years is given by

$$A(t) = P\left(1 + \frac{0.01r}{n}\right)^{nt}.$$

Suppose that \$1000 is deposited and draws interest at the nominal rate of 4% per year. Find the amount on deposit at the end of 1 year if the interest is compounded in each of the following ways.
    a) Annually   b) Semiannually   c) Quarterly

18. Suppose that there are two cultures of bacteria, both of which contain N_0 bacteria at a given time t. If the number of bacteria in one culture triples every 8 hours and the number of bacteria in the other doubles every $4\frac{4}{5}$ hours, which culture will contain more bacteria at the end of 24 hours?

8.2 Rational Exponents

As you already know from earlier mathematics courses, if a is a real number and n is a positive integer, then a^n denotes the product of a with itself n times. That is,

$$a^n = \underbrace{a \cdot a \cdot a \cdot \ldots \cdot a}_{n \text{ factors}}.$$

The number a is called the **base** and the integer n, the **exponent** in this expression.

Using this definition of a^n and the laws of associativity and commutativity for multiplication, we can verify the following identities, where a and b are real numbers and n and m are positive integers.

    I) $a^m \cdot a^n = a^{m+n}$.
    II) $(a^m)^n = a^{nm}$.
    III) $(ab)^n = a^n \cdot b^n$.

These identities are valid for $a \in R$ and $m, n \in I^+$.*

For any given $a \in R$, we can think of the definition of a^n as defining a function on the set of positive integers into the set of real numbers. We will denote this function by $\exp_a$ (read "exponential to the base a"). Thus,

    A) $\exp_a(n) = a^n$,   where $n \in I^+$.

Usually, we will omit the parentheses in the expression $\exp_a(n)$ and simply write $\exp_a n$. For any given real number a, the function $\exp_a$ is the set of ordered pairs of the form (n, a^n), where n is an integer. Hence, equa-

* The symbol I denotes the set of all integers and I^+ the set of positive integers.

tion A can be thought of as defining a family of functions, with a different function obtained for each replacement of the parameter a.

Notice that Formula I above expresses the following property of the function $\exp_a$.

$$\text{I) } \exp_a m \cdot \exp_a n = \exp_a (m + n).$$

Formulas II and III express certain relationships between members of the one-parameter family of functions defined by equation A, where the parameter a has R as its domain. These relationships are expressed in functional notation below.

$$\text{II) } \exp_{(\exp_a m)} n = \exp_a (m \cdot n)$$
$$\text{III) } \exp_{ab} n = \exp_a n \cdot \exp_b n.$$

To extend each member of the family of functions defined by equation A to a function having the set of integers for its domain could be done in many ways. The extensions of these functions that we now make will be such that they continue to satisfy Formulas I, II, and III, where n is any integer. When we extend these functions, it is necessary to stipulate that a be a nonzero real number. Definition 8.2b on page 303 makes clear the need for this restriction.

It is not difficult to see that the requirement that Formulas I, II, and III be satisfied completely determines such an extension of $\exp_a$, where $a \neq 0$. For example, if m is a positive integer and $n = 0$, Formula I gives

$$\exp_a m \cdot \exp_a 0 = \exp_a (m + 0) = \exp_a m,$$

or

$$a^m \cdot a^0 = a^{m+0} = a^m, \quad \text{with } a \neq 0.$$

Thus, we must define $a^0 = \exp_a 0 = 1$.

Definition 8.2a If $a \neq 0$ and $a \in R$, then

$$a^0 = 1.$$

We also see that if n is a positive integer and a is a nonzero real number then, by Formula I, $\exp_a n \cdot \exp_a (-n) = \exp_a (n + (-n)) = \exp_a 0 = 1$. For this to be the case, it must be true that

$$a^n \cdot a^{-n} = 1, \quad \text{or } a^{-n} = \frac{1}{a^n}, \quad \text{with } a \neq 0.*$$

* Because the notation occasionally becomes rather cumbersome in this chapter and the next, we will *sometimes* write quotients horizontally, as $1/a^n$, rather than vertically, as $\frac{1}{a^n}$.

Hence, we have the following definition for negative integral exponents.

Definition 8.2b If $a \in R$, $a \neq 0$, and n is a positive integer, then

$$a^{-n} = \frac{1}{a^n}.$$

With these definitions of a^0 and a^{-n}, where n is a positive integer, we have defined an extension of $\exp_a$, with $a \neq 0$. This extension is defined on the set of all integers.

$$\exp_a n = a^n, \qquad \text{where } a \neq 0 \text{ and } n \in I.$$

We leave it for you to verify that Formulas I, II, and III are satisfied under this extension provided that $a \neq 0$ and $m, n \in I$. A specific example is given for Formula II to help convince you that the formulas do apply to the new definitions.

Example 1 In Formula II on page 301, let $m = -2$ and $n = -3$; by Definition 8.2b, we have $(a^{-2})^{-3} = \dfrac{1}{(a^{-2})^3} = \dfrac{1}{(1/a^2)^3} = \dfrac{1}{1/(a^2)^3} = \dfrac{1}{1/a^6} = a^6 = a^{(-2)(-3)}$.

8.2a EXERCISES

Evaluate each of the following expressions.

1. 21^0
2. 4^{-1}
3. $(\frac{3}{4})^{-2}$
4. $\dfrac{1}{5^{-3}}$
5. $(2^{-3})^{-2}$
6. $2^{-4} \cdot 4^2$
7. $\dfrac{3^{-2}}{6^{-2}}$
8. $(14^{-1} \cdot 7^{-2})^0$
9. $10^{-5} \cdot 10^{-3}$
10. 6.43×10^0
11. 7.19×10^4
12. 3.49×10^{-3}

Write each of the following as an equivalent expression in which all the exponents are positive. (Assume that $xy \neq 0$.)

13. $5x^{-3}$
14. $x^{-2}y^3$
15. $(5x)^{-3}$

16. $5x^{-1}y^2$

17. $\dfrac{5}{x^{-2}}$

18. $\dfrac{x^{-2}}{y^{-3}}$

19. $\dfrac{x^{-4}y}{xy^{-3}}$

20. $\dfrac{3x^{-4}}{5y^3}$

Find the set of solutions for each of the following conditions.

21. $x^{-2} = 9$.
22. $x^{-3} = \frac{1}{27}$.
23. $x^{-1} < 5$.
24. $x^{-2} \geq \frac{1}{4}$.

Scientists frequently use a method for expressing numbers that is called **scientific notation.** *In scientific notation, numbers are expressed in the form*

$$a \times 10^n$$

where $1 \leq |a| < 10$ and $n \in I$. For example, 43,000,000 is written as 4.3×10^7 and 0.0012 is written as 1.2×10^{-3}. Express each of the following numbers in scientific notation.

25. 75.3
26. 4.56
27. 0.146
28. 0.00055
29. 712,000
30. 6170
31. 0.0245
32. 0.000043

Exercises 33, 34, and 35 concern the verification that Formulas I, II, and III are satisfied by $\exp_a$, where $a \neq 0$, defined on the set of all integers.

33. Let a be a nonzero real number, and let m and n be positive integers. Prove each of the following theorems.

a) If $m > n$, then $\dfrac{a^m}{a^n} = a^{m-n}$. $\left(\text{Hint: } \dfrac{a^m}{a^n} = a^x \text{ if and only if } a^n \cdot a^x = a^m.\right)$

b) If $m < n$, then $\dfrac{a^m}{a^n} = a^{m-n}$. $\left(\text{Hint: First prove that } \dfrac{a^m}{a^n} = \dfrac{1}{a^{n-m}}.\right)$

34. Let a and b be real numbers, with $ab \neq 0$, and let m and n be integers. Prove each of the following theorems.
   a) If m or n is negative, $a^m \cdot a^n = a^{m+n}$. (Hint: Use Exercise 33.)
   b) If m or n is negative, $(a^m)^n = a^{mn}$.
   c) If n is negative, $(ab)^n = a^n b^n$.

35. If $ab \neq 0$ and m or n is zero, verify each of the following.
   a) $a^m \cdot a^n = a^{m+n}$.   b) $(a^m)^n = a^{mn}$.   c) $(ab)^n = a^n b^n$.

The next step is to extend the definition of the function $\exp_a$ so that its domain is the set of all *rational* numbers. To begin with, let us remember that we want this extension to be such that Formulas I, II, and III on page 301 are satisfied. We note that if p and q are integers, with $q \neq 0$, since $\frac{p}{q} = \left(\frac{1}{q}\right)p$, it follows that $\exp_a \frac{p}{q} = a^{p/q}$ must be equal to $(a^{1/q})^p$ so that Formula II still holds. Let us, therefore, consider what must be meant by

$$\exp_a \left(\frac{1}{q}\right) = a^{1/q}, \quad \text{with } q \in I^+.$$

Again, since Formula II must be satisfied and since $\left(\frac{1}{q}\right)q = 1$, we have

$$(a^{1/q})^q = a^1 = a.$$

But this means that $a^{1/q}$ is a number that, when raised to the qth power equals a. In other words, $a^{1/q}$ must be a qth root of a. We know that, if a is negative, the qth roots for certain values of q will be nonreal numbers. For example, if $a = -1$ and $q = 2$, the only square roots of -1 are the imaginary numbers i and $-i$. To satisfy Formula II, therefore, we would be required to define $(-1)^{1/2}$ as either i or $-i$. Since we want $\exp_a$ to be a function whose range is in R, we will limit ourselves to the functions $\exp_a$, where $a > 0$. (Of course, when a is negative, the qth roots of a will be real for certain values of q. For example, the fifth root of $a = -1$ is -1. To simplify the discussion, however, we will restrict ourselves to $a > 0$.)

We are still faced with the problem of defining what we mean by $a^{1/q}$, where $a > 0$ and q is a nonzero integer. Since any rational number can be written in the form $\frac{p}{q}$, where q is a positive integer, we will confine ourselves to the case where $a > 0$ and $q \in I^+$. It will then follow by Formula II that, if s is a negative integer, for example, if $s = -3$, then

$$a^{1/s} = a^{1/-3} = a^{(1/3)(-1)} = (a^{1/3})^{-1} = 1/a^{1/3}.$$

Hence, if we can define $a^{1/q}$, for q a positive integer, then $a^{1/s}$, where s is a negative integer, will be uniquely defined.

Even with these restrictions, we still have some problems. If $a > 0$ and $q \in I^+$, what will we mean by $a^{1/q}$? We have already noted that $a^{1/q}$ must be a qth root of a. That is, we must have $(a^{1/q})^q = a$. Another way of saying this is to say that $a^{1/q}$ must be a zero of the polynomial

$$x^q - a.$$

We know by Theorem 5.4b, page 218, which is a consequence of the Fundamental Theorem of Algebra, that a polynomial of degree q will have q zeros (real and complex) if multiplicities are counted. Which one of these numbers shall we choose as the defined value of $a^{1/q}$? We have already stipulated that the qth root of a must be a real number because we want the range of the function $\exp_a$ to be included in R. But even in as simple a case as $a = 1$ and $q = 2$, we see that there are two real numbers 1 and -1 which could be used as the value of $1^{1/2}$, since both 1 and -1 are zeros of $x^2 - 1$.

The property of polynomials expressed in the following theorem helps us out of our difficulties.

Theorem 8.2a *If $a > 0$ and q is a positive integer, then there is one and only one positive real number that is a zero of the polynomial*

$$x^q - a.$$

Assuming this theorem for the moment, we can then give the following definition of $a^{1/q}$.

Definition 8.2c *If $a \in R$, $a > 0$, and q is a positive integer, then $a^{1/q}$ is defined to be the unique positive zero of the polynomial $x^q - a$.*

The proof of Theorem 8.2a follows from the Intermediate Value Theorem (page 171) and the fact that the polynomial function f defined by $f(x) = x^q - a$, with $a > 0$, is a strictly increasing function for positive x. An outline of this proof is now given.

First, $f(0) = -a$, which is less than 0. Second, $f(x) > 0$ for x sufficiently large, say $x = a + 1$. Therefore, by the Intermediate Value Theorem, $f(x)$ must be zero for some positive real number b between 0 and $a + 1$. Now function f is strictly increasing for positive x, because the slope function of f, defined by $f'(x) = qx^{q-1}$, is greater than zero whenever $x > 0$. Hence, $f(x)$ can equal zero for at most one positive x, which means that $f(x)$ is zero for one and only one positive real number.

The value of $a^{1/q}$ given in Definition 8.2c is called the **principal qth root of a**, and is sometimes denoted by $\sqrt[q]{a}$.

There are $q - 1$ other distinct complex qth roots of a; that is, there are $q - 1$ complex numbers which, raised to the qth power, are equal to a.

Theorem 8.2a has shown us that there is precisely one positive real number with this property, and we have defined it as the *principal qth root of a*. (You will often see references simply to the qth root of a when the *principal qth root of a* is meant.)

8.2b EXERCISES

Evaluate each of the following.

1. $25^{1/2}$
2. $64^{1/3}$
3. $16^{0.25}$
4. $243^{0.2}$
5. $32^{-1/5}$
6. $64^{-0.5}$
7. $125^{-1/3}$
8. $81^{-1/4}$
9. $(\frac{4}{9})^{1/2}$
10. $(1\frac{17}{64})^{-1/2}$
11. $3.24^{0.5}$
12. $0.064^{-1/3}$

Verify that $(ab)^{1/q} = a^{1/q} \cdot b^{1/q}$ *in each of the following cases.*

13. $a = 9; b = 16; q = 2.$
14. $a = 8; b = 27; q = 3.$
15. $a = \frac{1}{4}; b = \frac{1}{36}; q = -2.$
16. $a = 5\frac{4}{9}; b = 1\frac{11}{25}; q = 2.$
17. $a = 0.125; b = 216; q = -3.$

Find the set of solutions for each of the following conditions.

18. $x^{1/2} = 0.4.$
19. $x^{-1/2} = 2.$
20. $x^{1/3} \geq 3.$
21. $x^{-1/5} < 2.$
22. By definition, if $a > 0$ and q is a positive integer, then $(a^{1/q})^q = a$. Prove that if $a > 0$ and q is a positive integer, then $(a^q)^{1/q} = a$. (Hint: Consider the polynomial $x^q - a^q$.)

8.3 More about Rational Exponents

We return now to the problem of extending the domain of the function $\exp_a$, $a > 0$, to the set of all rational numbers, so that Formulas I, II, and

III of section 8.2 are satisfied. It is clear from Theorem 8.2a that we can uniquely define $\exp_a \frac{1}{q}$ by

$$\exp_a \frac{1}{q} = a^{1/q},$$

where $\frac{1}{q}$ is the principal qth root of a.

If Formula II is to be satisfied, then it will be the case that

$$a^{p/q} = a^{p \cdot (1/q)} = (a^p)^{1/q} = (a^{1/q})^p,$$

where p is an integer. Hence, we must define $a^{p/q}$ as follows:

$$a^{p/q} = (a^{1/q})^p = (a^p)^{1/q},$$

Keep in mind that q must be a positive integer; but any rational number can be written with a positive denominator, so there is no problem. Furthermore, the expressions $(a^{1/q})^p$ and $(a^p)^{1/q}$ must be equal, as we now demonstrate. From Formula II as it applies to integers and from commutativity and associativity of multiplication, we have

$$((a^{1/q})^p)^q = (a^{1/q})^{pq} = (a^{1/q})^{qp} = ((a^{1/q})^q)^p = a^p.$$

But because $(a^{1/q})^p > 0$, the equality $((a^{1/q})^p)^q = a^p$ means that $(a^{1/q})^p$ is the principal qth root of a^p; that is, $(a^{1/q})^p = (a^p)^{1/q}$. Hence, we have the following definition.

Definition 8.3a If $a \in R$ and $a > 0$, then

$$a^{p/q} = (a^{1/q})^p = (a^p)^{1/q},$$

where q is a positive integer and p is any integer.

With this definition of $\exp_a \frac{p}{q}$ we have extended the function $\exp_a$ to one that is defined on all the rationals. The question now is whether this extension continues to satisfy Formulas I, II, and III when the exponents are rational numbers.

To verify that these three formulas do hold for rational exponents is simply a matter of computation and using some of the properties which we have derived in this section. We now restate Formulas I, II, and III for rational exponents and then show that Formula I is satisfied under this extension. You are asked to verify Formulas II and III in the exercises on page 310.

Theorem 8.3a *If a and b are positive real numbers and r_1 and r_2 are rational numbers, then*

I) $a^{r_1} \cdot a^{r_2} = a^{r_1+r_2}$;

II) $(a^{r_1})^{r_2} = a^{r_1 r_2}$;

III) $(ab)^{r_1} = a^{r_1} b^{r_1}$.

Proof To verify that $a^{r_1} \cdot a^{r_2} = a^{r_1+r_2}$, we let $r_1 = \dfrac{p_1}{q_1}$ and $r_2 = \dfrac{p_2}{q_2}$. Then

$$a^{p_1/q_1} \cdot a^{p_2/q_2} = a^{p_1 q_2/q_1 q_2} \cdot a^{p_2 q_1/q_1 q_2}$$
$$= (a^{1/q_1 q_2})^{p_1 q_2} \cdot (a^{1/q_1 q_2})^{p_2 q_1}$$
$$= (a^{1/q_1 q_2})^{(p_1 q_2 + p_2 q_1)}$$
$$= a^{(p_1 q_2 + p_2 q_1)/q_1 q_2}$$
$$= a^{p_1/q_1 + p_2/q_2}.$$

You should be able to justify each of the successive equalities in the preceding chain. If we equate the first and last steps of this chain, we obtain

$$a^{p_1/q_1} \cdot a^{p_2/q_2} = a^{p_1/q_1 + p_2/q_2},$$

which is equivalent to

$$a^{r_1} \cdot a^{r_2} = a^{r_1+r_2}.$$

This completes the proof of Part I of the theorem. □

Up to this point, we have shown that there is a natural extension of the function $\exp_a$ with $a > 0$, whose domain is the set of integers, to a function whose domain is the set of rationals. Out next goal is to extend this function to one having R for its domain.

Before we are able to make this further extension, we need another theorem.

Theorem 8.3b *The function $\exp_a$ defined on the rationals is a strictly increasing function if $a > 1$ and a strictly decreasing function if $0 < a < 1$. If $a = 1$, then $\exp_a$ is the constant function f defined by $f(x) = 1$, where x is a rational number.*

The empirical evidence gained by plotting portions of the graphs of such functions should convince you of the validity of this theorem, and we will not give a formal proof. (Although we have all the properties needed to prove this theorem, the proof is long and involved and would interrupt our development.) Partial graphs of the functions $\exp_2$ and

$\exp_{\frac{1}{2}}$, which correspond to the examples discussed earlier, will resemble the graphs in Figures 8-5 and 8-6, respectively.

Figure 8-5

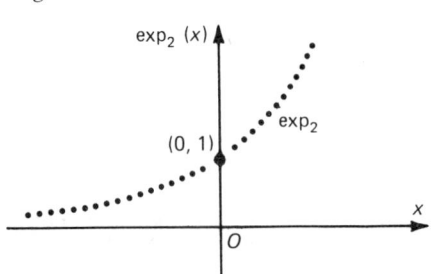

Figure 8-6

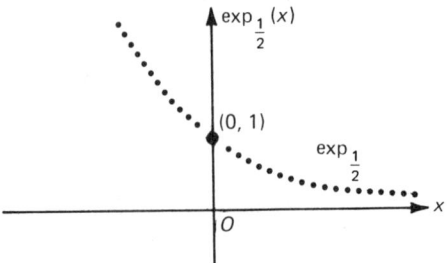

8.3 EXERCISES

Evaluate each of the following.

1. $25^{3/2}$
2. $16^{3/4}$
3. $64^{3/2}$
4. $625^{3/4}$
5. $0.16^{5/2}$
6. $0.027^{2/3}$
7. $0.0001^{3/4}$
8. $3(\frac{1}{9})^{3/2}$
9. $\frac{1}{2}(16)^{3/2}$
10. $49^{-3/2}$
11. $0.008^{-4/3}$
12. $0.01^{5/2}$

13. a) Indicate the points on the graph of $\exp_2$ with abscissas which are multiples of $\frac{1}{2}$ in the interval $-2 \leq x \leq 2$ (that is, for $x = -2$, $-1\frac{1}{2}$, -1, and so on).
    b) On the same set of axes used in part a, indicate the points on the graph of $\exp_4$ with abscissas corresponding to those used in part a.
14. Repeat Exercise 13 for $\exp_{1/2}$ and $\exp_{1/4}$.
15. Prove that the graph of $\exp_{1/a}$, where $a > 1$, is the reflection in the y-axis of the graph of $\exp_a$.
16. Let a and b be positive real numbers and let r_1 and r_2 be rational numbers, Verify each of the following statements.
    a) $(a^{r_1})^{r_2} = a^{r_1 r_2}$.   b) $(ab)^{r_1} = a^{r_1} \cdot b^{r_1}$.
17. Prove each of the following theorems:
    a) If $a > 1$, the function $\exp_a$ defined on the integers is strictly increasing. That is, prove that, if $a > 1$ and m and n are integers such that $m > n$, then $a^m > a^n$.

b) If $0 < a < 1$, the function $\exp_a$ defined on the integers is decreasing.
18. Develop an indirect proof for each of the following theorems:
   a) If $a > 1$, the function $\exp_a$ defined on the rational numbers is strictly increasing. (Hint: Assume that $a > 1$ and $p_1/q_1 > p_2/q_2$, where p_1, q_1, p_2, q_2 are integers and q_1 and q_2 are positive. Then show that $a^{p_1/q_1} \leq a^{p_2/q_2}$ leads to a contradiction of the theorem given in Exercise 17a.)
   b) If $0 < a < 1$, the function $\exp_a$ defined on the rational numbers is decreasing.

8.4 The Function $\exp_a$ Defined on R

To extend the domain of $\exp_a$ to R, we will make use of Theorem 8.3b and certain properties of sequences discussed in Chapter 1. Let us consider a specific case.

You know that $\sqrt{2}$ corresponds to a nonterminating, nonrepeating decimal. The decimal expansion of $\sqrt{2}$ to 19 places is

$$1.4142135623730950488 \ldots$$

Now consider the sequence $\langle r_n \rangle$ whose terms

$$1, 1.4, 1.41, 1.414, 1.4142, 1.41421, \ldots$$

are the successive truncations of the decimal expansion of $\sqrt{2}$. Given any positive ϵ, we can find a positive integer N such that if $n > N$, then $|r_n - \sqrt{2}| < \epsilon$, so $\lim_{n \to \infty} r_n = \sqrt{2}$.

Now if $a \in R$ and $a \neq 1$, the sequence $\langle a^{r_n} \rangle$, whose terms are

$$a^1, a^{1.4}, a^{1.41}, a^{1.414}, a^{1.4142}, a^{1.41421}, \ldots$$

is a restriction of the function $\exp_a$ defined on the rationals. If $a > 1$, $\exp_a$ is an increasing function and, correspondingly, $\langle a^{r_n} \rangle$ is an increasing sequence. Further, $a^{r_n} < a^2$ for all n, so the sequence is bounded above. Hence, by Theorem 1.8a, $\langle a^{r_n} \rangle$ converges to its least upper bound. Since $r_n \to \sqrt{2}$, we define this least upper bound to be $a^{\sqrt{2}}$.

Although we did not prove it in Chapter 1, any nonincreasing sequence bounded below converges to its greatest lower bound. Now, if $0 < a < 1$, the sequence $\langle a^{r_n} \rangle$ is a restriction of the decreasing function $\exp_a$ and hence is nonincreasing. Further, $a^{r_n} > 0$ for all n, so the sequence is bounded below and hence converges to its greatest lower bound, which we define to be $a^{\sqrt{2}}$.

If $a = 1$, the sequence $\langle a^{r_n} \rangle = \langle 1 \rangle$, so $\langle a^{r_n} \rangle$ converges to its least upper bound, 1. Hence we define $1^{\sqrt{2}}$ to be 1.

Every real number x corresponds to some nonterminating decimal such that the successive truncations of this decimal

$$r_1, r_2, r_3, \ldots, r_n, \ldots$$

form an increasing sequence that converges to x.*

If $a \geq 1$, the sequence $\langle a^{r_n} \rangle$ is a nondecreasing sequence that converges to its least upper bound. If $0 < a < 1$, $\langle a^{r_n} \rangle$ is a nonincreasing sequence that converges to its greatest lower bound. In either case, we define this bound to be a^x.

Definition 8.4a If $a \in R$, $x \in R$, and $\langle r_n \rangle$ is an increasing sequence of rational numbers that converges to x, then

$$\exp_a x = \lim_{n \to \infty} a^{r_n}.$$

This definition does define a function on R that is decreasing when $0 < a < 1$ and increasing when $a > 1$. We will accept without proof that the function $\exp_a$ with domain R is a continuous function and that this extension of $\exp_a$ satisfies the formulas for exponents which were given for positive integral exponents on page 301 and for rational exponents on page 309.

We now summarize what we have learned in the last several sections.

1. If a is a positive real number, then there is a function $\exp_a$, defined by $\exp_a x = a^x$, and whose domain is R, that is an extension of the function $\exp_a$, defined by $\exp_a n = a^n$, whose domain is the set of integers.
2. This extension $\exp_a$ is a continuous, increasing function if $a > 1$ and a continuous, decreasing function if $0 < a < 1$.
3. If x and y are any real numbers and a and b are positive real numbers, then the function $\exp_a x = a^x$ satisfies the following formulas:

   I) $a^x \cdot a^y = a^{x+y}$.
   II) $(a^x)^y = a^{xy}$.
   III) $(ab)^x = a^x b^x$.

Since the base a is restricted to positive real numbers, another obvious property is that $\exp_a x > 0$ for all $x \in R$. Furthermore, if $a > 0$ and $a \neq 1$, the range of $\exp_a$ is R^+, the set of all positive reals. If $a = 1$, the range of $\exp_a$ is, of course, $\{1\}$. Thus, in addition to $\exp_a$ being continuous, strictly increasing for $a > 1$, and strictly decreasing for $0 < a < 1$, the range of $\exp_a$ is not bounded above; and 0 is the greatest lower bound of the range of the function. We state the following theorem without proof.

* The decimal representation of a real number is not necessarily unique. To satisfy this statement in the case of $\frac{1}{2}$, for example, we would use the decimal .49999 . . . rather than the customary .5000 . . . .

Theorem 8.4a *If a is a positive real number and $a \neq 1$, then $\exp_a x$, where $x \in R$, is not bounded above, and the greatest lower bound for $\exp_a$ is zero.*

8.4 EXERCISES

1. Evaluate each of the following.
    a) $3^{1.7} \cdot 3^{2.3}$
    b) $(4^\pi)^{2/\pi}$
    c) $(10^{-\sqrt{2}})^{-3\sqrt{2}}$
    d) $(\sqrt{3})^{3/4} \cdot (\sqrt{3})^{5/4}$

Indicate whether each of the following statements is true or false and justify your answer.

2. $3^{3\sqrt{2}} > 2^{\sqrt{3}}$.
3. $2^{\pi\sqrt{3}} > 2 \cdot 2^{\sqrt{3}}$.
4. $(\frac{3}{4})^{2\sqrt{5}} > (\frac{3}{4})^{\sqrt{19}}$.
5. $(0.5)^\pi > (0.5)^{22/7}$.

The fact that $\exp_a$, where $a > 1$, is strictly increasing and $\exp_a$, where $0 < a < 1$, is strictly decreasing enables us to make the following statements:

> If $a > 1$, then $a^x < a^y$ if and only if $x < y$.
> If $0 < a < 1$, then $a^x < a^y$ if and only if $x > y$.

Use this information to solve each of the following inequalities.

6. $2^{2x} < 2^{5x-4}$.
7. $(0.24)^x > (0.24)^{1-x}$.
8. $10^{x+5} > 100^x$.
9. $9^{2x} < 3^{5x-1}$.
10. $2^{1-x} > 8^x$.
11. $125^{3x} > 25^{2x+1}$.
12. a) Show that, if a is a positive number, then $\exp_a x = \exp_{1/a}(-x)$.
    b) Use part a to prove that, if $0 < a < 1$, then $\exp_a$ defined on the real numbers is not bounded above and has 0 as its greatest lower bound.

8.5 The Graph of $\exp_a$

The drawing of some reasonably good characterizations of the graphs of functions of the form $\exp_a$ is simplified if we first investigate the composite function $f \circ l_a$, where $l_a(x) = ax$, which is defined by

$$(f \circ l_a)(x) = f(ax), \quad \text{for } x \in R.$$

Of course, if $a = 1$, then $f(ax) = f(x)$, and the graph of $f \circ l_1$ is the same as the graph of f. Now consider all cases where $a \neq 1$. We see that

$f(ax)$ will take on all the values that $f(x)$ takes on as x ranges over the set of real numbers. Therefore, the ranges of f and $f \circ l_a$ are the same. The difference is that $f(ax)$ takes on the specific value $f(x_0)$ when $x = \dfrac{x_0}{a}$ rather than when $x = x_0$, as $f(x)$ does.

Example 1 Let us consider the special case where $a = 2$ and see what we can determine about the graph of $f \circ l_2$, which is defined by $(f \circ l_2)(x) = f(2x)$. It follows that

$$(f \circ l_2)(0) = f(2 \cdot 0) = f(0);$$
$$(f \circ l_2)(1) = f(2 \cdot 1) = f(2);$$
$$(f \circ l_2)(2) = f(2 \cdot 2) = f(4);$$
$$(f \circ l_2)(-1) = f(2 \cdot -1) = f(-2);$$

and so on. Thus, we can see that $(f \circ l_2)(x)$ takes on the same values as $f(x)$ but that each occurs at *one half* the distance from the y-axis on the graph of $f \circ l_2$ as on the graph of f. Hence, the graph of $f \circ l_2$ is the graph of f compressed *towards* the y-axis. Figure 8-7 illustrates this situation.

Figure 8-7

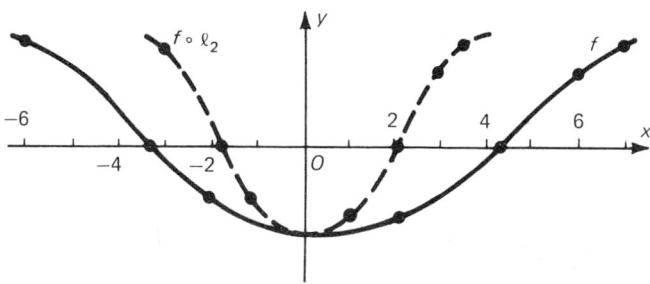

Example 2 Consider the case where $a = \frac{1}{2}$ so that $(f \circ l_{\frac{1}{2}})(x) = f(\frac{1}{2}x)$. From values such as

$$(f \circ l_{\frac{1}{2}})(2) = f(\tfrac{1}{2} \cdot 2) = f(1),$$
$$(f \circ l_{\frac{1}{2}})(-2) = f(\tfrac{1}{2} \cdot -2) = f(-1),$$

and so on, it should be clear that $(f \circ l_{\frac{1}{2}})(x)$ takes on the same values as $f(x)$, but that each occurs on the graph of $f \circ l_{\frac{1}{2}}$ at *twice* the distance from the y-axis as on the graph of f. Thus, the graph of $f \circ l_{\frac{1}{2}}$ is the graph of f stretched away from the y-axis. The graph in Figure 8-8 illustrates this situation.

Figure 8-8

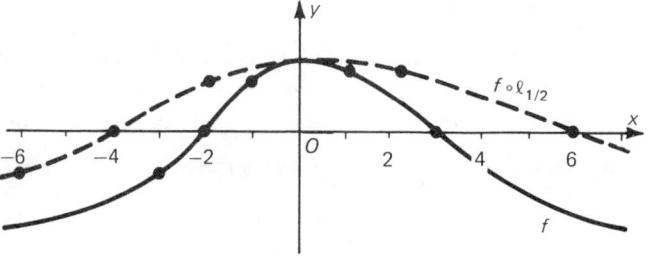

In general, corresponding to every point in the graph of f having coordinates $(x_0, f(x_0))$, there is a point in the graph of $f \circ l_a$, defined by $(f \circ l_a)(x) = f(ax)$, having coordinates $\left(\dfrac{x_0}{a}, f(x_0)\right)$, and conversely.

Now, given a point in the plane having coordinates (x_0, y_0), what is the effect of multiplying the abscissa by a positive constant, $\dfrac{1}{a} = c$? Well, (cx_0, y_0) is a point in the horizontal line $h(x) = y_0$, just as (x_0, y_0) is. If $c > 1$ and $x_0 > 0$, then (cx_0, y_0) is c times as far to the right of the y-axis as is the point (x_0, y_0). If $c > 1$ and $x_1 < 0$, then (cx_1, y_0) is c times as far to the left of the y-axis as is (x_1, y_0). We say that the point (cx_0, y_0) is a *horizontal* "shifting" of the point (x_0, y_0). These two cases are illustrated in Figure 8-9.

Figure 8-9

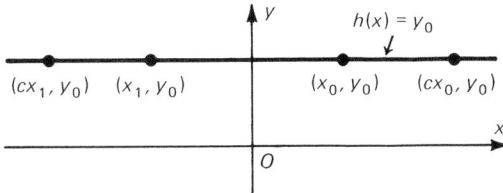

Similarly, if $0 < c < 1$ and $x_0 > 0$, then (cx_0, y_0) is c times as far from the y-axis as is (x_0, y_0) and is a shifting to the left (towards the y-axis) of the point (x_0, y_0). If $0 < c < 1$ and $x_1 < 0$, then (cx_1, y_0) is a shifting to the right (towards the y-axis) of the point (x_1, y_0). These two cases are illustrated in Figure 8-10.

Figure 8-10

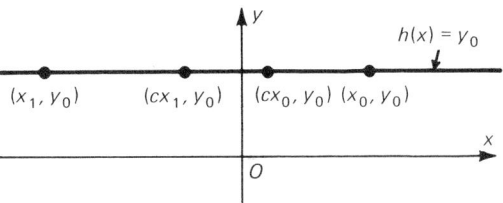

Notice that, in Example 1 on page 314, $a = 2$ so that $c = \dfrac{1}{a}$ is less than 1. In Example 2, $a = \tfrac{1}{2}$ so that $c > 1$. Hence, the results for those examples agree with this general discussion for any positive real number a.

Now consider what happens to the graph of a function f if *every* point (x_0, y_0) in its graph is shifted to the point (cx_0, y_0), where $c > 0$. From what we have just observed about individual points in the graph, it should be apparent that, if $c > 1$, the graph of f is compressed about the y-axis. Also, if $0 < c < 1$, the graph of f is stretched about the y-axis.

We can summarize the cases when $a > 0$ as follows: If the graph of f consists of the points having coordinates (x, y), then the graph of $f \circ l_a$ consists of the points having coordinates $\left(\dfrac{x}{a}, y\right)$ and the graph of

$f \circ l_a$ is the graph of f compressed about the y-axis when $a > 1$. (Remember that, if $a > 1$, then $0 < \frac{1}{a} < 1$.) If $0 < a < 1$, then the graph of $f \circ l_a$ is the graph of f stretched about the y-axis. (Remember that, if $0 < a < 1$, then $\frac{1}{a} > 1$.)

Now suppose that $a < 0$; then $-a > 0$. We know that the graph of $f \circ l_a$ is the reflection in the y-axis of the graph of $f \circ l_{-a}$ because the graph of $f(ax) = f \circ l_a(x)$ is the reflection in the y-axis of the graph of $f(-ax) = f \circ l_{-a}(x)$. Since we have already discussed the graph of $f \circ l_b$, where $b > 0$, we know the properties of its reflection in the y-axis.

8.5a EXERCISES

Copy each graph of a given function f that is shown. Then, on the same set of axes, sketch that portion of the graph of the indicated function which can be determined.

1. Sketch $y = f(\tfrac{1}{3}x)$.

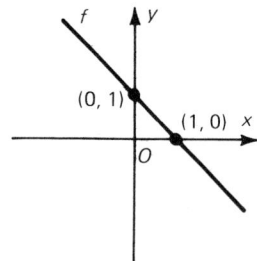

2. Sketch $y = f(3x)$.

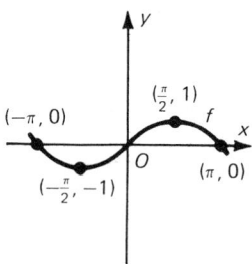

3. Sketch $y = f(-\tfrac{1}{2}x)$.

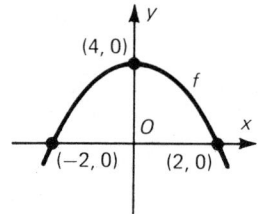

4. Sketch $y = f(-2x)$.

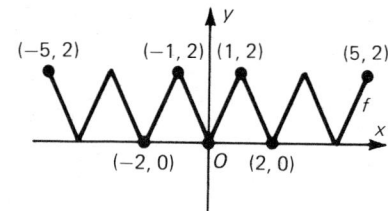

In each of Exercises 5–12, sketch the graph of the given function g. Then, on the same set of axes, sketch the graph of $g \circ l$ and write an equation that defines $g \circ l$.

5. $g(x) = 5 - 6x$; $l(x) = \tfrac{1}{2}x$.
6. $g(x) = x^2$; $l(x) = 3x$.
7. $g(x) = |x|$; $l(x) = 4x$.
8. $g(x) = |x|$; $l(x) = \tfrac{1}{3}x$.
9. $g(x) = \sin x$; $l(x) = 2x$.
10. $g(x) = \cos x$; $l(x) = \tfrac{1}{2}x$.
11. $g(x) = 2\cos x$; $l(x) = -2x$.
12. $g(x) = \tfrac{1}{2}\sin x$; $l(x) = -\tfrac{1}{2}x$.

From this point on, unless we specify otherwise, the function referred to as "$\exp_a$" is the function defined by

$$\exp_a x = a^x, \quad \text{where } a \in R^+, a \neq 1, \text{ and } x \in R.$$

Because $\exp_a$ is a continuous function which has 0 for its greatest lower bound and is not bounded above, the Intermediate Value Theorem (Theorem 4.3b, page 171) can be used to show that $\exp_a x$ takes on all positive real-number values as x ranges over R. For example, the function $\exp_2$ has this property, and, therefore, given any positive number a, there is a real number α such that $2^\alpha = a$. From this, we see that

$$\exp_a x = a^x = (2^\alpha)^x = 2^{\alpha x} = \exp_2 \alpha x.$$

This means that every function $\exp_a$ can be related to the function $\exp_2$ by choosing the proper α such that $2^\alpha = a$.

We know that, if $\alpha > 0$, then the graph of the function defined by $f(\alpha x)$ is a stretching or a compressing of the graph of f about the vertical axis. If $\alpha < 1$, the graph is stretched; if $\alpha > 1$, the graph is compressed. Thus, if $\alpha > 0$, the graph of $\exp_a$ will be a stretching or a compressing of the graph of $\exp_2$ about the vertical axis. If $\alpha < 0$, then the graph will be reflected in the vertical axis, as well as stretched or compressed.

On the basis of our discussion to this point, we can now make some reasonably good characterizations of the graphs of functions of the form $\exp_a$. First of all, we sketch the graph of $\exp_2$ (see Figure 8–11), about which we have the following information: (1) by Theorem 8.3a, $\exp_2$ is strictly increasing; (2) by Theorem 8.5a, the range of $\exp_2$ is not bounded above and has 0 as its greatest lower bound; and (3) from the graph in Figure 8–5 (page 310), the locations of certain points are known. We can now use the graph of $\exp_2$ in Figure 8–11, and the fact that $\exp_a x = \exp_2 \alpha x$ to construct several other graphs, as shown in Figure 8–12.

Figure 8-11

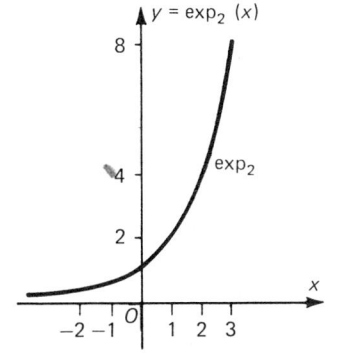

Figure 8-12

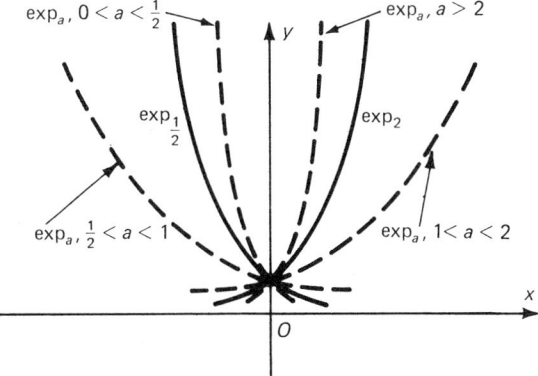

8.5b EXERCISES

Sketch the graph of each of the following functions.

1. $\exp_3$
2. $\exp_5$
3. $\exp_{1.1}$
4. $\exp_{1/3}$
5. $\exp_{2/3}$
6. $\exp_{0.1}$

As the graphs in Figure 8–12 indicate, $\exp_a$ is a one-to-one function. This means that if $a \in R^+$ and $a \neq 1$, then $a^x = a^y$ if and only if $x = y$. Use this fact to solve each of the following equations.

7. $3^x = 27$.
8. $5^x = 625$.
9. $4^x = \frac{1}{64}$.
10. $10^x = 0.001$.
11. $10^{x-2} = 10$.
12. $9^{2x} = 27^{x+1}$.
13. $25^{3x} = 125^{x-1}$.
14. $9^{x+1} = 27^{2x}$.

As the graphs in Figure 8–12 illustrate, the graphs of $\exp_a$ and $\exp_b$ intersect only in the point (0, 1) if $a \neq b$. This means that if $x \neq 0$, $a^x = b^x$ if and only if $a = b$. Use this fact to solve the following equations.

15. $x^{3/2} = 27$.
16. $x^{-2/3} = 16$.
17. $x^{3/4} = \frac{1}{8}$.
18. $x^{-3/4} = \frac{1}{27}$.
19. $(x - 2)^{2/3} = 25$.
20. $(x + 3)^{3/2} = 64$.
21. $(2x + 3)^{4/5} = 16$.
22. $2(x + 3)^{-1/2} = 128$.

8.6 The Slope Function of $\exp_a$ *(Optional)*

On the basis of these graphs in Figure 8–12, and our discussion so far, a natural assumption would be that the graphs of all the exponential functions of the form $\exp_a$, where $a > 0$, are concave upward. One way of verifying this assumption would be to show that $\exp_a$ has a slope function, which in turn has a slope function that is positive for all values of x. By

Theorem 4.6a, page 194, the fact that the slope function of the slope function is positive for all x would mean that the graph of the function is concave upward.

With this goal in mind, we will now attempt to determine if $\exp_a$ has a slope function and, if so, what that function is. First of all, we will take a specific example and investigate the possibility of a slope function for the function $\exp_2$.

Let $(x_0, 2^{x_0})$ be an arbitrary point on the graph of $\exp_2$. By Theorem 3.10a, if $\lim\limits_{x \to x_0} \dfrac{2^x - 2^{x_0}}{x - x_0}$ exists, then there is a tangent line to the graph of $\exp_2$ at $(x_0, 2^{x_0})$ that has a slope equal to this limit.

If x is close to x_0, but $x \neq x_0$, then $x = x_0 + \epsilon$, where ϵ is a small number, positive or negative. Hence,

$$\frac{2^x - 2^{x_0}}{x - x_0} = \frac{2^{x_0+\epsilon} - 2^{x_0}}{(x_0 + \epsilon) - x_0} = \frac{2^{x_0}(2^\epsilon - 1)}{\epsilon}$$

as $x \to x_0$, $\epsilon \to 0$.

Thus,
$$\lim_{x \to x_0} \frac{2^x - 2^{x_0}}{x - x_0} = \lim_{\epsilon \to 0} \left[\frac{2^{x_0}(2^\epsilon - 1)}{\epsilon} \right]$$
$$= \lim_{\epsilon \to 0} [2^{x_0}] \cdot \lim_{\epsilon \to 0} \left[\frac{2^\epsilon - 1}{\epsilon} \right]$$
$$= 2^{x_0} \cdot \lim_{\epsilon \to 0} \left[\frac{2^\epsilon - 1}{\epsilon} \right]$$

The following table gives some values of $\dfrac{2^\epsilon - 1}{\epsilon}$ for values of ϵ such that $|\epsilon| \leq 1$. The symbol "$\approx$" indicates an approximate value.

ϵ	$\dfrac{2^\epsilon - 1}{\epsilon}$
1	1.000
.5	≈ 0.828
.2	≈ 0.743
.1	≈ 0.718
.01	≈ 0.696
$-.01$	≈ 0.690
$-.1$	≈ 0.669
$-.2$	≈ 0.647
$-.5$	≈ 0.586
-1	0.5000

From this table it appears that, as ϵ approaches 0, $\dfrac{2^\epsilon - 1}{\epsilon}$ approaches some number between 0.690 and 0.696. It is a fact, which we cannot prove here, that the correct value of this number to six decimal places is 0.693147.

Hence, we have that the slope of the tangent line to the graph of $\exp_2$ at the point $(x_0, 2^{x_0})$ is equal to

$$K_{x_0} = 2^{x_0} \cdot k,$$

where k is a fixed constant which has a decimal approximation, to six places, of 0.693147. Therefore, $\exp_2$ has a slope function $\exp_2'$, which is defined for $x \in R$ by

$$\exp_2' x = 2^x \cdot k, \quad \text{with } k \approx 0.693147.$$

Notice that the slope function of this exponential function is also an exponential function.

Provided that we make one quite natural assumption, it is now an easy matter to determine the slope function $\exp_a'$ for the exponential function $\exp_a$, where $a > 0$.

As we saw at the beginning of this section,

$$\exp_a x = a^x = (2^\alpha)^x = 2^{\alpha x} = \exp_2 \alpha x,$$

or, equivalently,

$$\exp_a x = \exp_2 \alpha x, \quad \text{where } 2^\alpha = a.$$

This means, of course, that the graph of $\exp_2 \alpha x$ is the graph of $\exp_2$ stretched or compressed about the vertical axis, or reflected in the vertical axis and then stretched or compressed about it, depending upon the value of α (see page 317). If l is a tangent line to some point on the graph of $\exp_2$, then l—under this same stretching or compressing, or reflection followed by stretching or compressing—is translated to line l_{e^*}, as shown in Figures 8–13 and 8–14. [This follows easily since, if l is a linear function, $l(\alpha x)$ defines a linear function for any real α, $\alpha \neq 0$.] It would now seem reasonable to assume that l_{e^*} is a tangent line to the graph of $\exp_2 \alpha x$. This agrees with an assumption we made in connection with tangents to graphs of polynomial functions (see page 177), and we will make this assumption for $\exp_a x = \exp_2 \alpha x$.

Figure 8-13

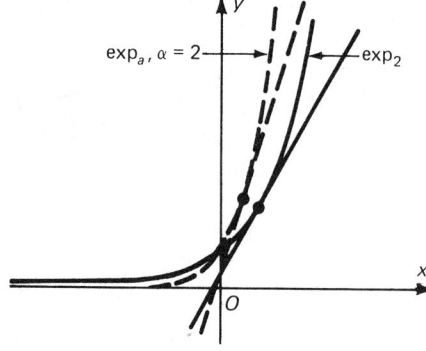

Figure 8-14

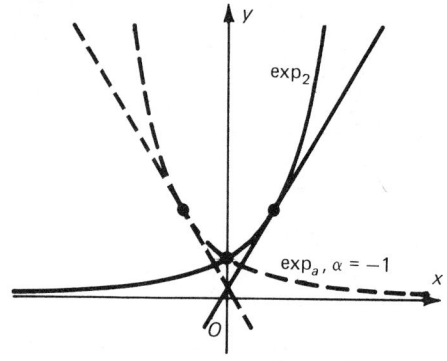

Since the slope of the tangent line to the graph of $\exp_2$ at $(x_0, 2^{x_0})$ is $2^{x_0} \cdot k$, where $k \approx 0.693147$, the equation of the tangent line to the graph of $\exp_2$ at this point is

$$l(x) = 2^{x_0} \cdot k(x - x_0) + 2^{x_0}, \qquad \text{where } k \approx 0.693147.$$

Now the graph of $\exp_2 \alpha x$ is a stretching or compressing of the graph of $\exp_2$ as every point $(u_0, 2^{u_0})$ of that graph is translated to $\left(\dfrac{u_0}{\alpha}, 2^{u_0}\right)$. Assuming that the tangent lines undergo the same stretching or compressing, we see that the equation of the tangent line to the graph of $\exp_2 \alpha x$ at $\left(\dfrac{u_0}{\alpha}, 2^{u_0}\right)$ is

$$l(\alpha x) = k \cdot 2^{u_0}(\alpha x - u_0) + 2^{u_0}.$$

If we now let $\dfrac{u_0}{\alpha} = x_0$, then $2^{u_0} = 2^{\alpha x_0}$, and the equation of the tangent line to $\exp_2 \alpha x$ at the point $(x_0, 2^{\alpha x_0})$ is

$$l(\alpha x) = k \cdot 2^{\alpha x_0} \alpha (x - x_0) + 2^{\alpha x_0}.$$

This means that the slope of the tangent line to $\exp_2 \alpha x$ at $(x_0, 2^{\alpha x_0})$ is $k \cdot \alpha \cdot 2^{\alpha x_0}$. Therefore, the slope function of $\exp_2 \alpha x$ is defined by

$$\exp_2' \alpha x = k \cdot \alpha \cdot 2^{\alpha x}, \qquad \text{where } k \approx 0.693147.$$

However, since $2^{\alpha} = a$, $2^{\alpha x_0} = a^{x_0}$. Hence, the equation of the tangent line to the graph of $\exp_2 \alpha x$ at $(x_0, 2^{\alpha x_0})$ is

$$l(\alpha x) = \alpha k \cdot a^{x_0}(x - x_0) + a^{x_0}.$$

Because $\exp_2 \alpha x = \exp_a x$, where $a = 2^{\alpha}$, we see that the slope of the tangent line to the graph of $\exp_a$ at (x_0, a^{x_0}) is $\alpha k \cdot a^{x_0}$. Hence, the slope function of $\exp_a$ is defined by

$$\exp_a' x = \alpha k \cdot a^x = \alpha k \cdot \exp_a x, \qquad \text{where } 2^{\alpha} = a, \text{ and } k \approx 0.693147.$$

If we now choose the particular value $\alpha = \dfrac{1}{k}$, or $a = 2^{1/k}$, where $k \approx 0.693147$, we have

$$\exp_a' x = (\exp_{2^{1/k}})'(x) = \dfrac{1}{k} \cdot k \cdot (2^{1/k})^x = (2^{1/k})^x.$$

The number $2^{1/k}$ is a very useful number in mathematics, and it is denoted by the letter e.

Definition 8.6a $e = 2^{1/k}$, where k is the number that $\dfrac{2^{\alpha} - 1}{\alpha}$ approaches as α approaches 0.

The number e is irrational and since $k \approx 0.693147$, $2^{1/k} \approx 2^{1/.693} \approx 2^{1.443}$. This number, $e \approx 2^{1.443}$, is approximated, to the first fifteen places, by the decimal 2.7 1828 1828 45 90 45.

From our work above, since $e = 2^{1/k}$, we see that

$$\exp'_e x = e^x = \exp_e x.$$

That is, the slope function of the function $\exp_e$ is the function $\exp_e$.

Thus, $\exp_e$ is its own slope function, and the slope of the tangent line at each point of its graph is equal to the ordinate of that point of the graph. (See Figure 8–15.)

Figure 8-15

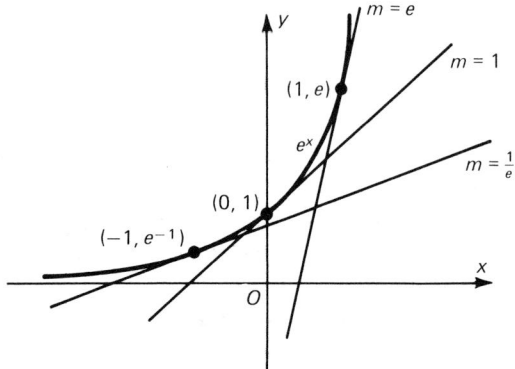

The number e is not usually defined as we have defined it, but is more often defined as the limit of the sequence

$$\left\langle \left(1 + \frac{1}{n}\right)^n \right\rangle.$$

8.6 EXERCISES

For each of the following exercises, give the slope of the tangent line drawn at the point on the graph of $\exp_2$ with the given abscissa. Express the slope in terms of k, where 0.693147 is an approximation for k.

1. $x = -2.0$.
2. $x = -1.0$.
3. $x = -0.5$.
4. $x = 0$.
5. $x = 0.5$.
6. $x = 1.0$.

7. $x = 1.5$.
8. $x = 2.0$.
9. $x = 2.5$.
10. It is possible to obtain an idea of the shape of the graph of a function by simply drawing a number of its tangents. To illustrate this technique, use 0.7 for k and 1.4 for $\sqrt{2}$ and graph all the tangent lines at the points given in exercise 1. Do *not* draw the graph of $\exp_2$.
11. Prove that the graph of $\exp_2$ does not have a horizontal tangent line at any of its points.

Determine the slope function of each of the functions defined below.

12. $f: f(x) = 4^x$.
13. $g: g(x) = (\frac{1}{2})^x$.
14. $F: F(x) = 8^x$.
15. $G: G(x) = 0.25^x$.
16. Evaluate $\left(1 + \dfrac{1}{n}\right)^n$ to the nearest hundredth for $n = 1, 2, 3, 4, 5$.

Recall that, if f is a function, a is any real number, and the slope function f' is defined, then the slope function $(af)'$ is defined, and $(af)'(x) = a \cdot f'(x)$. Find the slope function in each of the following exercises.

17. $f: f(x) = 5 \exp_2 x$.
18. $g: g(x) = -3 \cdot 2^x$.
19. $F: F(x) = \frac{1}{3} \cdot 2^x$.
20. $G: G(x) = -\frac{3}{4} \exp_2 x$.
21. Prove that if F is a function defined by
$$F(x) = A \cdot e^{cx}, \quad \text{with } x \in R,$$
where A and e are real numbers, then F has a slope function F' defined by
$$F'(x) = cA \cdot e^{cx}, \quad \text{with } x \in R.$$

8.7 Concavity of the Graph of $\exp_a$ (*Optional*)

As we have seen, the function $\exp_a$, where $a > 0$, has a slope function $\exp'_a$ defined by
$$\exp'_a x = ak \cdot \exp_a x,$$
where $2^\alpha = a$ and $k \approx 0.693147$.

If an increasing or decreasing function defined on the set of all rational numbers is extended to a continuous function whose domain is the set of all reals, then the extended function is also increasing or decreasing, respectively. Hence, we know that $\exp_2$ is increasing function because the function $f(x) = 2^x$, where x is rational, is increasing. (This is also verified by the fact that $\exp_2' x = k \cdot \exp_2 x > 0$, for $x \in R$, which, by Theorem 4.5a, means that $\exp_2$ is an increasing function.) In the case of the function $\exp_a$, if $0 < a < 1$, then $\alpha < 0$, and if $a > 1$, then $\alpha > 0$. Since $\exp_a x > 0$ for all $x \in R$, it follows that

a) $\exp_a' x = \alpha k \cdot \exp_a x$ is greater than 0 if $a > 1$; and
b) $\exp_a' x = \alpha k \cdot \exp_a x$ is less than 0 if $0 < a < 1$.

Therefore, by Theorem 4.5a, we know that, if $0 < a < 1$, the function $\exp_a$ is a decreasing function; and, if $a > 1$, the function $\exp_a$ is an increasing function.

Now that we have determined for which replacements of a $\exp_a$ is increasing and for which replacements it is decreasing, we will investigate the concavity of the graph of each such function. From the graphs already given, we are led to expect that $\exp_a$ is always concave upward. Furthermore, as we remarked in the preceding section (page 319), if we can show that $\exp_a'$ has a slope function $(\exp_a')' = \exp_a''$ that is positive for all x, then we can conclude that the graph of $\exp_a$ is concave upward (see Theorem 4.6a, page 194).

By Theorem 4.4a we know that, if a function f has a slope function f' and c is a constant, then the slope function $(cf)'$ of cf is equal to cf'. Thus, because

$$\exp_a' = \alpha k \cdot \exp_a,$$

it follows that

$$(\exp_a')' = \exp_a'' = \alpha k \cdot \exp_a' = (\alpha k)^2 \cdot \exp_a.$$

That is,

$$\exp_a'' = (\alpha k)^2 \cdot \exp_a.$$

Now $\exp_a x$ is positive for all values of x, $k > 0$, and $\alpha \neq 0$ if $a \neq 1$. Hence, $(\alpha k)^2 \cdot \exp_a = \exp_a''$ is positive for all values of x if $a \neq 1$. Thus, we have the following theorem concerning $\exp_a$.

Theorem 8.7a *If $a \neq 1$, then the graph of $\exp_a$ is concave upward.*

8.7 EXERCISES

Find the slope function of each of the following functions.

1. $f: f(x) = 10e^{2x}$.
2. $g: g(x) = 0.05e^{-3x}$.
3. $F: F(x) = -4e^{10x}$.
4. $G: G(x) = 100e^{0.1x}$.

In Exercises 5 and 6, let $f: f(x) = \frac{1}{2}e^x$ and $g: g(x) = \frac{1}{2}e^{-x}$.

5. Let c be a function such that $c(x) = f(x) + g(x)$.
   a) Prove that c is an even function.
   b) Prove that the graph of c is increasing for $x > 0$, decreasing for $x < 0$, and has a minimum point at $(0, 1)$.
   c) Prove that the graph of c is concave upward for all $x \in R$.
   d) Sketch the graphs of f, g, and c on the same set of axes.

6. Let s be a function such that $s(x) = f(x) - g(x)$.
   a) Prove that s is an odd function.
   b) Prove that the graph of s is increasing for all $x \in R$.
   c) Prove that the graph of s is concave upward for $x > 0$, concave downward for $x < 0$, and has an inflection point at $(0, 0)$.
   d) Sketch the graphs of f, g, and s on the same set of axes.

The functions c and s described in Exercises 5 and 6 are properly called the hyperbolic cosine function and the hyperbolic sine function, abbreviated cosh and sinh, respectively. They are so named because they have many properties and interrelationships similar to those of trigonometric functions, and because they are related to the geometric curve called the hyperbola in much the same way that the trigonometric functions are related to the circle. To illustrate the relationship between the cosh and sinh functions, verify each of the following properties.

7. $\cosh^2 x - \sinh^2 x = 1$.
8. $\sinh(2x) = 2 \sinh x \cdot \cosh x$.
9. $\cosh(2x) = \cosh^2 x + \sinh^2 x$.
10. $\cosh' x = \sinh x$.
11. $\sinh' x = \cosh x$.

Verify that $(f \circ g)'(x) = f'(g(x)) \cdot g'(x)$ in each of the following cases.

12. $f: f(x) = e^x$; $g: g(x) = x + a$, $a \in R$.
13. $f: f(x) = e^x$; $g: g(x) = cx$, $c \in R$.
14. $f: f(x) = e^x$; $g: g(x) = cx + a$, $a, c \in R$.

8.8 The Inverse of an Exponential Function

In this section we will show that, given any positive number x, there is a unique real number t such that $2^t = x$. The correspondence $x \to t$ defines the inverse function of the function $\exp_2$. In fact, we will show that if a is any positive number, with $a \neq 1$, then $\exp_a$ has an inverse function. We will then analyze this class of functions in terms of their properties and graphs in much the same way that we have analyzed the functions introduced in earlier chapters.

First of all, given the equation $2^t = x$, where x is a given positive number, how do we know that there is a unique real number t that satisfies this equation? From our study of the function $\exp_2$ we know that it is an increasing function and, thus, by Theorem 2.4a, page 55, it is a one-to-one function. Furthermore, the range of $\exp_2$ is R^+, the set of all positive real numbers. From these facts, we can immediately infer that, corresponding to any given positive number x, there is one and only one real number t such that $2^t = x$. Therefore, this correspondence defines the following function f on R^+ into R:

$$\{(x, f(x)) \mid x \in R^+ \text{ and } f(x) \text{ is the unique real number } t \text{ such that } 2^t = x\}.$$

This function is the set of ordered pairs of the form $(2^t, t)$, where $t \in R$, since it is defined by the correspondence that relates every real number $x = 2^t$ to t. That is, the function is defined by the correspondence

$$2^t \to t.$$

Example 1 This function contains the ordered pair $(8, 3)$ because $2^3 = 8$. It also contains $(\sqrt{2}, \frac{1}{2})$ since $2^{\frac{1}{2}} = \sqrt{2}$.

The function f just described is the inverse of the function $\exp_2$. That this is the case is clear, since the function $\exp_2$ is defined by the correspondence which relates every real number t to the unique positive number 2^t; that is, by the correspondence

$$t \to 2^t.$$

Therefore, $\exp_2$ is the set of ordered pairs of the form $(t, 2^t)$, where $t \in R$.

Example 2 The function $\exp_2$ contains the ordered pairs $(3, 8)$ and $(\frac{1}{2}, \sqrt{2})$ because, under the correspondence defining this function, $3 \to 2^3 = 8$ and $\frac{1}{2} \to 2^{\frac{1}{2}} = \sqrt{2}$.

Thus, f is the inverse function of $\exp_2$, which we could denote by $\exp_2^{-1}$, as we have denoted inverse functions in the past. It is such an important function, however, that we have a special name for it. This function is defined formally as follows.

Definition 8.8a The inverse function of the function $\exp_2$ is denoted by $\log_2$ and is called the **logarithm to the base 2.** This function is defined by $\log_2 (x) = t$, where t is the unique real number such that $2^t = x$, and where $x \in R^+$.

If there is no ambiguity about the meaning of an expression, we will usually omit the parentheses and write $\log_2 x$ or $\log_2 64$. In expressions like $\log_2 (x + y)$, we will use parentheses to make the meaning clear.

Since $\log_2$ is the inverse function of $\exp_2$, it follows from Theorem 2.5b, page 59, that, for all values of x for which both $\log_2(\exp_2 x)$ and $\exp_2(\log_2 x)$ are defined,

I) $\log_2(\exp_2 x) = x = \exp_2(\log_2 x)$.

Now the domain of $\exp_2$ is all of R and the range of $\exp_2$ is R^+. This means that $\log_2$, as the inverse function of $\exp_2$ has R^+ for its domain and R for its range. Hence, $\exp_2(\log_2 x) = x$ is defined only for $x \in R^+$, while $x = \log_2(\exp_2 x)$ is defined for all $x \in R$. Thus, Identity I holds for all $x \in R^+$.

Taking note of these observations and the fact that $\exp_2 x$ can be written as 2^x, we can now rewrite Identity I as

I') $\log_2 2^x = x$, if $x \in R$; $\quad x = 2^{\log_2 x}$ if $x \in R^+$.

This identity can be stated informally as follows. "If x is a real number, then the power to which 2 must be raised to get 2^x is x; if x is a positive number, then x is equal to 2 raised to the power to which 2 must be raised to get x." Thus, to emphasize the meaning of "$\log_2 x$," we can read this symbol as "the power to which 2 must be raised to get x."

8.8 EXERCISES

Rewrite each of the following exponential equations as a logarithmic equation.

1. $2^6 = 64$.
2. $2^0 = 1$.
3. $2^{1/2} = \sqrt{2}$.
4. $2^{-2} = 0.25$.

Write each of the following logarithms as a real number.

5. $\log_2 32$
6. $\log_2 0.5$
7. $\log_2 \frac{1}{128}$
8. $\log_2 0.0625$

Determine the value of x in each of the following exercises.

9. $\log_2 x = 3$.
10. $\log_2 x = -1$.
11. $\log_2 x = -4$.
12. $\log_2 x = 2.5$.

Determine the value of x in each of the following exercises.

13. $\log_2 2^{10} = x$.
14. $\log_2 \sqrt{2} = x$.
15. $\log_2 2^{1.5} + \log_2 2^{-1.2} = x$.

16. $\log_2 2^x = 5$.
17. $\log_2 2^x = -3.2$.
18. $\log_2 2^x - \log_2 2^7 = 5$.

Determine the value of x in each of the following exercises.

19. $2^{\log_2 10} = x$.
20. $2^{\log_2 x} = 6$.
21. $2^{\log_2 3} + 2^{\log_2 5} = x$.
22. $2^{\log_2 x} - 2^{\log_2 9} = 4$.
23. For what values of x is $\log_2 x$ positive? Negative? Justify your answers.
24. a) How is the graph of $\log_2$ related to the graph of $\exp_2$? Justify your answer.
    b) Sketch the graphs of $\exp_2$ and $\log_2$ on the same set of axes.
    c) For what values of x is the graph of $\log_2$ increasing? Decreasing?
    d) For what values of x is the graph of $\log_2$ concave upward? Concave downward?
25. Prove that if $b > 0$, then $\log_2\left(\dfrac{1}{b}\right) = -\log_2 b$. [Hint: Let $\log_2\left(\dfrac{1}{b}\right) = x$.]

8.9 Logarithms to the Base a

In the preceding section, we showed that the function $\exp_2$ has an inverse function, denoted by $\log_2$. Theorem 8.9a below follows directly from the fact that every function $\exp_a$, where $a > 0$ and $a \neq 1$, is either increasing or decreasing, from the definition of an inverse function, and from Theorem 2.5c, page 60. (You should be able to explain why the function $\exp_1$ does not have an inverse.)

Theorem 8.9a *Every exponential function of the form $\exp_a$, with $a > 0$ and $a \neq 1$, has an inverse function. The domain of the inverse of any such exponential function is R^+, the set of positive real numbers, and the value of this inverse at x, where $x \in R^+$, is the unique real number t that satisfies $a^t = x$.*

Definition 8.9a *The inverse function of $\exp_a$, where $a > 0$ and $a \neq 1$, is denoted by $\log_a$ and is called the **logarithm to the base a**. This function is defined as follows: $\log_a x$ is the unique real number t such that $a^t = x$, where $x \in R^+$.*

Just as the notation $\log_2 x$ means the power to which 2 must be raised to get x, the notation $\log_a x$ means the power to which a must be raised to get x.

In the rest of our discussion about logarithmic functions, we will assume that the base a is greater than 1. We have defined logarithms to any positive base a, with $a \neq 1$, but if $0 < b < 1$, then $\dfrac{1}{b} > 1$, and $\log_b x = -\log_{1/b} x$. To see that this is the case, we note that

$$b^t = (b^{-1})^{-t} = \left(\frac{1}{b}\right)^{-t}.$$

Therefore, if $t = \log_b x$, that is, if $b^t = x$, then $-t = \log_{1/b} x$, since $\left(\dfrac{1}{b}\right)^{-t} = x$. Thus, given any logarithm to a base between 0 and 1, we can express it in terms of a logarithm to a base greater than 1. For convenience, we shall therefore follow the usual custom and deal only with logarithms to bases greater than 1.

Example 1 Because 2 is the unique real number which satisfies the equation $3^t = 9$, $\log_3 9 = 2$. Similarly, $\log_5 625 = 4$ since $5^4 = 625$.

Because $\log_a$ is the inverse function of $\exp_a$, we have the following identities. These identities follow immediately from Theorem 2.5b, as do the identities for $\exp_2$ and $\log_2$ on page 327.

II) $\log_a (\exp_a x) = x$, with $x \in R$; $\quad x = \exp_a (\log_a x)$, with $x \in R^+$.

Since $\exp_a x$ can be written as a^x, these properties can also be expressed as

II') $\log_a a^x = x$, with $x \in R$; $\quad x = a^{\log_a x}$, with $x \in R^+$.

From the discussion in section 2.16 of Chapter 2, we know that the graph of the inverse of a given function f is the reflection in the line $y = x$ (which is the graph of the identity function I) of the graph of f. Hence, the graph of $\log_a$ is the reflection in the line $y = x$ of the graph of $\exp_a$. Since we already know a great deal about the nature of the graph of $\exp_a$, we can infer many properties of the graph of $\log_a$.

Figure 8-16

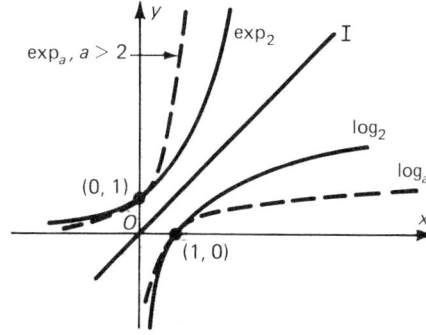

In Section 8.7, we learned that $\exp_a$, with $a > 1$, is a strictly increasing function whose graph is concave upward. As Figure 8-16 indicates, since the graph of $\log_a$ is the reflection in the line $y = x$ of the graph of $\exp_a$, it seems that $\log_a$ will be a strictly increasing function whose graph is concave downward. We note from Figure 8-16 that, if $a > 1$ and if $0 < x < 1$, then $\log_a x < 0$; if $x = 1$, $\log_a x = 0$; and if $x > 1$, $\log_a x > 0$.

8.9 EXERCISES

Write each of the following logarithms as a real number.

1. $\log_5 5$
2. $\log_{10} 1000$
3. $\log_4 2$
4. $\log_{10} 0.1$
5. $\log_3 243$
6. $\log_{10} 1$

Determine the value of x in each of the following exercises.

7. $\log_5 x = 3$.
8. $\log_{10} x = -3$.
9. $\log_4 x = 1.5$.
10. $\log_6 x = -1$.

Determine the value of a in each of the following exercises.

11. $\log_a 10{,}000 = 4$.
12. $\log_a \frac{1}{16} = -4$.
13. $\log_a \sqrt{3} = 0.25$.
14. $\log_a 27 = \frac{3}{2}$.
15. $\log_a \frac{1}{32} = -1.25$.
16. $\log_a \sqrt[3]{4} = \frac{2}{3}$.

Determine the value of x in each of the following exercises.

17. $\log_{10} 10^x = 5$.
18. $\log_7 (7^{3x+2}) = 14$.
19. $5^{\log_5 x} = 1.2$.
20. $4^{\log_4 (x-1)} = -\frac{1}{2}$.

Indicate whether each of the following statements is true or false and justify your answer.

21. $\log_{10} 10{,}000 - \log_{10} 1000 = \log_{10} 10$.
22. $(\log_3 9)(\log_9 3) = 1$.
23. $\log_7 7^3 = \log_3 3^7$.
24. $\log_{10} 10^{-1.5} = 10^{\log_{10} -1.5}$.

25. Consider the function $\exp_1$ and explain why 1 cannot be used as a base for a system of logarithms.
26. If $a > 1$, for what values of x is $\log_a x$ positive? Negative? Justify your answers by referring to the function $\exp_a$.
27. Let a be a positive number greater than 1, and let x_1 and x_2 be positive numbers. Prove that $\log_a x_1 < \log_a x_2$ if and only if $x_1 < x_2$. (Hint: Use the properties of $\exp_a$.)

8.10 Some Basic Properties of Logarithmic Functions

In our development, we obtained the logarithmic functions as inverses of exponential functions. Historically, the idea of the logarithm of a number arose when certain sequences of numbers were compared. This then led to the concept of logarithmic functions.

Consider the geometric sequence whose nth term is $\exp_2 n = 2^n$ and compare it term by term with the sequence of positive integers whose nth term is $I(n) = n$.

$$2, 2^2, 2^3, 2^4, \ldots, 2^n, \ldots$$
$$1, 2, 3, 4, \ldots, n, \ldots$$

We can multiply two terms of the geometric sequence by using the properties of exponents; for example, $2^3 \cdot 2^7 = 2^{3+7} = 2^{10}$.

Using the two sequences with term-by-term association, we could instead proceed in the following way: Consider the problem of multiplying 2^3 by 2^7. We first note that the term 2^3 in function $\exp_2$ is associated with 3 in function I and that the term 2^7 in $\exp_2$ is associated with 7 in I. We add 3 and 7 and locate their sum in the sequence of integers; this sum is, of course, 10. (We know that this sum is in sequence I because the positive integers are closed under addition.) The number 10 in $\langle n \rangle$ is associated with 2^{10} in $\langle 2^n \rangle$, which is the correct answer, as we already know. This procedure is shown diagrammatically below.

$$2, 2^2, 2^3, 2^4, 2^5, 2^6, 2^7, 2^8, 2^9, 2^{10}, 2^{11}, 2^{12}, \ldots$$
$$\downarrow \qquad \downarrow \qquad \uparrow$$
$$1, 2, 3, 4, 5, 6, 7, 8, 9, 10, 11, 12, \ldots$$
$$\underline{\qquad 3 + 7 \qquad}$$

Suppose now that, instead of writing the geometric sequence as indicated powers of 2, we write the actual numerical value of these powers.

$$2, 4, 8, 16, 32, 64, 128, 256, 512, 1024, \ldots$$
$$1, 2, 3, 4, 5, 6, 7, 8, 9, 10, \ldots$$

With an extensive listing (or table) of these values, we could use the process outlined above to multiply any two terms in the upper sequence. For example, from the table above, we see that 16 × 64 = 1024 because 4 + 6 = 10.

Suppose that we had been using such a table for multiplication (this obviously assumes we are poor arithmeticians) and then decided to multiply 3 by 5. Inasmuch as neither 3 nor 5 appears in the upper sequence, we might reason as follows: 3 lies between 2 and 4 in the upper sequence; so it should correspond to a number in the lower sequence halfway between 1 and 2, to which 2 and 4 correspond. Thus, 3 should correspond to 1.5. Similarly, 5 lies between 4 and 8 in the upper sequence, one-fourth of the way from 4 to 8; hence, by this reasoning, 5 should correspond to a number in the lower sequence one-fourth of the way from 2 to 3, which are the numbers that 4 and 8 correspond to, respectively. Thus, by this reasoning, 5 would correspond to 2.25. Adding 1.5 and 2.25, to which the numbers 3 and 5 correspond, respectively, we obtain 3.75 in the lower sequence.

Now we must determine to what number in the geometric sequence 3.75 corresponds. We see that 3.75 lies between 3 and 4, to which 8 and 16

$$
\begin{array}{cccccc}
2, & 3, & 4, & 5, & 8, & ?\quad 16, \ldots \\
& \downarrow & & \downarrow & & \uparrow \\
1, & 1.5, & 2, & 2.25, & 3, & 3.75,\ 4, \ldots \\
& \multicolumn{4}{c}{1.5\ +\ 2.25\ \longrightarrow}
\end{array}
$$

correspond, respectively. Because 3.75 is three-fourths of the way from 3 to 4, the number three-fourths of the way from 8 to 16 should correspond to 3.75. This number is, of course, 14. Now, as we all know, 3 × 5 is not 14. We have gotten an **approximate** answer to our problem by assuming correspondences for numbers lying between those in our two sequences. Such a process is called **linear interpolation.**

We might reasonably expect that, if there were more terms in each of these sequences and therefore less need to interpolate (or at least shorter intervals over which to interpolate), our results would be better.

Before we insert more terms in these sequences, let us examine the relationship between the corresponding terms. Each term in the geometric sequence $\langle 2^n \rangle$ corresponds to its exponent in the sequence of positive integers $\langle n \rangle$. For example, 2^1 corresponds to 1, 2^2 to 2, and so on; or, when we write the geometric sequence as 2, 4, 8, 16, . . . , each term corresponds to the positive integer in the sequence of positive integers that is the power to which 2 must be raised to obtain that term. That is, each term in the upper sequence corresponds in the lower sequence to its logarithm to the base 2. Thus, if we are to insert terms in the upper sequence, then the corresponding terms in the lower sequence should be their logarithms to the base 2 (or approximations thereto).

Suppose that we now insert the other positive integers into the upper sequence:

$$1, \quad 2, \quad 3, \quad 4, \quad 5, \quad 6, \quad 7, \quad 8, \quad 9, \quad 10, \quad \ldots$$
$$0, \quad 1, \quad 1.58, \quad 2, \quad 2.32, \quad 2.58, \quad 2.81, \quad 3, \quad 3.17, \quad 3.32, \quad \ldots$$

The number below each integer in the upper sequence is its logarithm to the base 2, correct to the nearest hundredth. What we have here is a primitive **logarithm table.** (We will not discuss methods of determining logarithms of numbers to various bases; we will simply accept them as given in an available table of logarithms.) By using the same procedure as before, we can determine approximations of the products of positive integers or, by interpolation, an approximation of the product of two positive numbers:

Example 1 Using our logarithm table to determine 2×3 leads to the following procedure:

$$1, \quad 2, \quad 3, \quad 4, \quad 5, \quad 6, \quad 7, \quad \ldots$$
$$\downarrow \quad \downarrow \quad\quad\quad\quad\quad\quad \uparrow$$
$$0, \quad 1, \quad 1.58, \quad 2, \quad 2.32, \quad 2.58, \quad 2.81, \quad \ldots$$
$$\swarrow \quad \searrow \quad\quad\quad \uparrow$$
$$1 + 1.58 \longrightarrow$$

No interpolation is needed since our table contains all the necessary entries.

Example 2 To find 2.5×2, we need to interpolate. Since 2.5 is halfway between 2 and 3, we want it to correspond to the number halfway between 1 and 1.58, which is $1 + \frac{1}{2}(1.58 - 1) = 1 + .29 = 1.29$. By interpolation between 4 and 5, we get 4.9, which is an approximation to the exact answer, 5.

$$1, \quad 2, \quad 2.5, \quad 3, \quad 4, \quad ?, \quad 5, \quad 6, \quad \ldots$$
$$\downarrow \quad \downarrow \quad\quad\quad\quad\quad\quad \uparrow$$
$$0, \quad 1, \quad 1.29, \quad 1.58, \quad 2, \quad 2.29, \quad 2.32, \quad 2.58, \quad \ldots$$
$$\swarrow \quad \searrow \quad\quad\quad \uparrow$$
$$1 + 1.29 = 2.29 \longrightarrow$$

Let us analyze the procedure for these two examples. To multiply two positive numbers a and b, we determine their logarithms (or approximations to them) to the base 2, add these logarithms, and then find the number (or approximation of it) which is equal to 2 raised to the sum of these two logarithms.

$$a \quad \cdot \quad b \quad = 2^{(\log_2 a + \log_2 b)}$$
$$\searrow \quad\quad \searrow \quad\quad\quad \uparrow$$
$$\log_2 a + \log_2 b \to (\log_2 a + \log_2 b)$$

From a property of the function $\exp_2$, we see that the equation $a \cdot b = 2^{(\log_2 a + \log_2 b)}$ is correct, since

$$2^{(\log_2 a + \log_2 b)} = 2^{\log_2 a} \cdot 2^{\log_2 b} = a \cdot b.$$

But since $a \cdot b = 2^{\log_2 (a \cdot b)}$, we have $2^{\log_2 a + \log_2 b} = 2^{\log_2 (a \cdot b)}$. Therefore, $\log_2 a + \log_2 b = \log_2 (a \cdot b)$. This is a basic property of $\log_2$ which corresponds to the following property of $\exp_2$:

$$2^x \cdot 2^y = 2^{x+y}.$$

In general, for any a, where $a > 0$ and $a \neq 1$,

A) $\qquad \log_a x_1 + \log_a x_2 = \log_a (x_1 \cdot x_2), \qquad$ with $x_1, x_2 \in R^+$.

This follows from the corresponding property of the function $\exp_a$:

$$\exp_a (x + y) = \exp_a x \cdot \exp_a y.$$

It is property A of the logarithmic functions which makes it possible to use logarithmic tables to determine approximations of the product of two positive real numbers. Of course, ours was a very primitive table to the base 2, and the examples we gave to demonstrate the procedure were very simple ones. By inserting many more elements into the two sequences, we could get better approximations and could use the table to solve more significant problems. In practical application, logarithmic tables are used which give logarithms to the base 10 instead of 2. Sometimes tables of logarithms to the base e are used, and tables to the base 2 are employed in a branch of communication theory.

Logarithms to the base e are often called **natural logarithms**, and those to the base 10, **common logarithms.** It is conventional practice to use the abbreviation "log" for "logarithm" and to refer to log tables rather than logarithmic tables.

8.10 EXERCISES

The table below represents an extension of the base-two log table on page 333. *Use this table to carry out the following computations.*

x	16	32	64	128	256	512	1024	2048	4096	8192
$\log_2 x$	4	5	6	7	8	9	10	11	12	13

1. 16×64
2. 32×256
3. 48×96
4. 20×112

Find the solution for each of the following conditions. (Remember that logarithms of nonpositive numbers are not defined.)

5. $\log_8 3 + \log_8 x = \log_8 21$.
6. $\log_7 9x + \log_7 x = \log_7 125$.
7. $\log_{10} 6 + \log_{10} (2x - 3) = \log_{10} (2x + 2)$.
8. $\log_4 (x - 3) + \log_4 (x + 3) = 2$.
9. $\log_3 (x + 2) + \log_3 (x - 4) = 3$.
10. $\log_2 (3x + 2) + \log_2 (2x - 2) = 2$.

11. Given that $\log_{10} 2 \approx 0.3010$, that $\log_{10} 3 \approx 0.4771$, and that $\log_{10} 7 \approx 0.8451$, where each logarithm is correct to the nearest ten-thousandth, complete a table of logarithms to the base 10 of the integers from 1 through 9. [Hint: Use the property that $\log_{10} (x_1 \cdot x_2) = \log_{10} x_1 + \log_{10} x_2$.]
12. Write each of the following numbers in scientific notation. (Remember that a number in scientific notation is in the form $a \times 10^n$, where $1 \leq |a| < 10$ and $n \in I$.) Then use formula A given on page 334 and the log table that you made for Exercise 3 to find the logarithm to the base 10 of the number.
    a) 20 b) 80,000 c) 400 d) 9000
    e) 7,000,000 f) 0.005 g) 0.3 h) 0.06

Exercise 12 illustrates an important property of base-ten logarithms. This property is that the base-ten logarithm of any positive number can be easily found if one knows the logarithms of the numbers between 1 and 10. Exercise 13 gives the precise relationship involved.

13. Let the positive number x be equal to $a \times 10^n$, where $1 \leq a < 10$ and $n \in I$. Prove that $\log_{10} x = n + \log_{10} a$. (Note: The integer n is called the **characteristic** $\log_{10} x$, and the number $\log_{10} a$ is called the **mantissa** of $\log_{10} x$.)
14. Since the characteristic of $\log_{10} x$ may be supplied by inspection once x has been expressed in scientific notation, tables of logarithms to the base 10 list only mantissas; that is, values of $\log_{10} a$, where $1 \leq a < 10$. The table of base-ten logarithms given in the back of the book lists the values of $\log_{10} a$ to the nearest ten-thousandth for certain values of a from 1.00 to 9.99. Use this table to find the approximate value of each of the following logarithms.
    a) $\log_{10} 3.87$; $\log_{10} 38.7$; $\log_{10} 0.0387$; $\log_{10} 3870$.
    b) $\log_{10} 7.32$; $\log_{10} 732$; $\log_{10} 0.732$; $\log_{10} 0.00732$.
    c) $\log_{10} 8.43$; $\log_{10} 84{,}300$; $\log_{10} 0.843$; $\log_{10} 84.3$.
    d) $\log_{10} 5.46$; $\log_{10} 0.000546$; $\log_{10} 5460$; $\log_{10} 0.0546$.

8.11 Further Properties of Logarithmic Functions

In section 8.10, we derived the following formula for logarithmic functions:

A) $\quad \log_a (x_1 \cdot x_2) = \log_a x_1 + \log_a x_2.$

We can easily derive an important special case of formula A; namely, when $x_2 = \dfrac{1}{x_1} = x_1^{-1}$. If $x_2 = x_1^{-1}$, then

$$\log_a x_1 + \log_a x_1^{-1} = \log_a (x_1 \cdot x_1^{-1}) = \log_a 1 = 0.$$

But $\log_a x_1 + \log_a x_1^{-1} = 0$ implies that

$$\log_a x_1^{-1} = -\log_a x_1.$$

This result leads us to another important result. Since division of a positive real number x_1 by a positive real number x_2 is equivalent to multiplying the positive real number x_1 by x_2^{-1}, we see that

$$\log_a \left(\frac{x_1}{x_2}\right) = \log_a (x_1 \cdot x_2^{-1}) = \log_a x_1 + \log_a x_2^{-1}$$
$$= \log_a x_1 - \log_a x_2.$$

This result, expressed as formula B below,

B) $\quad \log_a \left(\dfrac{x_1}{x_2}\right) = \log_a x_1 - \log_a x_2, \quad$ where $x_1, x_2 \in R^+,$

enables us to determine the quotient of two positive real numbers by subtracting their logs, obtained from a log table.

From the property of exponential functions given by the equation

$$(a^x)^y = a^{xy} = a^{yx},$$

where $a > 0$ and x and y are arbitrary real numbers, we can derive yet another important property of logarithmic functions.

Since the range of $\log_a$ is R and its domain is R^+, we know that if $x_1 \in R$, then there is an $x_2 \in R^+$ such that $x_2 = a^{x_1}$, or $x_1 = \log_a x_2$. Conversely, if $x_2 \in R^+$, then there is an $x_1 \in R$ such that $x_2 = a^{x_1}$. Now, using the above property of exponential functions, we obtain

$$x_2^y = (a^{x_1})^y = a^{yx_1} = a^{y \log_a x_2}.$$

Notice that the last equality in this chain results from the fact that $x_1 = \log_a x_2$. If we now take the logarithm to the base a of both sides of the equation $x_2^y = a^{y \log_a x_2}$, we get $\log_a (x_2^y) = \log_a (a^{y \log_a x_2})$.* Since $\log_a (a^{y \log_a x_2}) = y \log_a x_2$ and since x_2 is an arbitrary member of R^+, we have

C) $\quad \log_a (x^y) = y \log_a x, \quad$ where $x \in R^+$ and $y \in R.$

*Since $\log_a$ of a positive real number is unique, we *can* perform the operation of taking the $\log_a$ of both sides of an equation.

One use of formula C, which expresses another important property of logarithmic functions, is to provide a means of approximating the value of any power of a positive real number by using a log table.

Example 1 Suppose that we want to determine $9^{1/3}$. Using our $\log_2$ table, page 333, and the above formula, we see that

$$\log_2 (9^{1/3}) = \tfrac{1}{3} \log_2 9 \approx \tfrac{1}{3}(3.17) \approx 1.06.$$

In the lower sequence of the $\log_2$ table, the number 1.06 does not appear. But, because 1.06 is between 1 and 1.58, which are approximations for $\log_2 2$ and $\log_2 3$, respectively, we can interpolate. Since

$$1.06 = 1 + \tfrac{6}{58}(1.58 - 1),$$

we know that 1.06 must be an approximation for $\log_2$ of the number

$$2 + \tfrac{6}{58}(3 - 2) = 2.10.$$

Thus,
$$\log_2 (9^{1/3}) \approx 1.06 \approx \log_2 2.10,$$
or
$$9^{1/3} \approx 2.10.$$

The value of $9^{1/3}$, correct to hundredths, is 2.08^+, so even with our simple base-two log table, we have gotten a fairly good approximation. This technique allows us to approximate such quantities as $2^{\sqrt{2}}$, e^π, and so on, with a fair degree of accuracy.

8.11 EXERCISES

Let x and y be positive real numbers and let a be a positive number such that $a \neq 1$. Further, let $\log_a x = r$ and $\log_a y = s$. Use the properties of logarithmic functions to express each of the following in terms of r and s.

1. $\log_a (x \cdot y)$
2. $\log_a \left(\dfrac{x}{y}\right)$
3. $\log_a (x \cdot y^2)$
4. $\log_a (x \cdot y)^2$
5. $\log_a \sqrt{x \cdot y}$
6. $\log_a \sqrt[3]{\dfrac{x}{y}}$

Use common logarithms to find the value, correct to three significant digits, of each of the following.

7. $\dfrac{594}{237}$
8. $(1.04)^{20}$

9. $\dfrac{9.81}{0.746}$

10. $3^{1.92}$

11. $\sqrt[3]{59.3}$

12. $5^{3/4}$

Alternative forms of the common logarithm of a number may be obtained just by adding and subtracting the same integer. For example, $\log_{10} 2 \approx 0.3010$ may also be written as $1.3010 - 1$, $2.3010 - 2$, and so on. Similarly, if $\log_{10} x = -0.1000$, we may also write $\log_{10} x$ as $(-0.1000 + 1) - 1 = 0.9000 - 1$, as $1.9000 - 2$, and so on. You will find that such alternative forms are useful in the computations in Exercises 13–16.

13. $\dfrac{4.59}{84.1}$

14. $2^{-2.15}$

15. $\sqrt[3]{0.467}$

16. $(0.123)^{0.75}$

Use common logarithms to find the set of solutions for each of the following equations.

17. $7^x = 10$.
18. $3^{-x} = 10$.
19. $2^{x+2} = 9$.
20. $5^{1-3x} = 72$.
21. $4^{7x} = 10^{x+1}$.
22. $8^{2x+1} = 5^{4x}$.
23. Use common logarithms to determine which of the two numbers, e^π, or π^e, is greater.

8.12 Computing with Common Logarithms

So far in this chapter, we have used only a very simple base-two log table in discussing log functions. The base most commonly used for computing with logarithms is base ten. In this section, we will discuss certain computational procedures involving base-ten, or common, logarithms. These procedures, as you will see, are justified in terms of the properties of logarithmic functions discussed in sections 8.10 and 8.11.

If a computation with positive real numbers involves only the operations of multiplication, division, raising to a power, and extracting a root, the computation may be carried out by common logarithms for the following reasons.

a) A positive number x can be approximated rather closely from tables if its logarithm is known.

b) By using the properties of logarithmic functions, it is possible to obtain fairly easily a good approximation of the logarithm of a complicated expression from the logarithms of the individual numbers in the expression.

The following examples illustrate the procedure. Since all logarithms will be common logarithms, we will omit the subscript 10 when referring to the logarithm of a number.

Example 1 Determine x if $x = \dfrac{27.4 \times (76.1)^2}{(8.98)^3}$.

Solution From the properties of logarithmic functions, it follows that

$$\log x = \log (27.4 \times (76.1)^2) - \log (8.98)^3$$
$$= \log 27.4 + 2 \log 76.1 - 3 \log 8.98.$$

The logs of these numbers are obtained from the common log table in the back of the book by means of the procedure described in Exercises 13 and 14 of Section 8.10.

To speed the computation, it is suggested that the work be arranged in some tabular fashion such as the following.

$$\log 27.4 \approx 1.4378 \qquad\qquad \left[\begin{array}{l}\log 76.1 \approx 1.8814.\\ 2 \log 76.1 \approx 3.7628.\end{array}\right.$$
$$2 \log 76.1 \approx 3.7628 \longleftarrow$$
$$ 5.2006 \qquad\qquad \left[\begin{array}{l}\log 8.98 \approx 0.9533.\\ 3 \log 8.98 \approx 2.8599.\end{array}\right.$$
$$3 \log 8.98 \approx 2.8599 \longleftarrow$$
$$\log x \ \approx 2.3407, \text{ or } \log x \approx 0.3407 + 2.$$

Hence, $x \approx 2.19 \times 10^2$, or $x \approx 219$.

Example 2 Determine x if $x = \sqrt[3]{\dfrac{4.62 \times 0.487}{8.56}}$.

Solution From the properties of logarithmic functions, it follows that

$$\log x = \tfrac{1}{3}(\log 4.62 + \log 0.487 - \log 8.56).$$

$\log 4.62 \approx 0.6646$
$\log 0.487 \approx 0.6875 - 1$
$1.3521 - 1$
$\log 8.56 \approx 0.9325$
$0.4196 - 1 \longleftarrow$
$3)\overline{2.4196 - 3} \longleftarrow$

[Note the use of an alternative form of the logarithm to permit easier division by 3.

$\log x \approx 0.8065 - 1.$

Hence, $x \approx 6.40 \times 10^{-1}$, or $x \approx 0.640$.

8.12 EXERCISES

In each of the following exercises, use common logarithms to find the value of x correct to three significant digits.

1. $x = \dfrac{38.1\sqrt{512}}{347}.$

2. $x = \dfrac{(5.49)(86.7)}{\sqrt[3]{842}}.$

3. $x = \sqrt[4]{\dfrac{(28.3)(67.1)}{0.0612}}.$

4. $x = \sqrt{\dfrac{(14.7)(93.6)}{(28.7)(165)}}.$

5. The period t in seconds of the motion of a simple pendulum L feet in length is given by $t = 2\pi \sqrt{\dfrac{L}{g}}$, where g is the gravitational constant. Use $\pi \approx 3.14$ and $g \approx 32.2$ to determine the period of a pendulum 48.9 feet long.

6. If the measures of the sides of $\triangle ABC$ are a, b, and c, then the area of the triangle is given by $A = \sqrt{s(s-a)(s-b)(s-c)}$, where s is $\dfrac{a+b+c}{2}$, which is the semiperimeter of the triangle. Find the area of a triangle whose sides measure 18.6, 22.3, and 28.1.

7. The value at the end of t years of a $1000 bond which accumulates interest at the rate of 4% per year compounded semiannually is given by $A = 1000(1.02)^{2t}$.
   a) Find the value of the bond at the end of 20 years.
   b) How long does it take the bond to double in value?

8. If the half-life of a radioactive substance is k years, the amount of the material left after t years is given by $W(t) = W_0 \cdot 2^{-t/k}$, where W_0 is the original amount of the material. Find the half-life of a substance to the nearest year if 28.6 grams of the material decay to 27.1 grams in 2 years.

9. If a certain culture of bacteria doubles in number every 16 hours, the number of bacteria present at the end of t hours is given by $N(t) = N_0 \cdot 2^{0.0625t}$, where N_0 is the number of bacteria present at $t = 0$. If 100 bacteria are present in the culture at noon, how many may be expected to be present at 6 P.M. on the same day?

8.13 Logarithms to Different Bases

As we remarked earlier, the logarithms most frequently used are *binary*, *natural*, and *common* logarithms, which are logarithms to the bases 2, e, and 10, respectively. For computational purposes, common logarithms are used almost exclusively. As we will show presently, if a and b are any two real numbers greater than 0 which represent logarithmic bases, then there is a constant $C_{a,b}$ such that

$$\log_b x = C_{a,b} \log_a x, \qquad \text{where } x \in R^+.$$

The value of the constant $C_{a,b}$ is determined by the values of a and b.

We first observe that, if a and b are any two bases, then they are both positive numbers greater than 1. Therefore, there is a real number α such that $a = b^\alpha$. Since $a^{\log_a x} = x$ and $a = b^\alpha$,

$$(b^\alpha)^{\log_a x} = x.$$

If we now take the logarithm to the base b of each side of this equation, we have

$$\log_b x = \log_b \left((b^\alpha)^{\log_a x}\right).$$

By formula C, page 336, $\log_b \left((b^\alpha)^{\log_a x}\right) = \log_a x \cdot \log_b b^\alpha$. Hence,

$$\log_b x = \log_a x \cdot \log_b b^\alpha.$$

Since $a = b^\alpha$, $\log_b b^\alpha = \log_b a$. Thus, the above equation becomes

D) $\log_b x = \log_b a \cdot \log_a x.$

Since $a > 1$, $\log_b a > 0$. On the basis of this fact, one interpretation that we can make of formula D is that the graph of $\log_b$ is a stretching or compressing of the graph of $\log_a$ about the x-axis. If $\log_b a > 1$, it is a stretching of the graph of $\log_a$ about the x-axis; and if $0 < \log_b a < 1$, it is a compressing. (See Figure 8-17.)

Figure 8-17

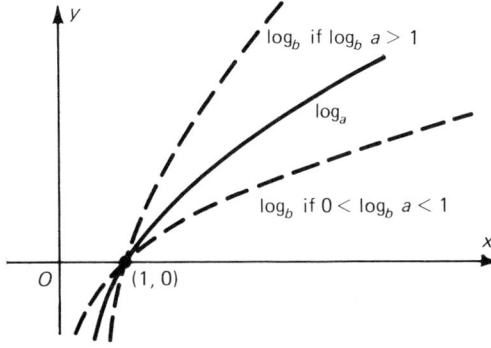

Formula D may be difficult to remember because you may forget whether the constant factor is $\log_b a$ or $\log_a b$. The following property of logarithmic functions should help you remember this formula.

Theorem 8.13a *The quotient of the logarithms of two numbers is independent of the base of the logarithmic function. Hence, if a and b are any two numbers greater than 0 that are not equal to 1 and $y \neq 1$, then*

$$\frac{\log_a x}{\log_a y} = \frac{\log_b x}{\log_b y}.$$

(We must specify that $y \neq 1$ since $\log_a 1 = 0$ for any base a.)

Proof From formula D, we see that $\log_b x = \log_b a \cdot \log_a x$ and $\log_b y = \log_b a \cdot \log_a y$. Forming the quotients of the left and right sides of these two equations, we have

$$\frac{\log_b x}{\log_b y} = \frac{\log_b a \, \log_a x}{\log_b a \, \log_a y} = \frac{\log_a x}{\log_a y},$$

which is what we wanted to prove. □

Now suppose that we want to write the relationship between $\log_a x$ and $\log_b x$ expressed by formula D. Remembering the above property, we write

$$\frac{\log_a x}{\log_a y} = \frac{\log_b x}{\log_b y}.$$

If $y = a$, we have $\log_a y = \log_a a = 1$, and

$$\log_a x = \frac{\log_b x}{\log_b a}.$$

This is equivalent to
$$\log_b x = \log_b a \cdot \log_a x.$$

If $y = b$, we have
$$\frac{\log_a x}{\log_a b} = \log_b x,$$

which is equivalent to
$$\log_b x = \frac{1}{\log_a b} \cdot \log_a x.$$

The equations $\log_b x = \log_b a \, \log_a x$ and $\log_b x = \frac{1}{\log_a b} \log_a x$ show that
$$\log_b a = \frac{1}{\log_a b}.$$

In applying formula D to specific examples, the following constants are useful.

$$\log_e 10 \approx 2.303.$$
$$\log_{10} e \approx 0.434.$$
$$\log_e 2 \approx 0.693.$$

From formula D we can also derive the result
$$\log_{1/a} x = -\log_a x$$
which we mentioned on page 329. Although we have chosen to work with logarithms whose bases are greater than 1, formula D, as well as the other formulas in this chapter, holds for logarithms to any base b where $b > 0$ and $b \neq 1$.

It is an increasingly common practice to denote the natural logarithmic function $\log_e$ by ln; thus, for example, we can write
$$\ln 10 \approx 2.303,$$
$$\ln 2 \approx 0.693.$$

8.13 EXERCISES

Use the table of common logarithms to find the value of each of the following.

1. $\log_2 38$
2. $\log_3 71$
3. $\log_5 0.271$
4. $\log_7 0.0132$
5. $\ln 127$
6. $\ln 8.76$
7. $\ln 0.302$
8. $\ln 0.0871$

Use the table of common logarithms to find the value of x in each of the following exercises.

9. $\log_2 x = 1.27$.
10. $\ln x = 3.31$.
11. $\log_5 x = 0.715$.
12. $\ln x = -0.259$.
13. Prove that, if a and b are positive numbers greater than 1, then
$$\log_b a = \frac{1}{\log_a b}.$$
14. Let $f : f(x) = \ln x$ and $g : g(x) = \log_a x$, where $a > 1$. Prove that $g = (\log_a e) \cdot f$.
15. Find the solutions for $(\ln x)^2 = \ln x^2$.
16. Let a, b, c, d be positive numbers greater than 1. Prove that
$$\log_a b \cdot \log_b c \cdot \log_c d = \log_a d.$$

9

The Rational Functions

Introduction

In Chapter 4 we noted that any polynomial function p, defined by

$$p(x) = a_n x^n + a_{n-1} x^{n-1} + \ldots + a_1 x + a_0; \quad x \in R,$$

where n is a positive integer, could be thought of as

$$p = a_n I^n + a_{n-1} I^{n-1} + \ldots + a_1 I + a_0,$$

where I is the identity function, and I^n is the repeated product in which I appears n times as a factor. Further, you learned that since $a_k I^k$, k a positive integer, can be thought of as the product of the constant function $f : f(x) = a_k$, $x \in R$, with the function I^k, the set of all polynomial functions can be generated from the constant functions and the identity function by repeated use of the binary operations of addition and multiplication. In this chapter we will consider the class of functions that can be generated from polynomials by use of the operation of division.

9.1 The Rational Functions

If we apply the binary operation of division to the set of all polynomial functions we obtain a new set of functions called the rational functions.

Definition 9.1a A **rational function** is a function h, defined by $h = \dfrac{p_1}{p_2}$ where p_1 and p_2 are polynomial functions. That is,

$$h(x) = \frac{p_1(x)}{p_2(x)}, \quad x \in R,$$

such that $p_2(x) \neq 0$, where p_1 and p_2 are polynomial functions.

Some examples of rational functions are given by f, g, and h, where

$$f(x) = \frac{2x}{x^2 + 1}, \quad g(x) = \frac{x^5 - 3x + 2}{x^2 - 5}, \quad h(x) = \frac{1}{x}, \quad x \in R,$$

it being understood that the function is not defined if the denominator is equal to zero.

Since $u(x) = 1$, $x \in R$, is a polynomial of degree 0, it follows that the set of polynomial functions is a subset of the rational functions, since any polynomial function p can be written as $p = \dfrac{p}{u}$, where $u(x) = 1$.

In arithmetic we say that a fraction $\dfrac{a}{b}$ is a proper fraction if $a < b$, and is an improper fraction if $a \geq b$. We say that a rational function $\dfrac{p_1}{p_2}$ is a **proper rational function** if the degree of p_1 *is less than* the degree of p_2 and is an **improper rational function** if the degree of p_1 *is greater than or equal to* the degree of p_2. Thus in the examples given above f and h are proper rational functions and g is an improper rational function.

By direct division one can always write an improper rational function as the sum of a polynomial and a proper rational function. For example consider the function g, defined above by

$$g(x) = \frac{x^5 - 3x^2 + 2}{x^2 - 5}.$$

Dividing $x^2 - 5$ into $x^5 - 3x^2 + 2$, we have

$$\require{enclose}
\begin{array}{r}
x^3 + 5x - 3 \\
x^2 - 5 \enclose{longdiv}{x^5 - 3x^2 + 2} \\
\underline{x^5 - 5x^3 } \\
5x^3 - 3x^2 \\
\underline{5x^3 - 25x } \\
-3x^2 + 25x + 2 \\
\underline{-3x^2 + 15} \\
+25x - 13
\end{array}$$

Thus $\dfrac{x^5 - 3x^2 + 2}{x^2 - 5} = x^3 + 5x - 3 + \dfrac{25x - 13}{x^2 - 5}.$

Therefore $g(x) = p_1(x) + \dfrac{p_2(x)}{p_3(x)},$

where $p_1(x) = x^3 + 5x - 3,$

$p_2(x) = 25x - 13,$

and $p_3(x) = x^2 - 5.$

Thus g is the sum of polynomial p_1 and the proper rational function $\dfrac{p_2}{p_3}$.

The operations of addition, subtraction, multiplication, division and composition applied to rational functions produce rational functions.

9.1 EXERCISES

In each of the following exercises, tabulate the values of x for which the given rational function is not defined.

1. $f : f(x) = \dfrac{x^2 - 1}{x^2 - 1}.$

2. $g : g(x) = \dfrac{x + 1}{x^2 + 1}.$

3. $h : h(x) = \dfrac{3x^3 - 4x^2 + 2}{x}.$

4. $t : t(x) = \dfrac{x^2 - 2x + 1}{x^2 + x - 2}.$

5. $F : F(x) = \dfrac{123}{x^3 + 3x}.$

6. $G : G(x) = \dfrac{x}{x^3 - 2x}.$

7. $H : H(x) = \dfrac{x^2 + 2x + 1}{x^3 + 1}.$

8. $T : T(x) = \dfrac{1 - x}{x^3 - 1}.$

In each of the following exercises, state whether the function is a proper rational function or an improper rational function. Express each improper rational function as a sum of a polynomial function and a proper rational function.

9. $f : f(x) = \dfrac{x^2 + 1}{x}.$

10. $g : g(x) = \dfrac{x^2}{x^2 - 1}.$

11. $h : h(x) = \dfrac{x^3 - 9}{x - 2}.$

12. $r : r(x) = \dfrac{x - 1}{x^2 - 1}.$

13. $F : F(x) = \dfrac{6x^2 + 5x - 15}{2x + 3}.$

14. $G : G(x) = \dfrac{10x^4 + x^2 - 5}{5x^2 + 3}.$

15. $H : H(x) = \dfrac{2x^2 - 11x + 3}{4x^2 + 2x - 3}.$

16. $T : T(x) = \dfrac{2x - 1}{8x^3 - 1}.$

In each of the following exercises, find $f+g$, $f-g$, $f \cdot g$, and $\dfrac{f}{g}$.

17. $f: f(x) = \dfrac{1}{x}$; $g: g(x) = \dfrac{2}{x-1}$.

18. $f: f(x) = \dfrac{4x}{2x-5}$; $g: g(x) = \dfrac{10}{5-2x}$.

19. $f: f(x) = \dfrac{x+3}{x-3}$; $g: g(x) = \dfrac{x-3}{x+3}$.

20. $f: f(x) = \dfrac{2}{x+1}$; $g: g(x) = \dfrac{4x}{x^2-1}$.

9.2 Graphs of Rational Functions

It is a bit more difficult to determine the character and the graph of a rational function than of a polynomial function. The basic reason is that if $r = \dfrac{p}{q}$ where p and q are polynomial functions and $q(x_0) = 0$, then $r(x_0)$ is not defined and the behavior of $r(x)$ for values near x_0 must be investigated. We will consider this problem shortly. We will now consider how to determine the values of x for which a rational function is positive, zero or negative.

If $r = \dfrac{p}{q}$ is a rational function, then for any value of x for which $r(x)$ is defined, $r(x)$ is a quotient of two numbers, $\dfrac{p(x)}{q(x)}$, $q(x) \neq 0$. Since a quotient of two numbers can be zero only when the numerator is zero we have the following theorem.

Theorem 9.2a *A rational function $r = \dfrac{p}{q}$, where p and q are polynomial functions, has for its zeros those zeros of p which are not zeros of q.*

We will be concerned only with the real zeros of a rational function. From Theorem 9.2a we see that the problem of finding the real zeros of a rational function consists of finding the real zeros of its numerator and denominator which are polynomials. For example if r is defined by

$$r(x) = \dfrac{x^2-1}{x^2-3x+2} = \dfrac{p(x)}{q(x)}.$$

we easily determine that the zeros of p are 1 and -1 and the zeros of q are 1 and 2. Thus the rational function r has only one zero, -1, which is the only zero of p which is not a zero of q. The reader will note that since $x^2 - 1 = (x - 1)(x + 1)$ and $x^2 - 3x + 2 = (x - 1)(x - 2)$ that

$$r(x) = \frac{(x-1)(x+1)}{(x-1)(x+2)}.$$

It is tempting to say that r is equal to the function f defined by

$$f(x) = \frac{x+1}{x+2}.$$

This is *not* the case however. For any value of $x \neq 1$, $x \neq -2$, $r(x) = f(x)$, but $f(1)$ is defined and $r(1)$ is not, since division by zero is not possible. (Neither r nor f is defined for $x = -2$). That is r is a restriction of f, f having one number, 1, in its domain which is not in the domain of r. This is a subtle point, but an important one. Practically we can determine the properties of r and the graph of r by determining the properties and graph of f and then remembering that r is not defined for $x = 1$. Thus the graph of r will be the graph of f with one point removed. That point will have coordinates $(1, f(1)) = (1, \frac{2}{3})$.

Let us now consider a procedure for determining those half-lines and open intervals of the x-axis for which a given rational function is positive or negative. If $r = \frac{p}{q}$, where p and q are polynomial functions, then the only values of x for which $r(x) < 0$ are those for which one of the two values $p(x)$ and $q(x)$ is negative and the other is positive. Since there are definite procedures for determining when a polynomial is negative, zero, or positive it follows that there is a procedure for determining when $r(x)$ is negative, zero, or positive.

For a specific example let us consider the function r defined by

$$r(x) = \frac{p(x)}{q(x)} = \frac{x^2 - x - 2}{2x^3 - 6x^2 + 2x - 6}, \quad x \in R, \, q(x) \neq 0.$$

Since $p(x) = (x - 2)(x + 1)$, $q(x) = 2(x - 3)(x^2 + 1)$, and $x^2 + 1 > 0$ for all x, the *sign diagrams* for the factors of $p(x)$ and $q(x)$ are shown in

Figure 9-1. From these diagrams it follows that $p(x) < 0$ if and only if $-1 < x < 2$ and $q(x)$ is less than zero if and only if $x < 3$.

Figure 9-1

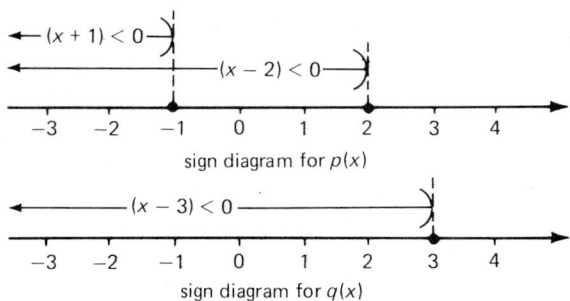

From the sign diagram in Figure 9-2, we conclude that $r(x) < 0$ if and only if $x < -1$ or $2 < x < 3$. Note that for $x < -1$, precisely 3 (an odd number) of all the factors of $p(x)$ and $q(x)$ are negative and the remaining factors are all positive. Similarly, between $x = 2$ and $x = 3$, precisely 1 (an odd number) of all the factors of $p(x)$ and $q(x)$ are negative and the remaining factors are positive. In general, if both $p(x)$ and $q(x)$ can be completely factored into products of constant, linear, and irreducible quadratic factors, then $r(x) = \dfrac{p(x)}{q(x)}$ will be negative if and only if an odd number of these factors are negative and the other factors are positive.

Figure 9-2

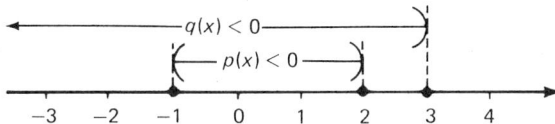

On the basis of the above analysis, we can say that the graph of

$$r(x) = \frac{p(x)}{q(x)} = \frac{x^2 - x - 2}{2x^3 - 6x^2 + 2x - 6}$$

must lie in the screened regions of the plane in Figure 9-3. From the factors of $p(x)$ and $q(x)$, it can also be determined that -1 and 2 are zeros of r, and that $r(3)$ is not defined, as indicated by the dashed line $x = 3$ in Figure 9-3.

Figure 9-3

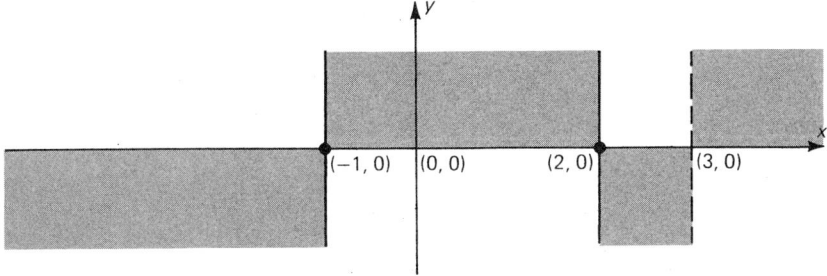

It is the case that every rational function r has a slope function r' which is also a rational function. In the next section we will give a rule for determining r' given r. Thus, given a rational function r, we will be able to determine the rational functions r' and r''. We have just seen that there are definite procedures for determining when a rational function is positive, zero, or negative. This means that given a rational function r we can determine when each of the functions r, r' and r'' is positive, zero, or negative. As we know from earlier work, this knowledge will give us most of the basic properties of r and will aid us in constructing a good representation of its graph.

9.2 EXERCISES

Find the real zeros of each of the following rational functions.

1. $f: f(x) = \dfrac{x^2 - 1}{x^2 + 2x + 1}$.

2. $g: g(x) = \dfrac{3x^2 - 9x}{x^2 - 9}$.

3. $h: h(x) = \dfrac{x^3 - x^2 + x}{x^4 + x^3 + 2x^2}$.

4. $r: r(x) = \dfrac{2x^2 - 2x - 40}{x^2 - 2x - 3}$.

5. $F: F(x) = \dfrac{6x^2 + x - 2}{2x^2 + 9x - 5}$.

6. $G: G(x) = \dfrac{6x^2 + 5x - 6}{6x^2 - 5x - 6}$.

7. $H: H(x) = \dfrac{x^3 + 1}{x^2 + 2x + 1}$.

8. $T: T(x) = \dfrac{x^2 - 2x + 1}{x^3 - 1}$.

Solve each of the following inequalities.

9. $\dfrac{x - 1}{x} > 0$.

10. $\dfrac{x}{x + 2} < 0$.

11. $\dfrac{x^2 - 1}{x^2 + 1} < 0$.

12. $\dfrac{x^3 - 1}{x - 4} > 0$.

13. $\dfrac{x^2 - 4x + 3}{x + 4} \geq 0$.

14. $\dfrac{x - 1}{x^2 - 3x - 10} \leq 0$.

15. $\dfrac{2x^2 + 7x}{3x^2 + 14x - 5} \leq 0$.

16. $\dfrac{x + 1}{x^3 + 1} \geq 0$.

In each of the following exercises, shade the regions of the plane in which the graph of r lies. Then indicate any x- or y-intercepts of the graph of r.

17. $r(x) = \dfrac{1}{x + 3}$.

18. $r(x) = \dfrac{2}{x^2 - 4}$.

19. $r(x) = \dfrac{x - 6}{x - 2}$.

20. $r(x) = \dfrac{x + 4}{2 - x}$.

21. $r(x) = \dfrac{x}{x - 1}$.

22. $r(x) = \dfrac{x + 2}{x}$.

23. $r(x) = \dfrac{x^2 + 4}{x^2 - 4}$.

24. $r(x) = \dfrac{x^2 - 9}{x^3 - 27}$.

Each of the following rational functions is a restriction of some polynomial function. Find this polynomial function and use its graph to determine the graph of the rational functions.

25. $f: f(x) = \dfrac{2x - 5}{2x - 5}$.

26. $g: g(x) = \dfrac{x^2 - 3x}{x}$.

27. $F: F(x) = \dfrac{x^3 - 1}{x - 1}$.

28. $G: G(x) = \dfrac{x^2 - x - 6}{3 - x}$.

9.3 Derivatives of Products, Quotients and Composites of Functions

In calculus courses it is proved that the product, quotient, and composite of functions whose derivatives (i.e., slope functions) exist also have derivatives, and formulas for finding these derivatives are developed and proved. Since it is beyond the scope of this book to do this in a rigorous way, we will merely state the results without proof.

Theorem 9.3a *If f and g are functions having derivatives f' and g', then the function $f \cdot g$ has a derivative $(f \cdot g)'$, where*

$$(f \cdot g)'(x) = f'(x)g(x) + f(x)g'(x),$$

which is defined for all values x such that $f'(x)$ and $g'(x)$ are defined.

Example 1 Let $f(x) = x$ and $g(x) = \sin x$, so $(f \cdot g)(x) = x \sin x$. We use the fact that $f'(x) = 1$ and $g'(x) = \cos x$.

$$(f \cdot g)'(x) = f(x) \cdot g'(x) + f'(x) \cdot g(x)$$
$$(f \cdot g)'(x) = x \cdot \cos x + 1 \cdot \sin x$$
$$= x \cos x + \sin x$$

Example 2 Let $f(x) = x^2$ and $g(x) = x^3$ so that $f'(x) = 2x$ and $g'(x) = 3x^2$.

$$(f \cdot g)'(x) = f(x) \cdot g'(x) + f'(x) \cdot g(x)$$
$$(f \cdot g)'(x) = (x^2)(3x^2) + (2x)(x^3)$$
$$= 3x^4 + 2x^4 = 5x^4$$

Notice that $(f \cdot g)(x) = x^2 \cdot x^3 = x^5$, so the result above is consistent with that obtained by using the formula for the slope function of a polynomial function you learned earlier.

Instead of giving the general formula for the derivative of the function $\dfrac{f}{g}$, where f and g are functions having derivatives, we first consider the special case where f is the unit function, defined by $f(x) = 1$, $x \in R$.

Theorem 9.3b *If g is a function having a derivative g', then the function $\dfrac{1}{g}$ has a derivative $\left(\dfrac{1}{g}\right)'$, where*

$$\left(\frac{1}{g}\right)'(x) = -\frac{g'(x)}{(g(x))^2},$$

which is defined for all x such that $g'(x)$ is defined and $g(x) \neq 0$.

Example 3 If $f(x) = \dfrac{1}{p(x)} = \dfrac{1}{x^2+1}$, then

$$f'(x) = -\frac{p'(x)}{(p(x))^2} = -\frac{2x}{(x^2+1)^2}.$$

Example 4 If $g(x) = \dfrac{1}{\sin x}$, then

$$g'(x) = -\frac{\sin' x}{(\sin x)^2} = -\frac{\cos x}{\sin^2 x}, \qquad \text{where } \sin x \ne 0.$$

Since the quotient $\dfrac{f}{g}$ can be thought of as the product $f \cdot \dfrac{1}{g}$ we can use Theorems 9.3a and 9.3b to get $\left(\dfrac{f}{g}\right)'$, that is

$$\left(\frac{f}{g}\right)' = \left(f \cdot \frac{1}{g}\right)' = f' \cdot \frac{1}{g} + f \cdot \left(\frac{1}{g}\right)'$$

$$= f' \cdot \frac{1}{g} + f\left(-\frac{g'}{g^2}\right)$$

$$= \frac{f'g - fg'}{g^2}.$$

We state this result as a theorem.

Theorem 9.3c *If f and g are functions having derivatives f' and g', then the function $\dfrac{f}{g}$ has a derivative $\left(\dfrac{f}{g}\right)'$, where*

$$\left(\frac{f}{g}\right)'(x) = \frac{f'(x)g(x) - f(x)g'(x)}{g^2(x)},$$

which is defined for all x such that f' and g' are defined and $g(x) \ne 0$.

This theorem allows us to find the derivative for any rational function $r = \dfrac{p}{q}$ as well as the derivatives of other quotients.

Example 5 If $r(x) = \dfrac{x^2 - 2}{x^3 + 3x - 1}$, then

$$r'(x) = \frac{\overset{f'(x)}{\downarrow}\ \cdot\ \overset{g(x)}{\downarrow}\ -\ \overset{f(x)}{\downarrow}\ \cdot\ \overset{g'(x)}{\downarrow}}{(x^3 + 3x - 1)^2}$$

$$r'(x) = \frac{(2x)(x^3 + 3x - 1) - (x^2 - 2)(3x^2 + 3)}{(x^3 + 3x - 1)^2}$$
$$\phantom{r'(x) = \frac{(2x)(x^3 + 3x - 1) - (x^2 - 2)(3x^2 + 3)}{(x^3 + 3x - 1)^2}}\ \uparrow$$
$$\phantom{r'(x) = \frac{(2x)(x^3 + 3x - 1) - (x^2 - 2)(3x^2 + 3)}{xxxxxx}}[g(x)]^2$$

$$= \frac{-x^4 + 9x^2 - 2x + 6}{(x^3 + 3x - 1)^2}.$$

Example 6 Suppose $g(x) = \dfrac{\sin x}{\cos x} = \tan x$. Then

$$g'(x) = \frac{\cos x \sin' x - \sin x \cos' x}{\cos^2 x}$$

$$= \frac{\cos x \cos x - \sin x (-\sin x)}{\cos^2 x}$$

$$= \frac{\cos^2 x + \sin^2 x}{\cos^2 x}$$

$$= \frac{1}{\cos^2 x},$$

since $\sin^2 x + \cos^2 x = 1$. Thus $\tan' x = \dfrac{1}{\cos^2 x}$, $\cos x \neq 0$. This agrees with the derivation of $\tan' x$ given in Section 7.2.

Theorem 9.3d *If f and g are functions having derivatives f' and g', then the function $f \circ g$ has a derivative $(f \circ g)'$, where*

$$(f \circ g)'(x) = f'(g(x))g'(x),$$

which is defined for all x such that $f' \circ g$ and g' are defined.

Example 7 The function $h : h(x) = \cos^3 x$ may be thought of as the composition of the functions f and g, where $f(x) = x^3$ and $g(x) = \cos x$. Hence

$$h'(x) = (f \circ g)'(x) = f'(g(x)) \cdot g'(x)$$

$$= 3(\cos x)^2 \cdot (-\sin x)$$

$$= -3 \cos^2 x \sin x.$$

Example 8 The function $h : h(x) = \sin\left(2x - \dfrac{\pi}{2}\right)$ may be thought of as $f \circ g$, where

$$f(x) = \sin x \quad \text{and} \quad g(x) = 2x - \frac{\pi}{2}.$$

In this case,

$$h'(x) = f'(g(x)) \cdot g'(x)$$

$$= \cos\left(2x - \frac{\pi}{2}\right) \cdot (2)$$

$$= 2 \cos\left(2x - \frac{\pi}{2}\right).$$

As a special case of Theorem 9.3d, suppose $f(x) = x^n$ and g is any function having a derivative. Then

$$f'(x) = nx^{n-1}$$

so
$$(f \circ g)'(x) = f'(g(x)) \cdot g'(x)$$
becomes
$$(f \circ g)'(x) = n(g(x))^{n-1} \cdot g'(x).$$

Example 9 The function $h: h(x) = (x^3 - 1)^7$ may be thought of as $f \circ g$, where

$$f(x) = x^7 \quad \text{and} \quad g(x) = x^3 - 1.$$

Hence
$$\begin{aligned} h'(x) &= n(g(x))^{n-1} \cdot g'(x) \\ &= 7(x^3 - 1)^6 \cdot 3x^2 \\ &= 21x^2(x^3 - 1)^6. \end{aligned}$$

Example 10 Given that $h(x) = (x^2 + 2x + 1)^3$, then

$$\begin{aligned} h'(x) &= 3(x^2 + 2x + 1)^2 \cdot (2x + 2) \\ &= (6x + 6)(x^2 + 2x + 1)^2. \end{aligned}$$

9.3 EXERCISES

In each of Exercises 1–6, use Theorem 9.3a to find f'. Then check your result by expressing $f(x)$ as a polynomial and using the methods you learned earlier to find f'.

1. $f: f(x) = (3x + 5)(7x + 3)$.
2. $f: f(x) = (2x - 7)(4x + 1)$.
3. $f: f(x) = 6x(x^4 - x)$.
4. $f: f(x) = x^3(x^2 - 2x + 1)$.
5. $f: f(x) = 6x(x - 1) + x(x + 1)$.
6. $f: f(x) = (x^2 + x - 1)(x^2 - x + 1)$.

For each of Exercises 7–14, use Theorems 9.3b or 9.3c to find f'.

7. $f: f(x) = \dfrac{1}{x}$.
8. $f: f(x) = \dfrac{1}{x^2}$.
9. $f: f(x) = \dfrac{1}{2x + 3}$.
10. $f: f(x) = \dfrac{1}{3 - x}$.
11. $f: f(x) = \dfrac{2x - 1}{3x + 1}$.
12. $f: f(x) = \dfrac{x^2}{2x + 1}$.
13. $f: f(x) = \dfrac{x^2}{x^2 - 16}$.
14. $f: f(x) = \dfrac{3 - x^2}{3 + x^2}$.

Find h' by two different methods. Show that the results are equal.

15. $h(x) = (2x + 3)^4 = 16x^4 + 96x^3 + 216x^2 + 216x + 81$.
16. $h(x) = (x^2 + 1)^5 = x^{10} + 5x^8 + 10x^6 + 10x^4 + 5x^2 + 1$.
17. $h(x) = (3x^2 - 4x)^2$.
18. $h(x) = (1 - x^2)^3$.

Find h' in each of the following exercises.

19. $h(x) = (x+1)^2(x^2+1)$.
20. $h(x) = (x-1)^3(x+2)^4$.
21. $h(x) = (2x^2+5)^7(4x^3+x)$.
22. $h(x) = \dfrac{1}{(x+2)^2}$.
23. $h(x) = \dfrac{-3}{(2x+1)^4}$.
24. $h(x) = \dfrac{x}{(x+3)^2}$.
25. $h(x) = \dfrac{10}{(x^2-1)^3}$.
26. $h(x) = \dfrac{2x+1}{(x-1)^2}$.

9.4 More about Graphs of Rational Functions

We have been assuming the continuity of rational functions where they are defined since they are the quotients of continuous functions, namely polynomial functions. (See page 93.) Also we can determine where rational functions are positive, negative or zero. Now that we have found how to determine the slope function of a rational function we can use earlier techniques to determine when a rational function r is increasing or decreasing, when its graph is concave up or down, and where its graph crosses the axes. We can also find relative maxima and minima of the function provided we can determine the real zeros of r, r' and r''. To illustrate, let us discover as much as we can about the graph of r, where

$$r(x) = \frac{x}{x^2+1}, \quad x \in R.$$

Now $r(x)$ is the quotient of the polynomials $p(x) = x$ and $q(x) = x^2 + 1$. The polynomial $p(x)$ has precisely one zero, namely 0, and $q(x)$ has no zeros, since, as we see from the discriminant of $q(x)$, $q(x)$ is an irreducible quadratic. It follows that $r(x)$ has precisely one zero, namely 0. Thus the only x-intercept of the graph of r is at $(0, r(0)) = (0, 0)$. This is also the y-intercept of the graph of r.

Since $p(x) = x$ is obviously positive for positive x and negative for negative x and $q(x) > 0$ for all x, we see that $r(x) > 0$ if $x > 0$ and $r(x) < 0$ if $x < 0$. Thus, except for the point $(0, 0)$, the graph of r lies entirely in the first and third quadrants. (That $q(x) > 0$ for all x is a consequence of the fact that an irreducible quadratic is either positive for all x or negative for all x (see page 222) and $q(0) = 1 > 0$.)

We also observe, in passing, that r is an odd function since

$$r(-x) = \frac{-x}{(-x)^2+1} = -\frac{x}{x^2+1} = -r(x).$$

Therefore the graph of r is symmetric with respect to the origin. (See page 102.)

The slope function of r, r' is defined by

$$r'(x) = \frac{(x^2+1) - x \cdot 2x}{(x^2+1)^2}$$
$$= -\frac{(x^2-1)}{(x^2+1)^2}$$
$$= -\frac{(x-1)(x+1)}{(x^2+1)^2}.$$

Since the numerator has precisely two zeros, -1 and 1, and the denominator has none we conclude that the zeros of $r'(x)$ are -1 and 1. It follows that the graph of r will have horizontal tangents at

$$(-1, r(-1)) = (-1, -\tfrac{1}{2}) \quad \text{and} \quad (1, r(1)) = (1, \tfrac{1}{2}).$$

The denominator, $(x^2+1)^2$, of $r'(x)$ is always positive and the three factors of the numerator -1, $(x-1)$ and $(x+1)$ are negative for all x, for all $x < 1$ and for all $x < -1$ respectively. Figure 9-4 represents the sign diagram of $r'(x)$. From the diagram we see that

$$r'(x) < 0 \quad \text{if } x < -1$$
$$r'(x) > 0 \quad \text{if } -1 < x < 1$$
$$r'(x) < 0 \quad \text{if } x > 1.$$

Figure 9-4

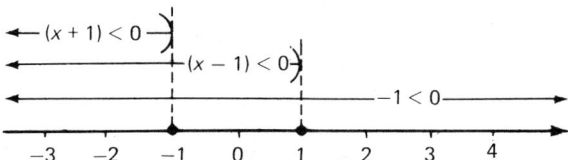

This means that r is a decreasing function for $x < -1$ and $x > 1$ and is an increasing function for $-1 < x < 1$. Since r is continuous, its graph must have a low point at $(-1, r(-1)) = (-1, -\tfrac{1}{2})$ and a high point at $(1, \tfrac{1}{2})$.

We could represent our present state of knowledge about the graph of r by the diagram in Figure 9-5.

Figure 9-5

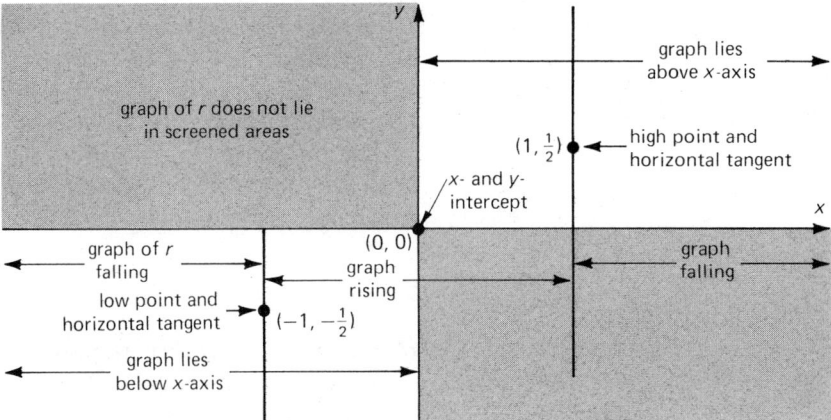

The slope function r'' of the function r' is defined by

$$r''(x) = -\left[\frac{(x^2+1)^2(x^2-1)' - (x^2-1)((x^2+1)^2)'}{((x^2+1)^2)^2}\right].$$

Now $(x^2 - 1)' = 2x$ and from recent work we know that

$$((x^2+1)^2)' = 2(x^2+1)(x^2+1)' = 2(x^2+1)(2x) = 4x(x^2+1).$$

Thus

$$\begin{aligned} r''(x) &= \frac{(x^2+1)^2(-2x) - (-(x^2-1))(4x)(x^2+1)}{(x^2+1)^4} \\ &= \frac{-(x^2+1)(2x) + 4x(x^2-1)}{(x^2+1)^3} \\ &= \frac{2x^3 - 6x}{(x^2+1)^3} \\ &= \frac{2x(x^2-3)}{(x^2+1)^3} \\ &= \frac{2x(x-\sqrt{3})(x+\sqrt{3})}{(x^2+1)^3}. \end{aligned}$$

Since the denominator $(x^2+1)^3$ has no zeros, the zeros of $r''(x)$ are those of the numerator $2x(x^2 - 3)$ which are easily seen to be 0, $\sqrt{3}$ and $-\sqrt{3}$. Using our procedure to determine the sign of $r''(x)$ we obtain the sign diagram in Figure 9–6. From the diagram we see that

$$\begin{aligned} r''(x) &< 0 &&\text{if } x < -\sqrt{3} \\ r''(x) &> 0 &&\text{if } -\sqrt{3} < x < 0 \\ r''(x) &< 0 &&\text{if } 0 < x < \sqrt{3} \\ r''(x) &> 0 &&\text{if } \sqrt{3} < x. \end{aligned}$$

Figure 9-6

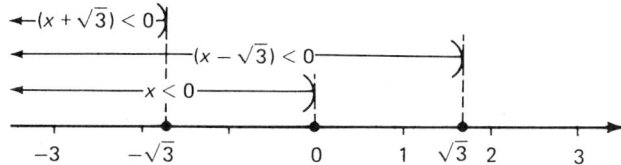

From Theorem 4.6a, p. 194, we can conclude that the graph of r is

$$\begin{aligned} &\text{concave down} &&\text{if } x < -\sqrt{3} \\ &\text{concave up} &&\text{if } -\sqrt{3} < x < 0 \\ &\text{concave down} &&\text{if } 0 < x < \sqrt{3} \\ &\text{concave up} &&\text{if } \sqrt{3} < x. \end{aligned}$$

On the basis of all the information we have at this point, we can give an approximation of the graph of r as in Figure 9–7 on page 358.

Figure 9-7

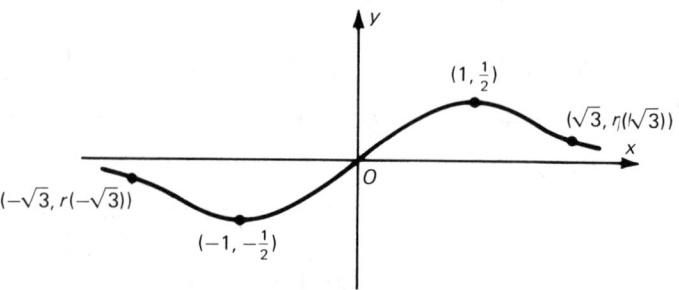

9.4 EXERCISES

For each of Exercises 1–3, find the values of x for which $r(x)$, $r'(x)$, and $r''(x)$ are positive, zero, and negative. Then use this information to draw a characterization of the graph of r.

1. $r(x) = \dfrac{1}{x^2 + 1}$.   2. $r(x) = \dfrac{x^2}{x^2 + 1}$.   3. $r(x) = \dfrac{x^3}{x^2 + 1}$.

Use the relationship between the graphs of f and af, Figure 9–7, and your graph for Exercise 1 to draw graphs of the functions in Exercises 4–6.

4. $r(x) = \dfrac{-1}{x^2 + 1}$.   5. $r(x) = \dfrac{4}{x^2 + 1}$.   6. $r(x) = \dfrac{-2}{x^2 + 1}$.

7. Given $r(x) = \dfrac{1}{x^2 + 1}$ and $f(x) = \dfrac{a}{x^2 + b^2}$, where $a \neq 0$ and $b > 0$, show that $f(x) = \dfrac{a}{b^2} r\left(\dfrac{x}{b}\right)$.

Use the results of Exercises 1 and 7 to draw graphs of the functions in Exercises 8–10.

8. $f(x) = \dfrac{12}{x^2 + 4}$.   9. $f(x) = \dfrac{18}{x^2 + 9}$.   10. $f(x) = \dfrac{-1}{x^2 + 4}$.

11. Given $r(x) = \dfrac{x}{x^2 + 1}$ and $f(x) = \dfrac{ax}{x^2 + b^2}$, where $a \neq 0$ and $b > 0$, show that $f(x) = \dfrac{a}{b} r\left(\dfrac{x}{b}\right)$.

Use Figure 9–7 and the result of Exercise 11 to draw graphs of the functions in Exercises 12–14.

12. $f(x) = \dfrac{-4x}{x^2 + 4}$.   13. $f(x) = \dfrac{3x}{x^2 + 9}$.   14. $f(x) = \dfrac{2x}{x^2 + 16}$.

9.5 Horizontal Asymptotes

In working with the rational function, $r(x) = \dfrac{x}{x^2 + 1}$ in Section 9.4, it was shown that $r(x)$ was decreasing for $x > 1$, and that $r(x)$ had a maximum value of $\frac{1}{2}$ at $x = 1$. It was also shown that $r(x) > 0$ for $x > 0$. It follows that as x gets large, $r(x)$ takes on smaller and smaller positive values. In fact, the graph of r gets arbitrarily close to the positive x-axis as x gets sufficiently large. For example, if we draw the horizontal lines $l_{1/10}$ and $l_{-1/10}$ defined by

$$l_{1/10}(x) = \frac{1}{10} \quad \text{and} \quad l_{-1/10}(x) = -\frac{1}{10},$$

all of the points of the graph of r (in Figure 9-8) having abscissas greater than 10 lie inside the horizontal strip bounded by these two lines since if $x > 10$,

$$\frac{x}{x^2 + 1} < \frac{x}{x^2} = \frac{1}{x} < \frac{1}{10}.$$

Figure 9-8

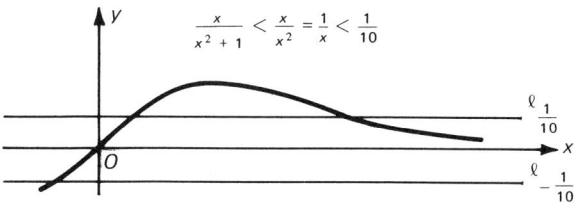

The choice of $\frac{1}{10}$ was quite arbitrary. If we let ϵ be any positive number and draw the horizontal lines l_ϵ and $l_{-\epsilon}$,

$$l_\epsilon(x) = \epsilon \qquad l_{-\epsilon}(x) = -\epsilon,$$

we see that if $x > \dfrac{1}{\epsilon}$ the graph of r will lie between these horizontal lines for x sufficiently large since

$$\frac{x}{x^2 + 1} < \frac{x}{x^2} < \frac{1}{x} < \frac{1}{\frac{1}{\epsilon}} = \epsilon \quad \text{if } x > \frac{1}{\epsilon} > 0.$$

We say that the graph of r is *asymptotic* to the x-axis and that the x-axis is a *horizontal asymptote* of the graph of r.

Another way of expressing this situation is to say that $r(x)$ approaches 0 as x gets arbitrarily large (or approaches ∞). As we did with sequences in Chapter 1, we express this as

$$\lim_{x \to \infty} r(x) = 0.$$

The notion of a limit of a function $f(x)$ as x approaches ∞ is a natural extension of the notion of a limit of a sequence.

Definition 9.5a The statement $\lim\limits_{x\to\infty} f(x) = L$ means that given any positive number ϵ, there is a number X such that if $x > X$, then

$$|f(x) - L| < \epsilon.$$

Since the graph of r is symmetric with respect to the origin, if $x < 0$ the graph of r will lie between the lines l_ϵ and $l_{-\epsilon}$ provided $|x|$ is sufficiently large, namely $|x| > \dfrac{1}{\epsilon}$. Thus the graph of r is asymptotic to the x-axis to the left as well as to the right. This can also be stated as follows:

$r(x)$ approaches 0 as
$|x|$ approaches infinity and $x < 0$ (or x approaches $-\infty$).

We write
$$\lim_{x\to-\infty} r(x) = 0.$$

The general notion of the limit of a function as x approaches $-\infty$ is given by the following definition.

Definition 9.5b The statement $\lim\limits_{x\to-\infty} f(x) = L$ means that given any positive number ϵ, there is a number X such that if $x < X$, then

$$|f(x) - L| < \epsilon.$$

If we let s be a function defined by

$$s(x) = r(x) + 1$$

then we know that the graph of s (in Figure 9-9) is the graph of r translated up one unit. The graph of s is asymptotic to the horizontal line $l(x) = 1$, and $\lim\limits_{x\to\infty} s(x) = \lim\limits_{x\to-\infty} s(x) = 1$.

Figure 9-9

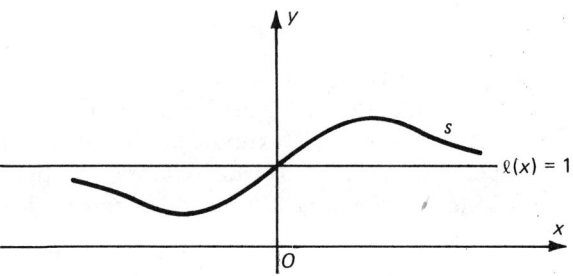

Definition 9.5c If f is a function, and $l(x) = L$ is a horizontal line such that either

1. $\lim\limits_{x \to \infty} f(x) = L$, or
2. $\lim\limits_{x \to -\infty} f(x) = L$,

then we say the graph of f is **asymptotic** to l and that l is a **horizontal asymptote** of the graph of f.

Note carefully that by this definition when we say a graph of a function is asymptotic to a horizontal line l we mean that the graph of f is asymptotic to the right *or* to the left *or* both to the right and left.

The graph of $f(x) = e^x$, for example, is asymptotic to the x-axis to the left. The graph of $g(x) = e^{-x}$ is asymptotic to the x-axis to the right. The graph of $r(x) = \dfrac{x}{x^2 + 1}$ is asymptotic to the x-axis both to the right and to the left. We would say that each of the graphs is asymptotic to the x-axis.

A special family of functions having graphs asymptotic to the x-axis is defined by:

$$f_n(x) = \frac{1}{x^n}, \qquad x \in R,$$

where n is any positive integer. This is clear since if $\epsilon > 0$, then if $x > 0$

$$\left| \frac{1}{x^n} \right| = \frac{1}{x^n} < \epsilon \text{ if } x > \frac{1}{\epsilon^{1/n}} = \epsilon^{-1/n}; \quad \text{that is } \lim_{x \to \infty} \frac{1}{x^n} = 0,$$

and if $x < 0$

$$\left| \frac{1}{x^n} \right| = \frac{1}{|x|^n} < \epsilon \text{ if } |x| > \epsilon^{1/n}; \quad \text{that is } \lim_{x \to -\infty} \frac{1}{x^n} = 0.$$

We make use of these results in considering the horizontal asymptotes for rational functions. Suppose $r = \dfrac{p}{q}$ is a rational function, where

$$p = a_n x^n + a_{n-1} x^{n-1} + \ldots + a_1 x + a_0$$

and

$$q = b_m x^m + b_{m-1} x^{m-1} + \ldots + b_1 x + b_0,$$

so that

$$r(x) = \frac{a_n x^n + a_{n-1} x^{n-1} + \ldots + a_1 x + a_0}{b_m x^m + b_{m-1} x^{m-1} + \ldots + b_1 x + b_0}.$$

If $x \neq 0$, we can factor x^n out of the numerator and x^m out of the denominator to obtain

$$r(x) = \frac{x^n}{x^m}\left(\frac{a_n + \dfrac{a_{n-1}}{x} + \cdots + \dfrac{a_1}{x^{n-1}} + \dfrac{a_0}{x^n}}{b_m + \dfrac{b_{m-1}}{x} + \cdots + \dfrac{b_1}{x^{m-1}} + \dfrac{b_0}{x^m}}\right)$$

Since each of the terms other than a_n and b_n in the numerator and denominator of the second factor are of the form $\dfrac{c}{x^k}$ where c is a constant and k is a positive integer, they all approach zero as x approaches ∞ or $-\infty$. That is, the limit of the second factor above as x approaches ∞ or $-\infty$ is $\dfrac{a_n}{b_m}$. The behavior of the first factor namely $\dfrac{x^n}{x^m}$ will depend on the relationship of the integers m and n. Let us consider the cases:

1. If $m > n$, then $\dfrac{x^n}{x^m} = \dfrac{1}{x^{m-n}}$, where $(m - n)$ is a positive integer and $\lim\limits_{x \to \infty} \dfrac{1}{x^{m-n}} = \lim\limits_{x \to -\infty} \dfrac{1}{x^{m-n}} = 0$. This means that if $m > n$, $r(x)$ is the product of two terms, one of which has the limit $\dfrac{a_n}{b_m}$, a constant, while the other term $\dfrac{1}{x^{m-n}}$ has the limit zero. Thus $\lim\limits_{x \to \infty} r(x) = \lim\limits_{x \to -\infty} r(x) = 0$. This means the graph of r is asymptotic to the x-axis both to the right and to the left.

2. If $m = n$, $\dfrac{x^n}{x^m} = 1$ for $x \neq 0$. In this case $\lim\limits_{x \to \infty} r(x) = \lim\limits_{x \to -\infty} r(x) = \dfrac{a_n}{b_m}$. This means the graph of r is asymptotic to the horizontal line l, $l(x) = \dfrac{a_n}{b_m}$, both to the right and to the left.

3. If $m < n$, $\dfrac{x^n}{x^m} = x^{n-m}$, $n - m$ a positive integer. In this case $r(x)$ is the product of a term which has the limit $\dfrac{a_n}{b_m}$ as $|x|$ approaches ∞, and another term which is a positive integer power of x. In this case $r(x)$ behaves for large values of $|x|$ in a manner similar to $\dfrac{a_n}{b_m} x^{n-m}$. That is it will have the same sign as $\dfrac{a_n}{b_m} x^{n-m}$ and will get large in absolute value as $\dfrac{a_n}{b_m} x^{n-m}$ gets large in absolute value.

Summing up we can give the following theorem.

Theorem 9.5a *If $r = \dfrac{p}{q}$ where p and q are polynomial functions then*

1. *if the degree of p is less than that of q, the graph of r is asymptotic to the x-axis.*
2. *if the degree of p equals the degree of q the graph of r is asymptotic to the horizontal line l, $l(x) = \dfrac{a_n}{b_m}$, where a_n and b_m are the leading coefficients of $p(x)$ and $q(x)$ respectively.*
3. *if the degree of p is n, and is greater than the degree of q, which is m, then the graph of r behaves for large values of $|x|$ similarly to the graph of $s(x) = \dfrac{a_n}{b_m} x^{n-m}$.*

We can actually give a better description of the behavior of an improper rational function for large values of $|x|$ than is given in case 3 of the theorem above. We earlier commented that any improper rational function r can be written as the sum of a polynomial function p and a proper rational function, s. Thus
$$r(x) = p(x) + s(x)$$
and
$$|r(x) - p(x)| = |s(x)|.$$

Since s is a proper rational function its graph is asymptotic to the x-axis and $|s(x)|$ gets arbitrarily close to 0 as $|x|$ gets sufficiently large. But this means that the vertical distance between the graphs of r and p, namely $|r(x) - p(x)|$, gets arbitrarily close to zero as $|x|$ gets sufficiently large. Said somewhat less rigorously, the graphs of r and p are essentially indistinguishable for large values of $|x|$.

Example 1 If $r(x) = x + \dfrac{1}{x^2}$, then the graphs of $r(x)$ and $I(x) = x$ are essentially the same for large values of x. See Figure 9-10.

Figure 9-10

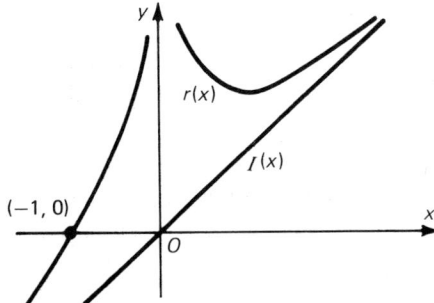

9.5 EXERCISES

For each of Exercises 1–8, determine whether the graph of the given function has a horizontal asymptote. If the graph does not have an asymptote, describe the behavior of the graph for large values of $|x|$.

1. $r(x) = \dfrac{4x + 5}{2x + 3}$.

2. $r(x) = \dfrac{x^2 + x - 3}{5x^2 + 10}$.

3. $r(x) = \dfrac{x^2 + 7x - 8}{2x - 5}$.

4. $r(x) = \dfrac{5x - x^2}{2x^2 - 3x}$.

5. $r(x) = \dfrac{5x^3 + 1}{20x^3 - 8000x}$.

6. $r(x) = \dfrac{x^4 - 4}{x^2 + 2}$.

7. $r(x) = \dfrac{x^2 - 7x + 11}{3x^2 + 10{,}000}$.

8. $r(x) = \dfrac{50x^{10} + 100}{x^{11} + x^6 + 1}$.

9. a) Evaluate $\lim\limits_{x \to \infty} \left(\dfrac{1}{x} \cdot \sin x\right)$. [Hint: $|\sin x| \leq 1$ for all $x \in R$.]

   b) What does this indicate about the graph of $f : f(x) = \dfrac{1}{x} \cdot \sin x$?

10. Evaluate $\lim\limits_{x \to \infty} (e^{-x} \cos x)$. What does this result indicate about the graph of $g : g(x) = e^{-x} \cos x$?

9.6 Vertical Asymptotes

In selecting the case of the rational function which we considered in some detail, namely r, where $r(x) = \dfrac{x}{x^2 + 1}$, we carefully chose the denominator to be a polynomial which did not have any real zeros. The reason for our doing this was to avoid temporarily the special problems which arise when the denominator of a rational function has real zeros. In this section we will consider these problems.

If r is a rational function, $r = \dfrac{p}{q}$, where the polynomial $q(x)$ has real zeros $x_1, x_2, \ldots, x_k$, and $x_1 < x_2 < \ldots < x_k$, then $r(x)$ is not defined for $x = x_j$, $j = 1, 2, \ldots, k$. The function r is, however, a continuous function on each of the following half-lines or intervals:

$$x < x_1,\ x_1 < x < x_2,\ x_2 < x < x_3,\ \ldots,\ x_{k-1} < x < x_k,\ x > x_k.$$

We are particularly interested in the behavior of $r(x)$ for values of x near the values $x_1, x_2, \ldots, x_k$. Let us consider the special case where r is defined by $r(x) = \dfrac{1}{x}$.

First it is fairly easy to describe the behavior of the graph of r for values of x *outside* some interval containing 0, say $-1 < x < 1$. We know that $r(x) > 0$ for $x > 0$ and $r(x) < 0$ for $x < 0$. This means the graph of r lies entirely in the first and third quadrants. We further know that the graph of r has no x- or y-intercept. Since the slope function of r is given by $r'(x) = -\dfrac{1}{x^2}$ we see that the graph of r is falling for $x \neq 0$. From Theorem 9.5a we know the graph of r is asymptotic to the x-axis. See Figure 9-11.

Figure 9-11

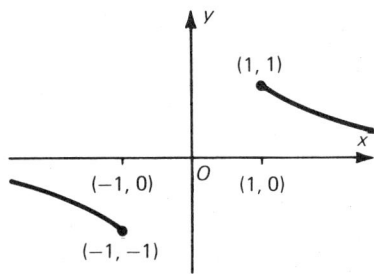

What happens as x approaches 0 from the right? It is apparent that if we take x sufficiently close to zero, but keep x positive, we can make $\dfrac{1}{x}$ larger than any given positive number. For example we see that if N is any positive number

$$0 < x < \frac{1}{N} \text{ implies that } \frac{1}{x} > N.$$

Said in another way: If we draw any horizontal line above the x-axis we can be sure that all points of the graph of $r(x)$ lie above this horizontal line provided x is sufficiently close to zero and positive. We say that the graph of r is asymptotic on the right to the vertical axis and the vertical axis is a *vertical asymptote* for the graph of r. See Figure 9-12.

Figure 9-12

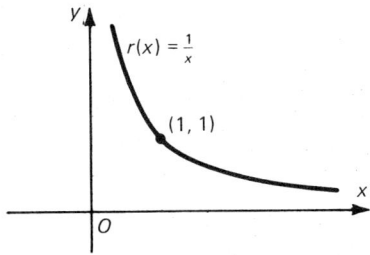

In a case we have already considered we saw that the graph of $\log_a$, $a > 1$, has the property that if L is any horizontal line lying below the x-axis then all points of the graph of $\log_a$, $(x, \log_a x)$, for which x is sufficiently small and positive will lie below this horizontal line. We say the graph of $\log_a$ is asymptotic on the right to the vertical axis. See Figure 9-13.

Figure 9-13

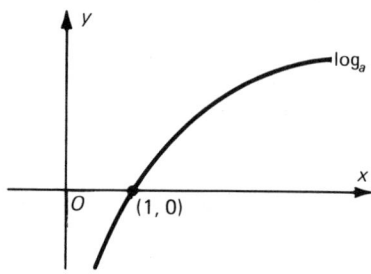

Returning to our discussion of $r(x)$ we see that since

$$r(-x) = \frac{1}{-x} = -r(x),$$

r is an odd function and its graph is symmetric with the origin. (Since r is its own inverse function $r(r(x)) = \dfrac{1}{\left(\dfrac{1}{x}\right)} = x$, we could have observed that the portion of the graph of r lying between the y-axis and the vertical line $x = 1$ is the reflection in the graph of I of that portion of its graph lying to the right of $x = 1$.) With these observations we graph r as in Figure 9-14. We say that graph of r is asymptotic to the line $x = 0$, i.e. the vertical axis, both on the right and left.

Figure 9-14

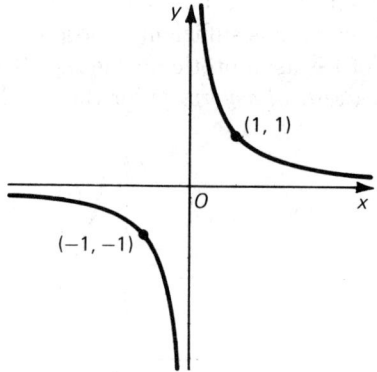

If we let the function s be defined by $s(x) = r(x + 1) = \dfrac{1}{x + 1}$, we know from previous work that the graph of s (in Figure 9-15) is the graph of r translated 1 unit to the left.

Figure 9-15

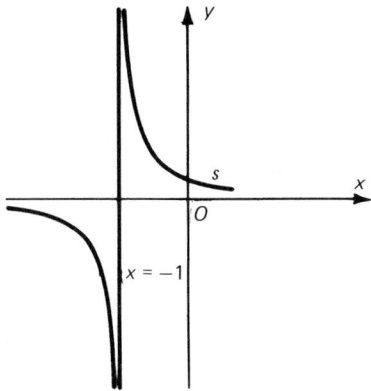

The graph of s is asymptotic to the vertical line $x = -1$, both on the right and left.

We now give a general definition of vertical asymptote. Where the expression $\begin{cases} \text{left} \\ \text{right} \end{cases}$ appears read either "left" or "right" consistently throughout the definition.

Definition 9.6a If L is the vertical line $x = a$ and f is a function defined in some interval directly to the $\begin{cases} \text{left} \\ \text{right} \end{cases}$ of a, (but not necessarily at a) and N is *any* positive number, and if either

1. all points of the graph lying to the $\begin{cases} \text{left} \\ \text{right} \end{cases}$ of L and sufficiently close to L will lie above the horizontal line defined by $l(x) = N$, or

2. all points lying to the $\begin{cases} \text{left} \\ \text{right} \end{cases}$ of L and sufficiently close to L will lie below the horizontal line defined by $l(x) = -N$,

then we say that L is a **vertical asymptote** for the graph of f and that the graph of f is **asymptotic to the line L on the** $\begin{cases} \text{left} \\ \text{right} \end{cases}$.

A graph which has a vertical asymptote L may be asymptotic to L on the right or the left or on both sides. The graph of $s(x) = \dfrac{1}{x+1}$ is asymptotic to the vertical line $x = -1$ on both the right and left. The graph of $\log_a$, $a > 1$, is asymptotic to the vertical line $x = 0$ on the right only.

9.6 VERTICAL ASYMPTOTES   367

If r is a rational function, $r = \dfrac{p}{q}$, and the polynomials p and q have a common linear factor $(x - x_0)$ occurring to the kth power in p and the mth power in q, then

$$r(x) = \frac{p(x)}{q(x)} = \frac{(x - x_0)^k s(x)}{(x - x_0)^m v(x)}, \qquad s(x_0) \neq 0 \text{ and } v(x_0) \neq 0,$$

by the Factor Theorem, p. 203. If $k \geq m$ then r is almost the same as the function R defined by

$$R(x) = \frac{(x - x_0)^{k-m} s(x)}{v(x)},$$

the only difference being that r is not defined at x_0 and $R(x_0) = 0$. Thus no vertical asymptote occurs at $x = x_0$. If $k < m$ then r is equally well defined by

$$r(x) = \frac{s(x)}{(x - x_0)^{m-k} v(x)}$$

as by

$$r(x) = \frac{p(x)}{q(x)},$$

since $\dfrac{s(x)}{(x - x_0)^{m-k} v(x)} = \dfrac{p(x)}{q(x)}$ for all values of x for which either side is defined. Neither side of the equation is defined for $x = x_0$.

Let us suppose, therefore, that r is a rational function, $r = \dfrac{p}{q}$, where q has a zero of multiplicity m, $m > 0$, at x_0 and $p(x_0) \neq 0$. Then by the Factor Theorem

$$r(x) = \frac{p(x)}{(x - x_0)^m v(x)}$$

$$= \frac{1}{(x - x_0)^m} \frac{p(x)}{v(x)}.$$

Since neither $p(x_0)$ nor $v(x_0)$ is equal to zero, and since both p and s are continuous functions, $\dfrac{p(x)}{v(x)}$ approaches the constant $\dfrac{p(x_0)}{v(x_0)}$ as x approaches x_0.

If $x > x_0$ but x is close to x_0, then $\dfrac{1}{(x - x_0)^m}$ is large. In fact $\dfrac{1}{(x - x_0)^m}$ can clearly be made larger than any positive number M by requiring $x > x_0$, but sufficiently close to x_0, i.e. by making $x - x_0$ sufficiently small. If $x - x_0 < \frac{1}{M^{1/m}}$, then

$$\frac{1}{x - x_0} > M^{1/m}$$

$$\frac{1}{(x - x_0)^m} > M.$$

Thus if $x > x_0$ and x approaches x_0, $r(x)$ will have the sign of $\frac{p(x_0)}{v(x_0)}$ and become arbitrarily large in absolute value. But this means the graph of r is asymptotic on the right to the vertical line $x = x_0$. See Figure 9–16.

Figure 9-16

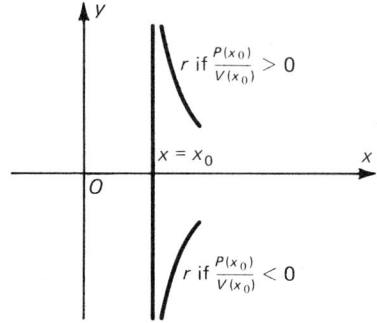

If $x < x_0$ but x is close to x_0, the analysis is quite similar. Since $(x - x_0) < 0$, $\frac{1}{(x - x_0)^m}$ will be positive if m is even and negative if m is odd. The factor $\frac{p(x)}{v(x)}$ will get close to $\frac{p(x_0)}{v(x_0)}$ as x approaches x_0 from the left and the factor $\frac{1}{(x - x_0)^m}$ will get arbitrarily large in absolute value. Thus $r(x)$ will get arbitrarily large in absolute value as x approaches x_0 from the left and will have the sign of $\frac{p(x_0)}{q(x_0)}$ if m is even and the negative of this if m is odd. But this means the graph of r is asymptotic to the line $x = x_0$ on the left. See Figure 9–17.

Figure 9-17

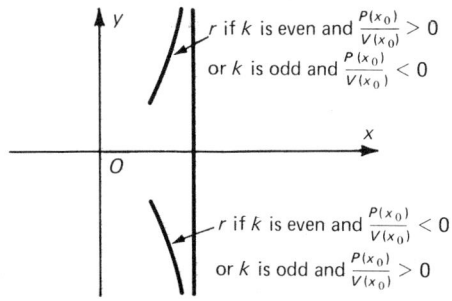

We consolidate the above analysis in the following theorem.

Theorem 9.6a *If r is a rational function, $r = \frac{p}{q}$, and x_0 is a zero of multiplicity m of $q(x)$, that is $q(x) = (x - x_0)^m v(x)$, and $p(x_0) \neq 0$, then the graph of r is asymptotic to the line $x = x_0$ on both the right and the left. The sign of $r(x)$ for x sufficiently near x_0, $x > x_0$, will be the sign of $\frac{p(x_0)}{v(x_0)}$ and the sign of $r(x)$ for $x < x_0$ and x sufficiently near x_0 will be the sign of $\frac{p(x_0)}{v(x_0)}$ if m is even, and the negative of this if m is odd.*

Example 1 $r(x) = \dfrac{x}{(x-1)^2(x+2)}$, $\dfrac{p(1)}{v(1)} = \dfrac{1}{3} > 0$ is shown in Figure 9-18.

Figure 9-18

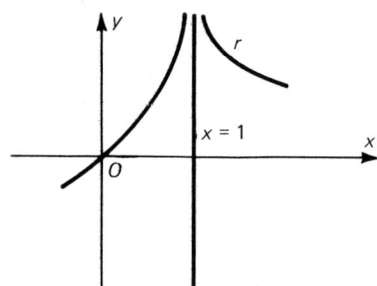

Example 2 $r(x) = \dfrac{-x^2}{(x-1)^3(x+2)}$, $\dfrac{p(1)}{v(1)} = -\dfrac{1}{3} < 0$ is shown in Figure 9-19.

Figure 9-19

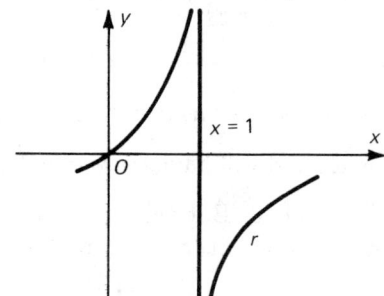

9.6a EXERCISES

For each of Exercises 1–10, determine whether the line $x = a$ is a vertical asymptote of the graph of r. Then sketch the graph of r for the interval $(a - 1, a + 1)$.

1. $r(x) = \dfrac{x+2}{x}$; $x = 0$.

2. $r(x) = \dfrac{x^2 - 7}{3x^3 + 2x}$; $x = 0$.

3. $r(x) = \dfrac{x^2 + 1}{x^4 - 1}$; $x = 1$.

4. $r(x) = \dfrac{x^3 - x^2}{x^3 + x}$; $x = 0$.

5. $r(x) = \dfrac{x^2 - 1}{x^2 - 2x + 1}$; $x = 1$.

6. $r(x) = \dfrac{x^2 + 2x}{x}$; $x = 0$.

7. $r(x) = \dfrac{x^2 + 2x - 15}{x + 5}$; $x = -5$.

8. $r(x) = \dfrac{3x^2 + 4x + 5}{x^2 + 8x - 20}$; $x = 2$.

9. $r(x) = \dfrac{x^2 - 5x + 6}{x^2 + x - 6}$; $x = 2$.

10. $r(x) = \dfrac{x^2 + 5x - 6}{x^3 + 6x^2}$; $x = -6$.

11. a) Sketch the graph of $r : r(x) = \dfrac{1}{x - 1}$.

    b) Given $f : f(x) = \dfrac{2x - 1}{x - 1}$, show that $f(x) = 2 + r(x)$.

    c) Use this information to sketch the graph of f.

Use the approach suggested by Exercise 11 to sketch the graph of each of the functions in Exercises 12–15.

12. $f : f(x) = \dfrac{3x + 7}{x + 2}$.

13. $f : f(x) = \dfrac{-2x - 1}{x + 1}$.

14. $f : f(x) = \dfrac{3x - 4}{x - 1}$.

15. $f : f(x) = \dfrac{x - 1}{x - 2}$.

We now have techniques for determining the basic properties of any rational function r and the nature of its graph provided we can determine the zeros of r, r', and r''. Since the determination of these zeros reduces to determining zeros of polynomials, we can approximate these zeros to any desired degree of accuracy if we cannot determine them exactly.

We conclude this section by applying some of our techniques to a rational function, which, although from its definition appears to be quite similar to the first rational function we considered, has a very different graph.

Consider the rational function r defined by

$$r(x) = \dfrac{x}{x^2 - 1}, \qquad x \in R,\ |x| \neq 1.$$

9.6 VERTICAL ASYMPTOTES

First the only zero of $r(x)$ is clearly 0. The sign diagram in Figure 9-20 shows that $r(x) < 0$ if $x < -1$ or $0 < x < 1$ and $r(x) > 0$ if $-1 < x < 0$ or $x > 1$.

Figure 9-20

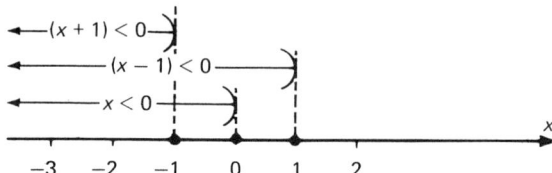

Since $r'(x) = \dfrac{(x^2 - 1) - x \cdot 2x}{(x^2 - 1)^2} = -\dfrac{x^2 + 1}{(x^2 - 1)^2}$, $r'(x)$ is negative whenever it is defined, and we conclude that r is a strictly decreasing function over the regions where it is defined, namely $x < -1$, $-1 < x < 1$ and $x > 1$. The slope of the tangent line to the graph of r at its x-intercept $(0, 0)$ is $r(0) = -1$. Now

$$r''(x) = -\left[\dfrac{(x^2 - 1)^2 2x - (x^2 + 1)2(x^2 - 1)2x}{((x^2 - 1)^2)^2}\right]$$

$$= \dfrac{2x(x^2 + 3)}{(x^2 - 1)^3}$$

upon simplification. The sign diagram for this function in Figure 9-21 indicates that $r''(x) < 0$ if $x < -1$ or $0 < x < 1$ and $r''(x) > 0$ if $-1 < x < 0$ or $x > 1$ (i.e., the sign of $r''(x)$ is the same as that of $r(x)$). Thus the graph of r is concave down if $x < -1$ or $0 < x < 1$ and is concave up if $-1 < x < 0$ or $x > 1$. (See Theorem 4.6a, p. 194.)

Figure 9-21

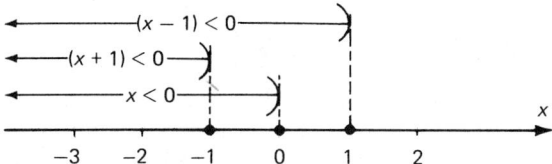

From Theorem 9.5c we can infer that the graph of r is asymptotic to the x-axis since r is a proper rational function.

From Theorem 9.6a we conclude that the graph is asymptotic to the vertical lines $x = -1$ and $x = 1$ on both the right and the left.

Using the above information we can draw an approximation of the graph of r as in Figure 9-22. We might have shortened the analysis somewhat by observing that r is an odd function and therefore the graph of r is symmetric with respect to the origin.

Figure 9-22

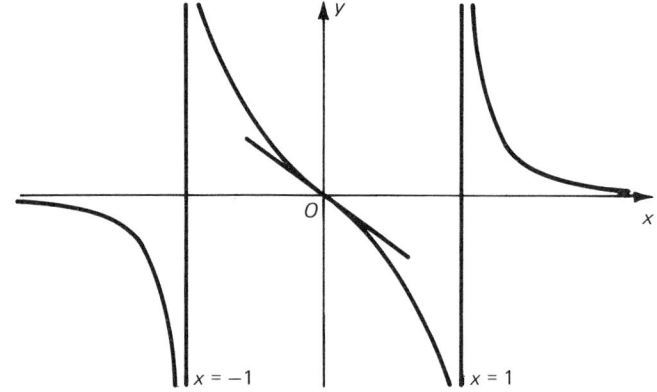

9.6b EXERCISES

For each of Exercises 1–8, determine the basic properties of the given function and draw its graph, including any horizontal or vertical asymptotes.

1. $r(x) = \dfrac{4}{x^2 - 4}.$

2. $r(x) = \dfrac{x^2}{x - 4}.$

3. $r(x) = \dfrac{x^2}{x^2 - 1}.$

4. $r(x) = \dfrac{x^3}{x - 2}.$

5. $r(x) = \dfrac{x + 1}{x^2 + 2}.$

6. $r(x) = \dfrac{6}{x^3 - x}.$

7. $r(x) = \dfrac{x^3}{x^2 - 1}.$

8. $r(x) = \dfrac{x}{(x - 1)^3}.$

10

Introduction to Polar Coordinates

Introduction

In this final chapter we give a brief introduction to a new type of coordinate system for the plane, called a *polar coordinate system*. For certain types of problems, the use of polar coordinates can greatly simplify their solution.

10.1 Polar Coordinates

To establish a polar coordinate system in the plane we select a point 0, called the **pole,** and a half-line, or ray, extending from the pole which determines the *zero direction*. This ray is called the **initial ray.** All other rays emanating from 0 are located in terms of the angle measured in radians (see Section 6.1) from the initial ray to the ray in question. Any point in the plane, at a positive distance from 0, can now be located by determining its distance r, from the pole, and the angle θ that the ray containing it makes with the initial ray. We then say that (r, θ) are **polar coordinates** of the point. For the pole 0, we say that $(0, \theta)$ is a set of polar coordinates for 0, for any value of θ. Some examples are given in Figure 10-1.

A ray from 0 which makes an angle of θ_0 with the initial ray has the equation $\theta = \theta_0$ in polar coordinates. A second (polar) coordinate of all points on this ray is θ_0. Of course, since the ray $\theta = \theta_0$ coincides with each of the rays $\theta = \theta_0 + 2n\pi$, where n is an integer, we see that the polar coordinates for a point in the plane are not unique. The point with polar coordinates (r, θ_0), also has any of the coordinates $(r, \theta_0 + 2n\pi)$, where n is an integer. The fact that a point in the plane does not have unique polar coordinates, unlike the situation with rectangular coordi-

Figure 10-1

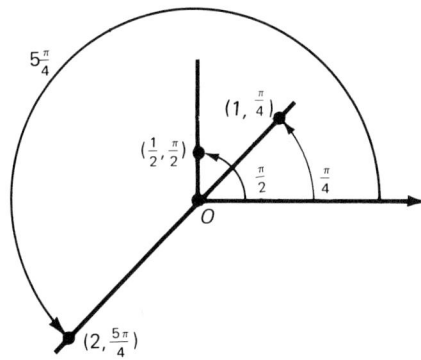

nates, can sometimes be inconvenient. This is often compensated for however, by the simplicity of the polar form of the equation of a curve. For example the polar form of the equation of a circle with radius 1 and center at the pole is $r = 1$. Since any point on this circle is at a unit distance from the origin it has polar coordinates of the form $(1, \theta)$ where θ is a real number. Conversely, any point with polar coordinates $(1, \theta)$ where θ is a real number, lies on this circle.

10.1 EXERCISES

Plot each of the following points and write two other pairs of polar coordinates for each point.

1. $\left(3, \dfrac{\pi}{4}\right)$
2. $\left(2, -\dfrac{\pi}{6}\right)$
3. $\left(\dfrac{1}{2}, \dfrac{2\pi}{3}\right)$
4. $\left(4, \dfrac{5\pi}{4}\right)$
5. $\left(1, -\dfrac{7\pi}{6}\right)$
6. $(3, \pi)$

Determine whether each of the following pairs of points is symmetric with respect to (a) the pole, (b) the line containing the initial ray, or (c) the line through the pole that is perpendicular to the initial ray.

7. $\left(8, \dfrac{\pi}{3}\right), \left(8, -\dfrac{\pi}{3}\right)$
8. $\left(6, \dfrac{\pi}{6}\right), \left(6, \dfrac{7\pi}{6}\right)$
9. $\left(2, \dfrac{3\pi}{4}\right), \left(2, -\dfrac{\pi}{4}\right)$
10. $(r, \theta), (r, -\theta)$
11. $(r, \theta), (r, \pi - \theta)$
12. $(r, \theta), (r, \pi + \theta)$
13. $(r, \pi + \theta), (r, \pi - \theta)$
14. $\left(r, \dfrac{\pi}{2} + \theta\right), \left(r, \dfrac{\pi}{2} - \theta\right)$
15. $\left(r, \dfrac{3\pi}{2} + \theta\right), \left(r, \dfrac{3\pi}{2} - \theta\right)$

Draw the graph of the points which satisfy each of the following conditions.

16. $r = 5$.
17. $\theta = 1$.
18. $\tan \theta = 1$.
19. Let l be a line that does not contain the pole, O. Draw $\overline{OA} \perp l$ and let the coordinates of A be (a, w). Let $P(r, \theta)$ be any other point on l.

a) Show that $r \cos (\theta - w) = a$.
b) Show that the coordinates of A also satisfy this equation.

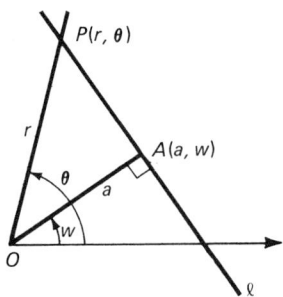

10.2 The Relationship of Polar and Rectangular Coordinates

Let us simultaneously consider a rectangular and a polar coordinate system, with the pole placed at the origin and the initial ray coinciding with the positive x-axis. Let (r, θ) be a set of polar coordinates for a point P with rectangular coordinates (x, y).

From our work with the trigonometric functions we know that the ray corresponding to the polar angle θ intercepts the unit circle about the origin in the point with rectangular coordinates $(\cos \theta, \sin \theta)$. By proportionality the same ray intercepts the circle of radius r about the origin in the point with rectangular coordinates $(r \cos \theta, r \sin \theta)$. (See Figure 10-2.) Since the point with rectangular coordinates $(r \cos \theta, r \sin \theta)$ lies on the circle of radius r about the origin and on the ray corresponding to the polar angle θ, it has polar coordinates (r, θ). This establishes a relationship between the rectangular and polar coordinates as shown in Figure 10-3. Namely, the point with polar coordinates (r, θ) has rectangular coordinates (x, y) where

$$x = r \cos \theta$$
$$y = r \sin \theta.$$

Figure 10-2

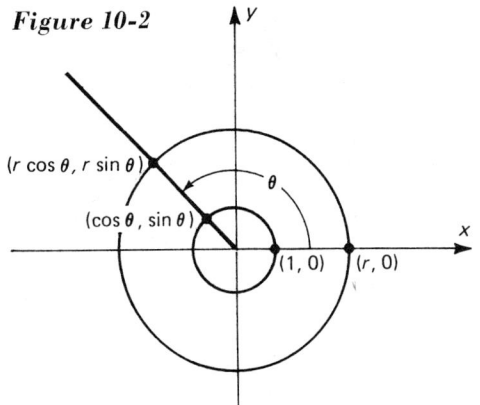

Figure 10-3

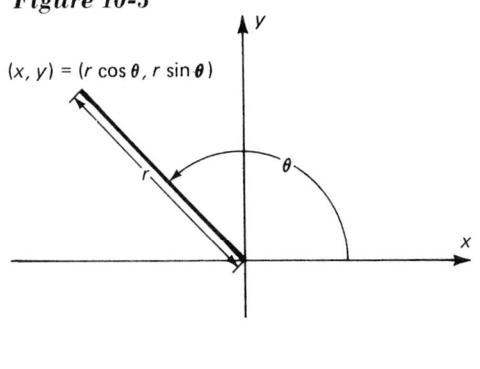

We can now write equations in polar coordinates for some familiar curves. For example the vertical line with equation $x = b$ in rectangular coordinates has the equation

$$r \cos \theta = b$$

in polar coordinates, or

$$r = \frac{b}{\cos \theta} = b \sec \theta.$$

This relationship is illustrated in Figure 10–4.

Figure 10-4

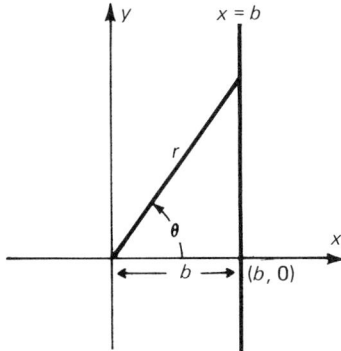

As another example consider the circle C of radius a with center at $x = a$, $y = 0$ in Figure 10–5. From geometry you should recall that $\measuredangle OPA$ is a right angle. Since the diameter $\overline{OA}$ has length $2a$, we see that

$$\frac{r}{2a} = \cos \theta$$

or
$$r = 2a \cos \theta.$$

Figure 10-5

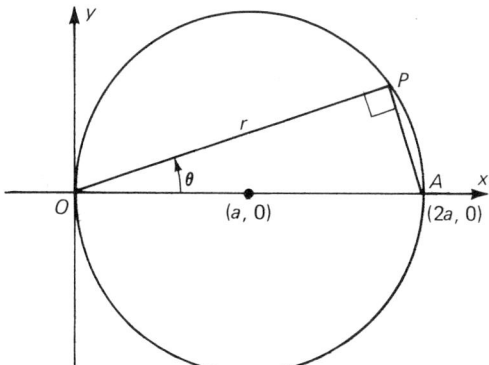

We could have derived the polar equation for this circle from its rectangular equation. Since C is the set of all points at a distance a from $(a, 0)$ we see that any point (x, y) on C satisfies the equation

$$\sqrt{(x-a)^2 + y^2} = a$$
or
$$(x-a)^2 + y^2 = a^2.$$

The latter equation is the standard rectangular form for the circle C. If we now substitute $x = r \cos \theta$ and $y = r \sin \theta$ into this equation, we obtain
$$(r \cos \theta - a)^2 + (r \sin \theta)^2 = a^2$$
or
$$r^2 \cos^2 \theta - 2ar \cos \theta + a^2 + r^2 \sin^2 \theta = a^2$$

Since $\cos^2 \theta + \sin^2 \theta = 1$, the last equation simplifies to
$$r^2 - 2ar \cos \theta = 0$$
or
$$r = 2a \cos \theta.$$

10.2 EXERCISES

Find rectangular coordinates for each of the following points.

1. $\left(10, \dfrac{11\pi}{6}\right)$
2. $\left(16, -\dfrac{\pi}{6}\right)$
3. $\left(14, \dfrac{2\pi}{3}\right)$
4. $(12, -\pi)$
5. $\left(6, \dfrac{5\pi}{4}\right)$
6. $\left(8, -\dfrac{3\pi}{4}\right)$

Find polar coordinates for each of the following points.

7. $(5\sqrt{3}, -5)$
8. $(-6, 6)$
9. $(8, 0)$
10. $(0, 10)$
11. $(10, 10)$
12. $(-8\sqrt{3}, -8)$

13. Show that if (r, θ) and (x, y) are polar and rectangular coordinate of the same point, then
   a) $r^2 = x^2 + y^2$,
   b) if $x \neq 0$, $\tan \theta = \dfrac{y}{x}$.

Sketch each of the following curves. Then identify the curve by transforming its equation to rectangular coordinates.

14. $r \sin \theta = 6$.
15. $r = 10 \sin \theta$.
16. $r = 6 \sin \theta + 8 \cos \theta$.

Sketch each of the following curves. Then transform its equation into polar coordinates.

17. $x^2 + y^2 - 10y = 0$.
18. $x^2 + y^2 - 12x = 0$.
19. $xy = 18$.
20. $y^2 - x^2 = 36$.
21. Show that if d is the distance between $P_1(r_1, \theta_1)$ and $P_2(r_2, \theta_2)$ then
$$d^2 = r_1^2 + r_2^2 - 2r_1 r_2 \cos(\theta_1 - \theta_2).$$

10.3 Conic Sections in Polar Coordinates

In Section 3.1 we briefly discussed the definition of the parabola as the plane curve which is the set of points P in the plane such that the distance of P from a fixed point F, called the focus, and its distance from a fixed line l, called the directrix, are equal.

$$d(P, F) = d(P, l).$$

This definition leads to simple polar equation for the parabola if we choose the origin as the focus and the line $x = -b$ as the directrix. From Figure 10-6 we see that the distance of the point P from the line $x = -b$ is $x + b$ and its distance from 0 is r. For P to be on the parabola we must have $r = x + b$, or since $x = r \cos \theta$, $r = r \cos \theta + b$. This simplifies to the polar equation of the parabola in the form

$$r = \frac{b}{1 - \cos \theta}, \quad \theta \neq 2n\pi, \; n \text{ an integer.}$$

Figure 10-6

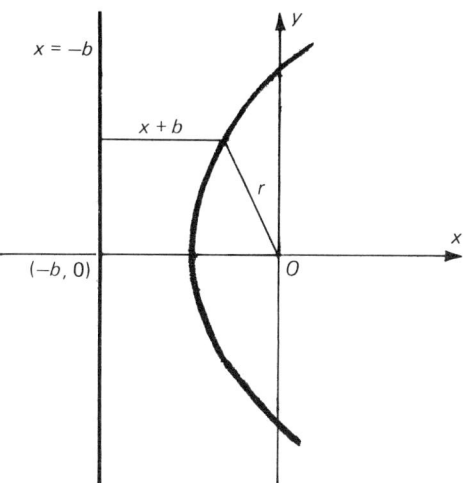

The **parabola** is one of three classical curves which are collectively called the *conic sections*. The other two are the **ellipse** and **hyperbola**. (See Figure 10-7 on page 380.) These curves are called conic sections because the intersection of a plane with a right circular cone of two nappes (i.e. a double right circular cone) is always one of these curves, or a straight line, a pair of intersecting straight lines, or a point.

It happens that both the ellipse and hyperbola can also be defined in terms of their relationship to a focus F and directrix l.

Consider the set S of points P in the plane such that

$$d(P, F) = e \, d(P, l),$$

where e is a positive number. The case where $e = 1$ we have already considered and S is a *parabola*. If $0 < e < 1$, S is an *ellipse*, and if $e > 1$, S is a *hyperbola*.

Figure 10-7

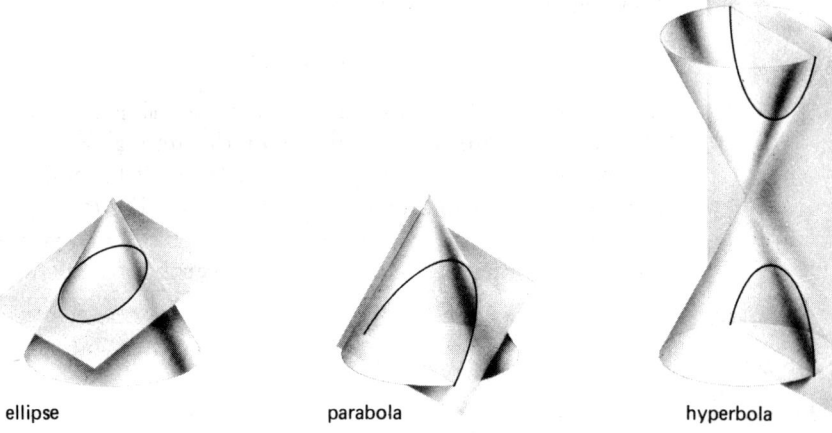

ellipse   parabola   hyperbola

Let us determine the polar form of the equation $d(P, F) = ed(P, l)$ where F is the origin and l is the line $x = -b$. As was the case in Figure 10-6 $d(P, F) = r$ and $d(P, l) = |x + b|$. If $x + b \geq 0$ (i.e. $x \geq -b$), then $|x + b| = x + b$ and our equation is

$$r = e(x + b) = e(r \cos \theta + b)$$

or

$$\text{A) } r = \frac{eb}{1 - e \cos \theta}.$$

If $x + b < 0$, (i.e. $x < -b$), then $|x + b| = -(x + b)$ and our equation is

$$r = -e(x + b) = -e(r \cos \theta + b)$$

or

$$\text{B) } r = \frac{-eb}{1 + e \cos \theta}.$$

Since r must be nonnegative, Equation B does not determine a curve if $1 + e \cos \theta \geq 0$. This leads to two conclusions.

1. For a parabola or an ellipse, $0 < e \leq 1$ and, consequently, $1 + e \cos \theta \geq 0$ for all θ. Therefore Equation B does not apply to these curves.

2. For the hyperbola, $e > 1$, so both Equations A and B are valid for certain values of θ. Equation A corresponds to the right-hand branch of the hyperbola and B to the left-hand branch. (See Figure 10–8.)

Figure 10-8

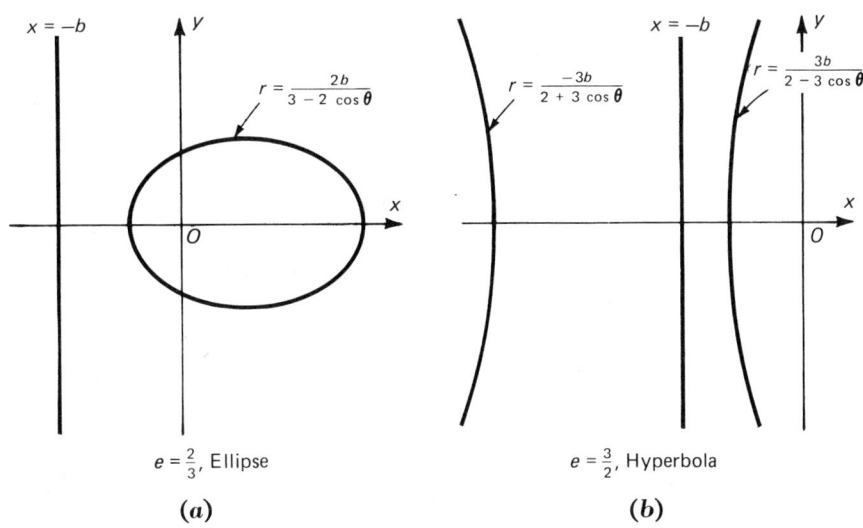

(a) $e = \frac{2}{3}$, Ellipse

(b) $e = \frac{3}{2}$, Hyperbola

10.3 EXERCISES

Transform each of the following equations of parabolas and ellipses to the standard polar form. Then find e and b and draw the curve and its directrix.

1. $r = \dfrac{12}{2 - \cos\theta}$.

2. $r = \dfrac{6}{1 - \cos\theta}$.

3. $r = \dfrac{16}{4 - 3\cos\theta}$.

4. $r = \dfrac{15}{3 - 2\cos\theta}$.

5. $r = \dfrac{-8}{\cos\theta - 1}$.

6. $r = \dfrac{15}{5 - 4\cos\theta}$.

Each of the following equations determines one branch of a hyperbola. Write an equation for the other branch and draw the hyperbola.

7. $r = \dfrac{8}{1 - 2\cos\theta}$.

8. $r = \dfrac{-20}{4 + 5\cos\theta}$.

9. $r = \dfrac{-12}{1 + 2\cos\theta}$.

10. $r = \dfrac{10}{2 - 3\cos\theta}$.

11. Show that the polar equation of a parabola with the origin as its focus and the line $y = -b$ as its directrix is $r = \dfrac{b}{1 - \sin\theta}$.

Find a polar equation or equations for each of the following conics.

12. Parabola with focus at $(0, 0)$ and directrix $x = b$.
13. Ellipse with focus at $(0, 0)$ and directrix $y = -b$.

14. Ellipse with focus at (0, 0) and directrix $x = b$.
15. Hyperbola with focus at (0, 0) and directrix $y = -b$.
16. Hyperbola with focus at (0, 0) and directrix $x = b$.

10.4 Other Polar Curves *(Optional)*

There are many other curves that are useful in solving practical problems. For example, James Watt developed a crossed parallelogram mechanism based on the curve known as the *lemniscate of Bernoulli* to approximate linear motion and thereby reduced the height of his steam engine housing by nine feet.

A curve whose equation is given in polar coordinates can be sketched by assigning values to θ, determining the corresponding values of r, plotting the points thus determined, and drawing a smooth curve through them.

However, we can frequently reduce the amount of labor by making use of knowledge of the symmetry of the curve and the range of values that can be assigned to θ.

Example 1 Consider the lemniscate $r^2 = 4 \cos 2\theta$. First we note that

$$\cos 2(-\theta) = \cos (-2\theta) = \cos 2\theta$$

and hence the curve is symmetric to the line containing the initial ray. Further,

$$\cos 2(\pi - \theta) = \cos (2\pi - 2\theta) = \cos (-2\theta) = \cos 2\theta,$$

so the curve is symmetric to the line through the pole that is perpendicular to the initial ray.

From these properties of symmetry, we need only know what the curve looks like for values of θ in the interval $0 \leq \theta \leq \frac{\pi}{2}$ in order to sketch the complete curve.

From the equation $r^2 = 4 \cos 2\theta$, it follows that r is defined only when $\cos 2\theta \geq 0$. Now if $\frac{\pi}{4} < \theta \leq \frac{\pi}{2}$, we have $\frac{\pi}{2} < 2\theta \leq \pi$ and $\cos 2\theta < 0$.

Hence we can further restrict the values of θ to the interval $0 \leq \theta \leq \frac{\pi}{4}$.

Now, as θ increases from 0 to $\frac{\pi}{4}$, 2θ increases from 0 to $\frac{\pi}{2}$ and $\cos 2\theta$

decreases from 1 to 0. Hence as θ increases from 0 to $\frac{\pi}{4}$, r^2 decreases from 4 to 0 or r decreases from 2 to 0. Thus, the desired portion of this lemniscate appears as in Figure 10-9. From the symmetry of the curve the complete graph can be sketched as in Figure 10-10.

Figure 10-10

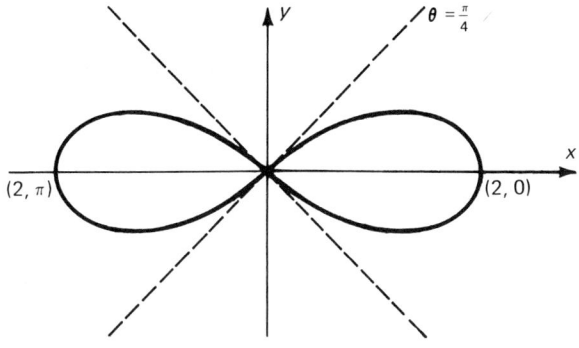

Figure 10-9

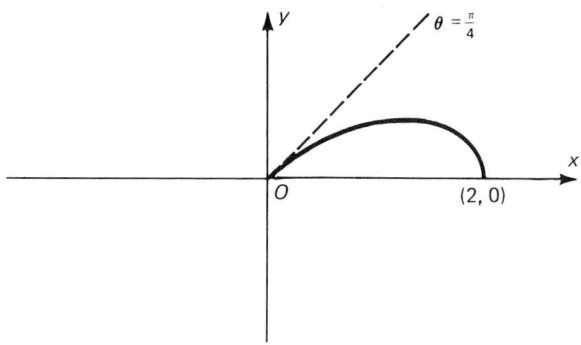

10.4 EXERCISES

Sketch the graph of each of the following curves.

1. $r = 4(1 - \cos \theta)$. (cardioid)
2. $r = 2(1 + \cos \theta)$. (cardioid)
3. $r = 3(1 - \sin \theta)$. (cardioid)
4. $r = 5(1 + \sin \theta)$. (cardioid)
5. $r = 4 + 2 \cos \theta$. (limaçon)
6. $r = 4|\sin 2\theta|$. (four-leaf rose)
7. $r = 6 \sin 3\theta$. (three-leaf rose)
8. $r^2 = 9 \sin 2\theta$. (lemniscate)
9. $r = 2\theta$. (spiral of Archimedes)
10. $r\theta = \pi$. (hyperbolic spiral)

Tables

*Common logarithms from 1.00 through 9.99**

n	0	1	2	3	4	5	6	7	8	9
10	0000	0043	0086	0128	0170	0212	0253	0294	0334	0374
11	0414	0453	0492	0531	0569	0607	0645	0682	0719	0755
12	0792	0828	0864	0899	0934	0969	1004	1038	1072	1106
13	1139	1173	1206	1239	1271	1303	1335	1367	1399	1430
14	1461	1492	1523	1553	1584	1614	1644	1673	1703	1732
15	1761	1790	1818	1847	1875	1903	1931	1959	1987	2014
16	2041	2068	2095	2122	2148	2175	2201	2227	2253	2279
17	2304	2330	2355	2380	2405	2430	2455	2480	2504	2529
18	2553	2577	2601	2625	2648	2672	2695	2718	2742	2765
19	2788	2810	2833	2856	2878	2900	2923	2945	2967	2989
20	3010	3032	3054	3075	3096	3118	3139	3160	3181	3201
21	3222	3243	3263	3284	3304	3324	3345	3365	3385	3404
22	3424	3444	3464	3483	3502	3522	3541	3560	3579	3598
23	3617	3636	3655	3674	3692	3711	3729	3747	3766	3784
24	3802	3820	3838	3856	3874	3892	3909	3927	3945	3962
25	3979	3997	4014	4031	4048	4065	4082	4099	4116	4133
26	4150	4166	4183	4200	4216	4232	4249	4265	4281	4298
27	4314	4330	4346	4362	4378	4393	4409	4425	4440	4456
28	4472	4487	4502	4518	4533	4548	4564	4579	4594	4609
29	4624	4639	4654	4669	4683	4698	4713	4728	4742	4757
30	4771	4786	4800	4814	4829	4843	4857	4871	4886	4900
31	4914	4928	4942	4955	4969	4983	4997	5011	5024	5038
32	5051	5065	5079	5092	5105	5119	5132	5145	5159	5172
33	5185	5198	5211	5224	5237	5250	5263	5276	5289	5302
34	5315	5328	5340	5353	5366	5378	5391	5403	5416	5428
35	5441	5453	5465	5478	5490	5502	5514	5527	5539	5551
36	5563	5575	5587	5599	5611	5623	5635	5647	5658	5670
37	5682	5694	5705	5717	5729	5740	5752	5763	5775	5786
38	5798	5809	5821	5832	5843	5855	5866	5877	5888	5899
39	5911	5922	5933	5944	5955	5966	5977	5988	5999	6010
40	6021	6031	6042	6053	6064	6075	6085	6096	6107	6117
41	6128	6138	6149	6160	6170	6180	6191	6201	6212	6222
42	6232	6243	6253	6263	6274	6284	6294	6304	6314	6325
43	6335	6345	6355	6365	6375	6385	6395	6405	6415	6425
44	6435	6444	6454	6464	6474	6484	6493	6503	6513	6522
45	6532	6542	6551	6561	6571	6580	6590	6599	6609	6618
46	6628	6637	6646	6656	6665	6675	6684	6693	6702	6712
47	6721	6730	6739	6749	6758	6767	6776	6785	6794	6803
48	6812	6821	6830	6839	6848	6857	6866	6875	6884	6893
49	6902	6911	6920	6928	6937	6946	6955	6964	6972	6981
50	6990	6998	7007	7016	7024	7033	7042	7050	7059	7067
51	7076	7084	7093	7101	7110	7118	7126	7135	7143	7152
52	7160	7168	7177	7185	7193	7202	7210	7218	7226	7235
53	7243	7251	7259	7267	7275	7284	7292	7300	7308	7316
54	7324	7332	7340	7348	7356	7364	7372	7380	7388	7396

*Mantissas; decimal points omitted

n	0	1	2	3	4	5	6	7	8	9
55	7404	7412	7419	7427	7435	7443	7451	7459	7466	7474
56	7482	7490	7497	7505	7513	7520	7528	7536	7543	7551
57	7559	7566	7574	7582	7589	7597	7604	7612	7619	7627
58	7634	7642	7649	7657	7664	7672	7679	7686	7694	7701
59	7709	7716	7723	7731	7738	7745	7752	7760	7767	7774
60	7782	7789	7796	7803	7810	7818	7825	7832	7839	7846
61	7853	7860	7868	7875	7882	7889	7896	7903	7910	7917
62	7924	7931	7938	7945	7952	7959	7966	7973	7980	7987
63	7993	8000	8007	8014	8021	8028	8035	8041	8048	8055
64	8062	8069	8075	8082	8089	8096	8102	8109	8116	8122
65	8129	8136	8142	8149	8156	8162	8169	8176	8182	8189
66	8195	8202	8209	8215	8222	8228	8235	8241	8248	8254
67	8261	8267	8274	8280	8287	8293	8299	8306	8312	8319
68	8325	8331	8338	8344	8351	8357	8363	8370	8376	8382
69	8388	8395	8401	8407	8414	8420	8426	8432	8439	8445
70	8451	8457	8463	8470	8476	8482	8488	8494	8500	8506
71	8513	8519	8525	8531	8537	8543	8549	8555	8561	8567
72	8573	8579	8585	8591	8597	8603	8609	8615	8621	8627
73	8633	8639	8645	8651	8657	8663	8669	8675	8681	8686
74	8692	8698	8704	8710	8716	8722	8727	8733	8739	8745
75	8751	8756	8762	8768	8774	8779	8785	8791	8797	8802
76	8808	8814	8820	8825	8831	8837	8842	8848	8854	8859
77	8865	8871	8876	8882	8887	8893	8899	8904	8910	8915
78	8921	8927	8932	8938	8943	8949	8954	8960	8965	8971
79	8976	8982	8987	8993	8998	9004	9009	9015	9020	9025
80	9031	9036	9042	9047	9053	9058	9063	9069	9074	9079
81	9085	9090	9096	9101	9106	9112	9117	9122	9128	9133
82	9138	9143	9149	9154	9159	9165	9170	9175	9180	9186
83	9191	9196	9201	9206	9212	9217	9222	9227	9232	9238
84	9243	9248	9253	9258	9263	9269	9274	9279	9284	9289
85	9294	9299	9304	9309	9315	9320	9325	9330	9335	9340
86	9345	9350	9355	9360	9365	9370	9375	9380	9385	9390
87	9395	9400	9405	9410	9415	9420	9425	9430	9435	9440
88	9445	9450	9455	9460	9465	9469	9474	9479	9484	9489
89	9494	9499	9504	9509	9513	9518	9523	9528	9533	9538
90	9542	9547	9552	9557	9562	9566	9571	9576	9581	9586
91	9590	9595	9600	9605	9609	9614	9619	9624	9628	9633
92	9638	9643	9647	9652	9657	9661	9666	9671	9675	9680
93	9685	9689	9694	9699	9703	9708	9713	9717	9722	9727
94	9731	9736	9741	9745	9750	9754	9759	9763	9768	9773
95	9777	9782	9786	9791	9795	9800	9805	9809	9814	9818
96	9823	9827	9832	9836	9841	9845	9850	9854	9859	9863
97	9868	9872	9877	9881	9886	9890	9894	9899	9903	9908
98	9912	9917	9921	9926	9930	9934	9939	9943	9948	9952
99	9956	9961	9965	9969	9974	9978	9983	9987	9991	9996

Values of trigonometric functions

Measure of ∡θ							
Degrees	Radians	sin θ	tan θ	cot θ	cos θ		
0	0.0000	0.0000	0.0000	—	1.0000	1.5708	90
1	.0175	.0175	.0175	57.290	.9998	1.5533	89
2	.0349	.0349	.0349	28.636	.9994	1.5359	88
3	.0524	.0523	.0524	19.081	.9986	1.5184	87
4	.0698	.0698	.0699	14.301	.9976	1.5010	86
5	.0873	.0872	.0875	11.430	.9962	1.4835	85
6	.1047	.1045	.1051	9.5144	.9945	1.4661	84
7	.1222	.1219	.1228	8.1443	.9925	1.4486	83
8	.1396	.1392	.1405	7.1154	.9903	1.4312	82
9	.1571	.1564	.1584	6.3138	.9877	1.4137	81
10	.1745	.1736	.1763	5.6713	.9848	1.3963	80
11	.1920	.1908	.1944	5.1446	.9816	1.3788	79
12	.2094	.2079	.2126	4.7046	.9781	1.3614	78
13	.2269	.2250	.2309	4.3315	.9744	1.3439	77
14	.2443	.2419	.2493	4.0108	.9703	1.3265	76
15	.2618	.2588	.2679	3.7321	.9659	1.3090	75
16	.2793	.2756	.2867	3.4874	.9613	1.2915	74
17	.2967	.2924	.3057	3.2709	.9563	1.2741	73
18	.3142	.3090	.3249	3.0777	.9511	1.2566	72
19	.3316	.3256	.3443	2.9042	.9455	1.2392	71
20	.3491	.3420	.3640	2.7475	.9397	1.2217	70
21	.3665	.3584	.3839	2.6051	.9336	1.2043	69
22	.3840	.3746	.4040	2.4751	.9272	1.1868	68
23	.4014	.3907	.4245	2.3559	.9205	1.1694	67
24	.4189	.4067	.4452	2.2460	.9135	1.1519	66
25	.4363	.4226	.4663	2.1445	.9063	1.1345	65
26	.4538	.4384	.4877	2.0503	.8988	1.1170	64
27	.4712	.4540	.5095	1.9626	.8910	1.0996	63
28	.4887	.4695	.5317	1.8807	.8829	1.0821	62
29	.5061	.4848	.5543	1.8040	.8746	1.0647	61
30	.5236	.5000	.5774	1.7321	.8660	1.0472	60
31	.5411	.5150	.6009	1.6643	.8572	1.0297	59
32	.5585	.5299	.6249	1.6003	.8480	1.0123	58
33	.5760	.5446	.6494	1.5399	.8387	.9948	57
34	.5934	.5592	.6745	1.4826	.8290	.9774	56
35	.6109	.5736	.7002	1.4281	.8192	.9599	55
36	.6283	.5878	.7265	1.3764	.8090	.9425	54
37	.6458	.6018	.7536	1.3270	.7986	.9250	53
38	.6632	.6157	.7813	1.2799	.7880	.9076	52
39	.6807	.6293	.8098	1.2349	.7771	.8901	51
40	.6981	.6428	.8391	1.1918	.7660	.8727	50
41	.7156	.6561	.8693	1.1504	.7547	.8552	49
42	.7330	.6691	.9004	1.1106	.7431	.8378	48
43	.7505	.6820	.9325	1.0724	.7314	.8203	47
44	.7679	.6947	.9657	1.0355	.7193	.8029	46
45	.7854	.7071	1.0000	1.0000	.7071	.7854	45
		cos θ	cot θ	tan θ	sin θ	Radians	Degrees
						Measure of ∡θ	

Answers for Selected Exercises

1.1 EXERCISES (page 3)

1. $\langle 1, \frac{1}{4}, \frac{1}{9}, \frac{1}{16}, \frac{1}{25}, \frac{1}{36}, \ldots \rangle$; neither   **3.** $\langle \frac{1}{2}, \frac{1}{4}, \frac{1}{8}, \frac{1}{16}, \frac{1}{32}, \frac{1}{64}, \ldots \rangle$; geometric   **5.** $\langle 1, \frac{5}{6}, \frac{7}{9}, \frac{3}{4}, \frac{11}{15}, \frac{13}{18}, \ldots \rangle$; neither   **7.** $\langle 0, 2, 6, 12, 20, 30, \ldots \rangle$; neither   **9.** $\langle 3, 0, -1, 0, 3, 8, \ldots \rangle$; neither   **11.** $\langle 2, -3, -8, -13, -18, -23, \ldots \rangle$; arithmetic   **13.** $\langle 1, 3, 6, 10, 15, 21, \ldots \rangle$; neither   **15.** $\langle 1, 2, 2, 4, 8, 32, \ldots \rangle$; neither   **17.** $\langle 4, 25, 46, 67, 88, 109, \ldots \rangle$ **19.** $\langle 4, 4.2, 4.4, 4.6, 4.8, 5, \ldots \rangle$ **21.** $\langle 3, 9, 27, 81, 243, 729, \ldots \rangle$ **23.** $\langle \sqrt{2}, 2, 2\sqrt{2}, 4, 4\sqrt{2}, 8, \ldots \rangle$ **25.** $x = \frac{1}{4}$

1.2 EXERCISES (page 6)

1. $n > 99$; $n > 999$ **3.** Limit is 0   **5.** No limit   **7.** No limit   **9.** Limit is 0
11. a) $n > 2$ **b)** $n > 5$ **c)** $n > 10$ **d)** $n > 100$ **13. a)** $n > 1$ **b)** $n > 2$ **c)** $n > 6$ **d)** $n > 7$

1.3 EXERCISES (page 10)

1. $\{x \mid x \geq 5\}$ **3.** $\{x \mid x \leq -5\}$ **5.** $\{x \mid -2 \leq x \leq 2\}$ **7.** $\{-4, 10\}$ **9.** $\{x \mid x < -2$ or $x > 8\}$ **11.** $\{x \mid x < -7$ or $x > -5\}$ **13.** $\{x \mid -12 \leq x \leq -8\}$
15. $\{x \mid 10 - \epsilon < x < 10 + \epsilon\}$ **17.** Union $= [-10, 10]$; Intersection $= (-5, 5)$
19. Union $= (-2, 5)$; Intersection $= (0, 3)$

1.4 EXERCISES (page 12)

1. a) 4 **b)** 100 **c)** 10,000 **d)** 1,000,000   **3. a)** 10 **b)** 100 **c)** 1000 **d)** 10,000   **5.** 18
7. 8

1.5 EXERCISES (page 15)

1. Bounded below only **3.** Bounded **5.** Bounded **7.** Unbounded **9.** Bounded
11. Bounded

1.7 EXERCISES (page 20)

1. glb $= -1$; lub $= \frac{1}{2}$; both in set **3.** glb $= 0$; lub $= 1$; lub in set **5.** glb $= -5$; lub $= 2.5$; both in set **7.** glb $= -3$; lub $= 3$; both in set **9.** Unbounded

1.8 EXERCISES (page 24)

1. $\frac{1}{7}$ **3.** 15 **5.** Nondecreasing **7.** Nondecreasing **9.** Nonincreasing **11.** $n + 3$
13. $7n(n + 1)(n + 2)$ **19.** Nonincreasing **21.** Nondecreasing

1.9 EXERCISES (page 28)

1. $\langle n^4 + 2n^3 - 7n^2 - 3n - 7 \rangle$; $\langle n^4 - 2n^3 - 3n^2 + 7n - 7 \rangle$ **3.** $\left\langle \dfrac{14n - 3}{10n} \right\rangle$; $\left\langle \dfrac{-6n + 7}{10n} \right\rangle$

5. $\left\langle \dfrac{3n - 5}{n^2 + 3n} \right\rangle$; $\left\langle \dfrac{n - 5}{n^2 + 3n} \right\rangle$ **7.** $\left\langle \dfrac{5n + 12}{(n + 3)(n + 2)^2} \right\rangle$; $\left\langle \dfrac{n}{(n + 3)(n + 2)^2} \right\rangle$ **9.** $\langle n + 5 \rangle$

11. $\left\langle \dfrac{n - 2}{3n} \right\rangle$ **13.** $\left\langle \dfrac{7(n^2 + n - 6)}{4} \right\rangle$

1.10 EXERCISES (page 32)

1. none; none; none; 0 **3.** $\frac{7}{5}$; $-\frac{3}{5}$; $\frac{2}{5}$; $\frac{2}{5}$ **5.** 0; 0; 0; 2 **7.** 0; 0; 0; $\frac{3}{2}$ **9.** none; none; none; 0
11. $\frac{4}{3}$; $-\frac{2}{3}$; $\frac{1}{3}$; $\frac{1}{3}$

2.1 EXERCISES (page 38)

1. Not a function. The set contains the ordered pairs (31, January) and (31, March). **3.** A function. **5.** A function. **7.** {(a, b), (e, f), (i, j), (o, p), (u, v)} **9.** Examples: $(-3, -6)$, $(0, -3)$, $(2, -1)$. Function. **11.** Examples: $(-5, 23)$, $(0, 32)$, $(10, 50)$. Function. **15.** (1, 1), (2, 7), (3, 13), (4, 19), (5, 25) **17.** (1, 4), (2, 10), (3, 18), (4, 28), (5, 40) **19.** (1, 2), (2, 4), (3, 6), (4, 8), (5, 10)

2.2 EXERCISES (page 42)

1. a) -3 **b)** -3 **c)** 4 **d)** $6 - 4a^2$ **e)** $-9a^2 + 6a + 5$ **3. a)** -8 **b)** $\frac{8}{27}$ **c)** $-\frac{1}{27}$ **d)** a^6
e) $1 - 3a + 3a^2 - a^3$ **5.** Domain $= R$; range $= R$ **7.** Domain $= \{x \mid x \neq 0\}$; range $= \{y \mid y \neq 0\}$ **13.** The set of negative integers. **17.** The set of positive odd integers.

2.3 EXERCISES (page 49)

1. a) 0, 3, −3   **b)** −6, 6, 27   **c)** No. The integers 1 and 2, for example, are not in the range of g.   **d)** Yes. Every real number y is the image of some real number $\frac{1}{3}y$, because R is closed under multiplication

3. 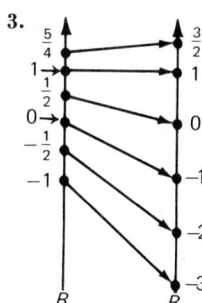   **5.** 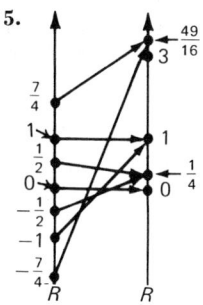   **7.**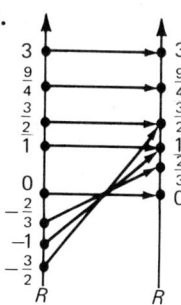

9. Not a function.   **11.** A function. Domain = $\{x \mid -4 \leq x \leq 4\}$; range = $\{y \mid 0 \leq y \leq 2\}$.

13. 3 units   **15.** $|4 - d|$ units   **17.** $|c - d|$ units   **19.** 3 units

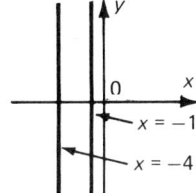   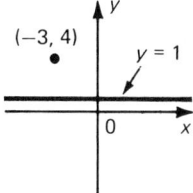

23. $|b - d|$ units

2.4 EXERCISES (page 55)

1. Increasing: $\{x \mid -3 \leq x \leq -1\}$; decreasing: $\{x \mid -1 \leq x \leq 1\}$ **3.** Increasing: $\{x \mid 3 \leq x \leq 3\frac{1}{2}\}$; decreasing: $\{x \mid -3 \leq x \leq -1\}$

5. One possible graph is shown below.

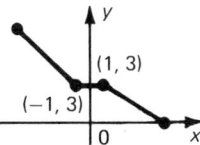

11. One possible graph is shown below.

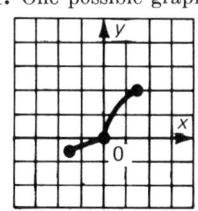

7. False. If this were the case then $g(0)$ would be less than $g(1)$, which would mean that g is not a nonincreasing function. Since $g(-1) = g(1) = 3$ and g is a nonincreasing function, it must also be true that $g(0) = 3$.

9. True. Since $g(0) = 3$ and g is a nonincreasing function, $x_1 \leq 0$ implies that $g(x_1) \geq 0$.

13. True. Since 4 is the greatest element of the range and 3 is the greatest element of the domain and h is increasing, $x_1 < 3$ implies that $h(x_1) < 4$.

15. True. Since $h(0) = 0$, $0 < x_2$ implies that $0 < h(x_2)$, or $h(x_2) > 0$.

2.5 EXERCISES (page 61)

1. Function f has no inverse because two of the elements obtained by reversing the components, $(1, -1)$ and $(1, 1)$, have the same first component.   **3.** Function h has an inverse. $h^{-1} = \{(-3, -3), (-1, -2), (0, 0), (2, 1), (3, 3)\}$.   **5.** The inverse is the set of all ordered pairs of real numbers in which the second component of each pair is greater by 5 than the first component.   **7.** The function has no inverse because all ordered pairs obtained by reversing the components would have 1 as their first component.   **9.** The function has no inverse. For example, both $(\pi, \sqrt{\pi})$ and $(\pi, -\sqrt{\pi})$ are contained in the set obtained by reversing the components.   **11.** $f^{-1}(x) = \dfrac{x-2}{-7}$. The domain and range of f^{-1} are R.   **13.** $h^{-1}(x) = \sqrt{x^2 + 3}$. The domain of h^{-1} is $\{x \mid x \in R \text{ and } x \geq 0\}$; the range of h^{-1} is $\{y \mid y \in R \text{ and } y \geq \sqrt{3}\}$.   **15.** $G^{-1}(x) = \dfrac{1 - 2x}{x}$. Domain of G^{-1}: $\{x \mid x \in R \text{ and } x < 0\}$; range of G^{-1}: $\{y \mid y \in R \text{ and } y < -2\}$.

2.6a EXERCISES (page 66)

1. -1 **3.** $-\sqrt{3}$ **5.** $-\sqrt{3}$ **7.** Collinear; $\dfrac{-4-2}{7-3} = \dfrac{8-(-4)}{-1-7}$.   **9.** Collinear; $\dfrac{-1-6}{1-6} = \dfrac{-8-(-1)}{-4-1}$.   **11.** Collinear; $\dfrac{\frac{2b+d}{3} - b}{\frac{2a+c}{3} - a} = \dfrac{d-b}{c-a}$.

13. 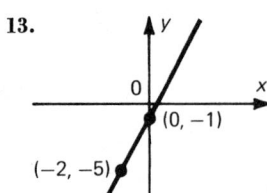   **15.** 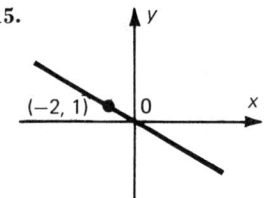   **17.**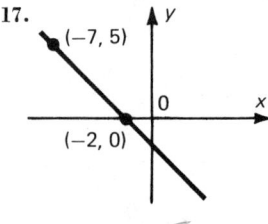

2.6b EXERCISES (page 68)

1. $0° < \alpha < 90°$ **3.** $0° < \alpha < 90°$ **5.** $90° < \alpha < 180°$

7. 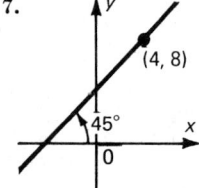   **9.** 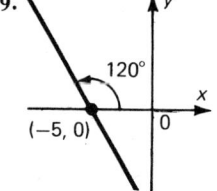   **11.**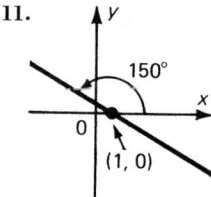

13. $45°$ **15.** $120°$ **17.** $72°$

2.6c EXERCISES (page 72)

1. $y = \frac{2}{3}x - 5$ **5.** $y = x + 1$ **11.** **15.**

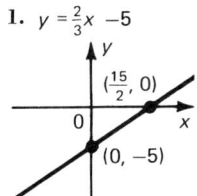 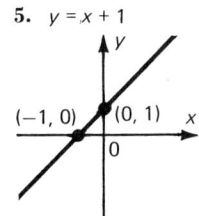 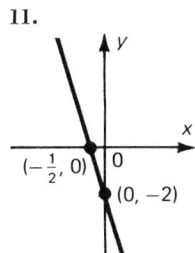 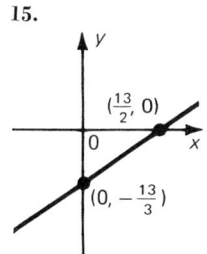

2.7 EXERCISES (page 76)

1. $y = \frac{1}{2}x + 2\frac{1}{2}$. **3.** $y = -\frac{1}{2}x$. **5.** $y = 4x + 8$.

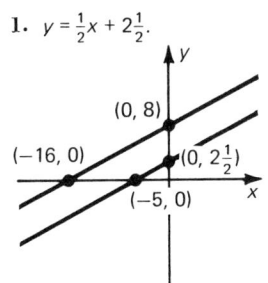 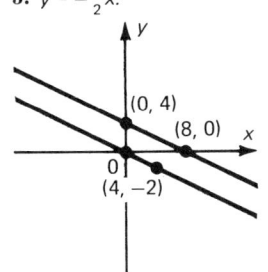 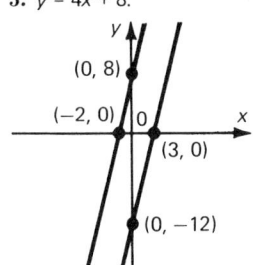

7. a) $y = \frac{3}{4}x$ and $y = -\frac{3}{2}x + 18$ are the equations of the lines containing the sides with slopes $\frac{3}{4}$ and $-\frac{3}{2}$. The coordinates of the third vertex are (8, 6).   **b)** $y = \frac{3}{4}x - 9$ and $y = -\frac{3}{2}x$. The coordinates of the third vertex are (4, −6).   **11. a)** The median to the side with endpoints (0, 0) and (8, 0) is contained in the line $y = 4x - 16$; the median to the side with endpoints (8, 0) and (6, 8) is contained in the line $y = \frac{4}{7}x$; and the median to the side with endpoints (6, 8) and (0, 0) is contained in the line $y = -\frac{4}{5}x + 6\frac{2}{5}$.   **b)** The intersection of lines $y = 4x - 16$ and $y = \frac{4}{7}x$ is the point $(4\frac{2}{3}, 2\frac{2}{3})$. Since $2\frac{2}{3} = -\frac{4}{5} \cdot 4\frac{2}{3} + 6\frac{2}{5}$, this point is also contained in the line $y = -\frac{4}{5}x + 6\frac{2}{5}$. Hence, the medians intersect in a common point.   **19. a)** The altitude to the side with endpoints (−6, 0) and (6, 0) is perpendicular to the x-axis; and it is contained in the line $x = 2$. The altitude to the side with endpoints (6, 0) and (2, 8) is contained in the line $y = \frac{1}{2}x + 3$. The altitude to the side with endpoints (2, 8) and (−6, 0) is contained in the line $y = -x + 6$.   **b)** The intersection of the lines $x = 2$ and $y = \frac{1}{2}x + 3$ is the point (2, 4). Since $4 = -2 + 6$, this point is also contained in the line $y = -x + 6$. Hence, the lines containing the altitudes intersect in a common point.

2.8 EXERCISES (page 78)

1. $\frac{9}{5}$ **3.** $5\sqrt{5}$ **5.** $\dfrac{43}{\sqrt{13}}$ **7.** A (10, 0); B (2, −4); C (4, 2); length of altitudes from A: $2\sqrt{10}$; from B: $2\sqrt{10}$; from C: $2\sqrt{5}$ **9.** $\dfrac{13}{\sqrt{5}}$

2.9 EXERCISES (page 81)

1. $\{x \mid x \leq 2\}$ **3.** $\{0\}$ **9.** $\{x \mid -18 < x < 18\}$ **11.** $\{x \mid x < -3 \text{ or } x > 3\}$
17. If $x < -1, f(x) = x + (-x - 1) = -1$. If $x \geq -1, f(x) = x + (x + 1) = 2x + 1$.

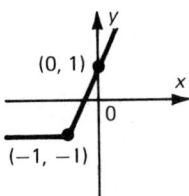

2.10 EXERCISES (page 85)

1. $x = -1$ **3.** $x = 0$ **5. a)** No **b)** All are 1 **c)** All are -1 **d)** No   **7.** 0.1   **9.** 0.0025
11. 0.0005   **13.** $p/3$ **15.** $p/2$

2.11 EXERCISES (page 88)

1. a) $\frac{11}{26}$ **b)** 1 **c)** 0 **d)** $\frac{21}{13}$ **3.** No; 1   **5.** No; 10   **7. a)** No **b)** Yes; 0   **9. a)** -2; Yes
b) Yes; -2

2.12 EXERCISES (page 91)

1. $(f + g)(x) = \dfrac{4x - 2}{x^2} \cdot (f - g)(x) = \dfrac{2x - 8}{x^2} \cdot D_f = D_g = D_{f+g} = \{x \mid x \neq 0\}$.

3. $(f + g)(x) = \dfrac{41}{14x} \cdot (f - g)(x) = \dfrac{-29}{14x} \cdot D_f = D_g = D_{f+g} = \{x \mid x \neq 0\}$.

5. $(f + g)(x) = \dfrac{-2x + 20}{15x^2} \cdot (f - g)(x) = \dfrac{8x - 20}{15x^2} \cdot D_f = D_g = D_{f+g} = \{x \mid x \neq 0\}$.

7. $(f \cdot g)(x) = x^2 + 2x - 8$ **9.** $(f \cdot g)(x) = x^3 + 1$ **11.** $g(x) = -2x^2 + 3x + 8$
13. $g(x) = 2x + 7$ **15.** $g(x) = x^2 + 2x - 1$ **17.** $\{5\}$ **19.** $\{3, -3\}$

2.14 EXERCISES (page 98)

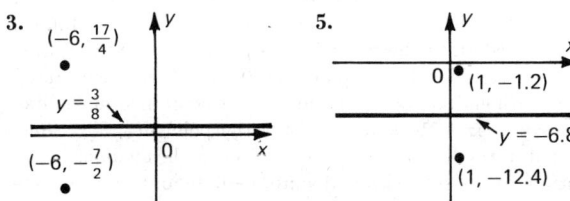

15. $(7, -.5)$ **17.** $(-1, -5.6)$ **19.** $(-\tfrac{5}{2}, \tfrac{5}{2})$ **21.** $(2, -3)$ **23.** $(-8, 5)$
25. $(-2 + \sqrt{3}, 0)$ **27.** x-axis   **29.** No symmetry   **31.** y-axis   **33.** No symmetry

2.15 EXERCISES (page 103)

1. $f(a) = -5; f(-a) = +5;$ and $-f(-a) - 5$. The function is even.   **3.** $f(a) = -\dfrac{1}{a^3};$

$f(-a) = -\dfrac{1}{(-a)^3} = \dfrac{1}{a^3};$ and $-f(-a) = -\dfrac{1}{a^3}$. The function is odd.   **5.** $f(a) = a^2 - 9;$

$f(-a) = a^2 - 9;$ and $-f(-a) = 9 - a^2$. The function is even.   **7.** $f(a) = a^2 - \dfrac{1}{a};$

$f(-a) = a^2 + \dfrac{1}{a};$ and $-f(-a) = -a^2 - \dfrac{1}{a}$. The function is neither even nor odd.

9. 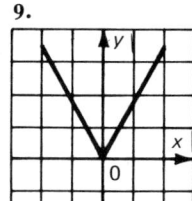 **11.** 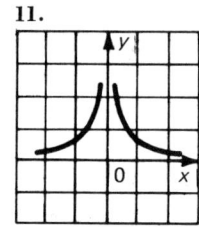 **15.** 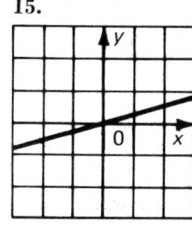 **17.**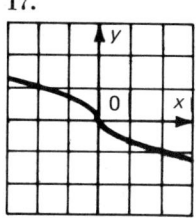

2.16 EXERCISES (page 108)

1. 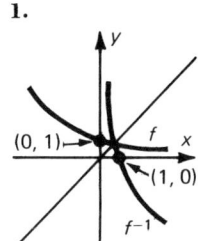 **3.** 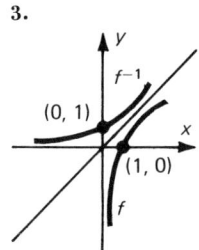 **5.** 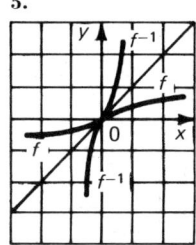 **7.** The function has no inverse. A restriction and its inverse are shown below.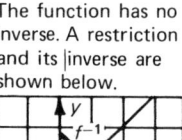

3.1 EXERCISES (page 116)

1. a) $\sqrt{(x_0)^2 + (y_0 - 4)^2}$ **b)** $|y_0 + 4|$ **c)** $\sqrt{(x_0)^2 + (y_0 - 4)^2} = |y_0 + 4|$.
e)

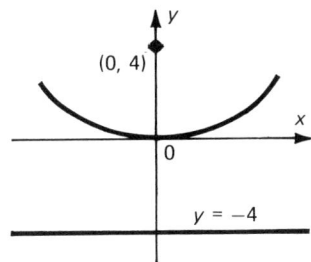

3. $\sqrt{(x_0 - 3)^2 + y_0^2} = |x_0 + 3|$; $x_0^2 - 6x_0 + 9 + y_0^2 = x_0^2 + 6x_0 + 9$; $y_0^2 = 12x_0$. $x = \dfrac{y^2}{12}$.

7. b) Yes; each vertical line in the plane intersects the graph in just one point.
c) If $p > 0$, the domain is R and the range is $\{y \mid y \geq 0\}$. If $p < 0$, the domain is R and the range is $\{y \mid y \leq 0\}$ **9.** Focus: $(0, 5)$; directrix: $y = -5$. **11.** Focus: $(\tfrac{1}{2}, 0)$; directrix: $x = -\tfrac{1}{2}$.
13. Focus: $(0, \tfrac{1}{8})$; directrix: $y = -\tfrac{1}{8}$. **15. a)** Coordinates of A: $(1, 1)$; coordinates of B: $(-1, 1)$
b) $d(0, A) = d(0, B) = \sqrt{2}$; $d(A, B) = 2$. **c)** Area: $\tfrac{1}{2}d(0, A) \cdot d(0, A) = \tfrac{1}{2}\sqrt{2} \cdot \sqrt{2} = 1$.

3.2 EXERCISES (page 121)

1. 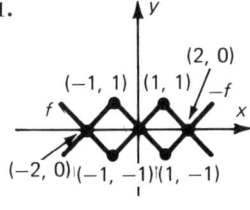 **3.**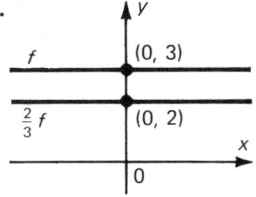

ANSWERS FOR SELECTED EXERCISES

5. Condition a **7.** Condition a **9.** Condition b

13. 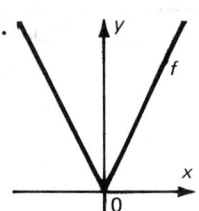 **15.** 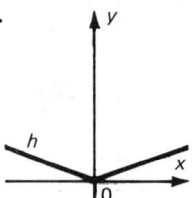 **17.**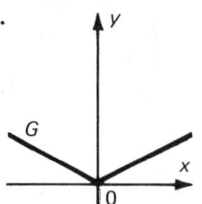

19. If $a = 1$, the graphs of f and af are identical. If $a = 0$, the graph of af is the graph of f compressed towards the x-axis so that every point lies in the x-axis.

3.3 EXERCISES (page 126)

1. Statement c is true.   **3.** Statement a is true.

5. 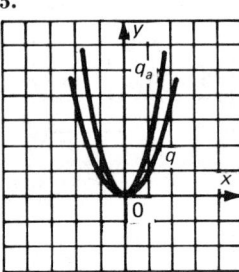   **7.** 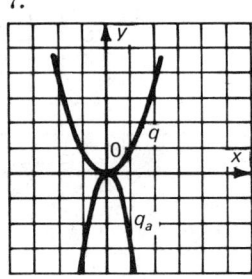   **9.**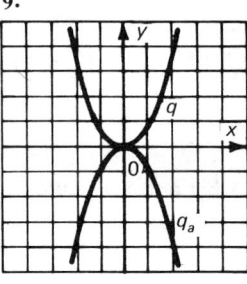

11. $\{x \mid -6 < x < 6\}$ **13.** $\{x \mid -3 < x < 1\}$ **15.** $\{x \mid x \leq -3 \text{ or } x \geq \frac{1}{2}\}$

17. a) $(4\sqrt{3}, 12)$ and $(-4\sqrt{3}, 12)$ **b)** $8\sqrt{3}$ **c)** $48\sqrt{3}$

3.4 EXERCISES (page 128)

1. 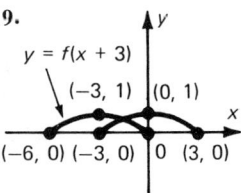   **5.** 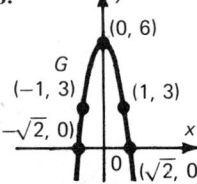   **7.**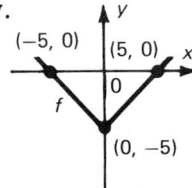

3.5 EXERCISES (page 133)

1. a) $(f \circ g)(1) = f(3) = 4$. **b)** $(g \circ f)(-2) = g(9) = 19$. **c)** $f(f(1)) = f(0) = 1$.
d) $g(g(-3.5)) = g(-6) = -11$.   **3.** $(f \circ g)(x) = f((x-1)^2) = 2(x-1)^2 - 3 = 2x^2 - 4x - 1$;
$(g \circ f)(x) = g(2x - 3) = (2x - 4)^2 = 4x^2 - 16x + 16$.   **5.** $(f \circ g)(x) = f(5x + 4) = -3$;
$(g \circ f)(x) = g(-3) = -11$.

9. 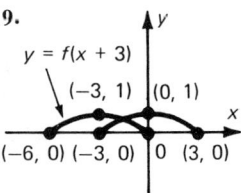   **11.** 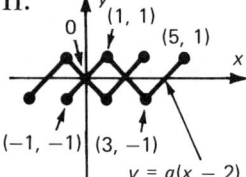   **13.**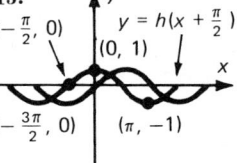

3.6 EXERCISES (page 138)

1. $g: y = -3x + 9$.   **3.** $r: y = -3x + 12$.   **5.** The graph of F is the graph of q_a translated 5 units to the left. $F: F(x) = x^2 + 10x + 25$. The graph of G is the graph of F translated 2 units upward; it is the graph of q_a translated 5 units to the left and 2 units upward. $G: G(x) = x^2 + 10x + 27$.

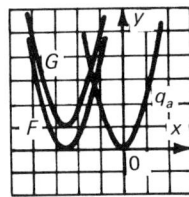

9. $Q_{1,-2,1}(x) = x^2 - 2x + 1 = (x^2 - 2x + 1) + 1 - 1 = (x - 1)^2$. Hence, the graph of Q is a 1 horizontal translate of the graph of $q(x) = x^2$.

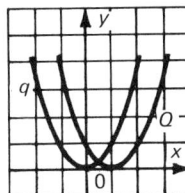

11. $Q_{1,-4,-1}(x) = x^2 - 4x - 1 = (x^2 - 4x + 4) - 1 - 4 = (x - 2)^2 - 5$. Hence, the graph of Q is a 2 horizontal and a -5 vertical translate of the graph of $q(x) = x^2$.

13. $Q_{-1,-2,-1}(x) = -x^2 - 2x - 1 = -(x^2 + 2x + 1) - 1 + 1 = -(x + 1)^2$. Hence, the graph of Q is a -1 horizontal translate of the graph of $q_{-1}(x) = -x^2$.

3.7 EXERCISES (page 142)

1. a) $x = 2$ **b)** $(2, -4)$ **c)** -4 is a minimum value. Since $a > 0$, the parabola is concave upward so that the vertex is the lowest point on the graph.   **d)** The y-intercept is $(0, 0)$. The image of the y-intercept is $(4, 0)$.

e)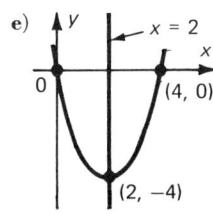

3. a) $x = 3$ **b)** $(3, 16)$ **c)** 16 is a maximum value. Since $a < 0$ the parabola is concave downward   **d)** The y-intercept is $(0, 7)$. The image of the y-intercept is $(6, 7)$

e)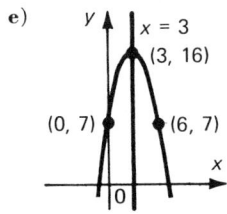

7. Hence, $a = -1$, $b = 6$, and $c = 0$. Thus, an equation that defines the parabola is $Q(x) = -x^2 + 6x$. The axis of symmetry is $x = 3$. The vertex is (3, 9).

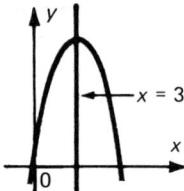

9. $8 = 16a - 4b + c$, $2 = 4a - 2b + c$, $8 = 16a + 4b + c$. Hence, $a = \frac{1}{2}$, $b = 0$, and $c = 0$. Thus, an equation that defines the parabola is $Q(x) = \frac{1}{2}x^2$. The axis of symmetry is $x = 0$, and the vertex is (0, 0).

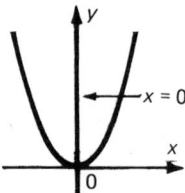

11. $x = y = 25$.   **13.** $A = x^2 + (20 - x)^2 = x^2 + 400 - 40x + x^2 = 2(x^2 - 20x) + 400 = 2(x^2 - 20x + 100) + 200$. Hence, the sum of the areas will be minimum, 200 sq. cm., when each piece is 10 cm. long.

15. Since $(d(0, P))^2 = x^2 + y^2$ and $y = -2x + 10$, we have $(d(0, P))^2 = x^2 + (-2x + 10)^2 = x^2 + 4x^2 - 40x + 100 = 5(x^2 - 8x) + 100 = 5(x^2 - 8x + 16) + 20 = 5(x - 4)^2 + 20$. Hence, the distance will be a minimum, $2\sqrt{5}$, when $x = 4$ and $y = 2$.

3.8 EXERCISES (page 149)

1. $\{0.8, -3.8\}$ **3.** $\{4.4, -0.4\}$ **5.** $\{2.3, 0.2\}$ **7.** $b^2 - 4ac = 9$. Hence, there are two real solutions, both rational.   **9.** $b^2 - 4ac = -7$. Hence, there are no real solutions.
11. $b^2 - 4ac = 57$. Hence, there are two real solutions, both irrational.   **15.** $\{x \mid 2 < x < 4\}$
21. $\{x \mid x \in R\}$ **25.** $0 < k < 3$

3.9 EXERCISES (page 154)

1.   **7.**   **13.**

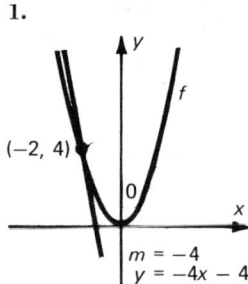
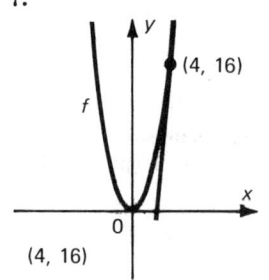
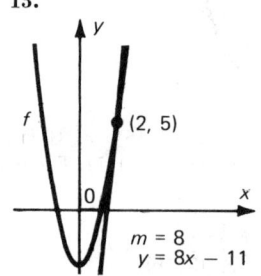

3.10 EXERCISES (page 156)

5. $y = 2x - \frac{7}{2}$.   **7.** $y = 9x + 10$.   **9.** $(-2, 1)$.   **11.** $(3, 6)$.   **13.** $(2, 0)$ and $(-2, 0)$. $y = -4x - 8$; $y = 4x - 8$.   **15.** $(5, -9)$ and $(3, -1)$. $y = -2x + 5$ and $y = -6x + 21$.

4.1 EXERCISES (page 161)

1. Yes. Function f is of degree 11.   **3.** No. The term $-2\sqrt{x} = -2x^{\frac{1}{2}}$ has a fractional exponent.
5. $g(-2) = -33$.   **7.** $G(2.5) = 25.75$.   **9.** $(f+g)(x) = 3x^3 + x^2 - x + 6$; $(f \cdot g)(x) = 3x^5 - 9x^4 + 17x^3 - 5x^2 + 7x + 5$. Degree of f is 3; degree of g is 2; degree of $f+g$ is 3; and degree of $f \cdot g$ is 5.   **11.** $a = 6$ and $b = 0$.   **13.** $a = 3$ and $b = 2$ or $a = -2$ and $b = -3$.
15. -1, $-\frac{2}{3}$, $\frac{2}{3}$ **17.** $-\frac{3}{2}$, $\frac{2}{3}$, 1 **19.** $P(x) = (x)(x-1)(x-2)$; 0, 1, 2
21. $P(x) = (x^2 + 9)(x - 3)(x + 3)$; -3, 3 **23.** $P(x) = (x-2)(x^2 + 2x + 4)$; 2
25. $P(x) = (x-1)(x^2 + x + 1)(x-2)(x^2 + 2x + 4)$; 1, 2

4.2 EXERCISES (page 166)

1. a) The domain is R. No. **b)** The range is R. **c)** $x = -2$, $x = 0$, or $x = 2$. $\{x \mid -2 < x < 0 \text{ or } x > 2\}$. $\{x \mid x < -2 \text{ or } 0 < x < 2\}$. **d)** As x becomes large, $|f(x)|$ becomes large, but $f(x) < 0$. As $|x|$ becomes large, but $x < 0$, $f(x)$ becomes large. **e)** $f'(x) = 0$ for $x = -1$ or $x = 1$. $f'(x) < 0$ for $\{x \mid x < -1 \text{ or } x > 1\}$. $f'(x) > 0$ for $\{x \mid -1 < x < 1\}$. **9.** As x becomes large, $P(x)$ becomes large. As $|x|$ becomes large, but $x < 0$, $|P(x)|$ becomes large, but $P(x) < 0$. **11.** As x becomes large, $|P(x)|$ becomes large, but $P(x) < 0$. As $|x|$ becomes large, but $x < 0$, $P(x)$ becomes large.
13. As x becomes large, $f(x)$ becomes large. As $|x|$ becomes large, but $x < 0$, $f(x)$ becomes large.
15. As x becomes large, $|f(x)|$ becomes large, but $f(x) < 0$. As $|x|$ becomes large, but $x < 0$, $|f(x)|$ becomes large, but $f(x) < 0$.

4.3 EXERCISES (page 171)

7. The graph of P has at least one x-intercept to the left of the origin if a_0 and a_n have the same sign.

4.4a EXERCISES (page 174)

1. $P'(x) = 6x^2 - 1$ **3.** $P'(x) = 2x^3 - 3x^2$ **5.** $P'(x) = -3x^3 + x^2 - 5x$ **7.** $m = 5$, $y = 5x - 9$ **9.** $m = 12$, $y = 12x + 10$ **11.** $(-1, 3)$, $(1, -1)$ **13. a)** $y = 1$ **b)** $y = 2$
15. a) $y = 51x - 145$ **b)** $y = -3x + 22$ **c)** $y = 48x - 123$

4.4b EXERCISES (page 180)

5. $y = -39x - 72$ **7. a)** $(2f)'(x) = 18x^2$ **b)** $(3g)'(x) = 42x$ **9.** $p(x) = x^2 + 5x + 3$
11. $p(x) = 2x^3 + 2x^2 + 2$

4.4c EXERCISES (page 182)

9. a) 102 **b)** 102.01; error $= 0.01$ **11. a)** $\dfrac{3{,}999{,}880}{3}\pi$ cc. **b)** 40π cc.

4.5 EXERCISES (page 188)

1. $f'(x) = 3x + 6$; hence, the graph of f is rising for $\{x \mid x > -3\}$ and falling for $\{x \mid x < -3\}$.
3. Rising: $\{x \mid x > 0\}$; falling: $\{x \mid x < 0\}$.   **5.** Rising: $\{x \mid x < -\frac{2}{3} \text{ or } x > \frac{2}{3}\}$; Falling: $\{x \mid -\frac{2}{3} < x < \frac{2}{3}\}$ **7.** Rising: $\{x \mid -4 < x < 2\}$; Falling: $\{x \mid x < -4 \text{ or } x > 2\}$
9. a) $x = -1$. **b)** $\{x \mid x \neq -1\}$ **c)** $\emptyset$ **11.** $f'(x) = 0$ at the point $(3, 27)$ on the graph of f. By theorem (1) $(3, 27)$ is a relative maximum point.

4.6 EXERCISES (page 195)

1. Concave upward: $\{x \mid x \in R\}$. The graph is never concave downward.   **3.** Concave upward: $\{x \mid x < \frac{3}{5}\}$; concave downward: $\{x \mid x > \frac{3}{5}\}$ **5.** Concave upward: $\{x \mid x < -\frac{3}{4} \text{ or } x > \frac{4}{3}\}$; concave downward: $\{x \mid -\frac{3}{4} < x < \frac{4}{3}\}$ **7.** $f''(x) = 0$ at $x = 0$. Since $f''(x) < 0$ for $x > 0$ and $f''(x) > 0$ for $x < 0$, $(0, 11)$ is a point at which the graph changes concavity.   **9.** $f''(x) = 0$ at $x = \frac{1}{2}$. The point $(\frac{1}{2}, 10\frac{3}{8})$ is not a point at which the graph changes concavity.

4.7 EXERCISES (page 201)

1. Positive: $\{x \mid -2 < x < 0 \text{ or } x > 3\}$; Negative: $\{x \mid x < -2 \text{ or } 0 < x < 3\}$ **3.** Positive: $\{x \mid -5 < x < 0 \text{ or } x > 2\frac{2}{3}\}$; Negative: $\{x \mid x < -5 \text{ or } 0 < x < 2\frac{2}{3}\}$
7. $P(x) = x(x + \sqrt{3})(x - \sqrt{3})$. Positive: $\{x \mid -\sqrt{3} < x < 0 \text{ or } x > \sqrt{3}\}$; zero: $\{-\sqrt{3}, 0, \sqrt{3}\}$; negative: $\{x \mid x < -\sqrt{3} \text{ or } 0 < x < \sqrt{3}\}$. $P'(x) = 3x^2 - 3 = 3(x-1)(x+1)$. Positive: $\{x \mid x < -1 \text{ or } x > 1\}$; zero: $\{-1, 1\}$; negative: $\{x \mid -1 < x < 1\}$. $P''(x) = 6x$. Positive: $\{x \mid x > 0\}$; zero: $\{0\}$; negative: $\{x \mid x < 0\}$.

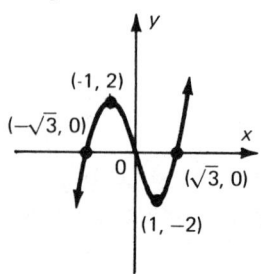

11. $P(x) = x(x-3)^2$. Positive: $\{x \mid x > 0, x \neq 3\}$; zero: $\{0, 3\}$; negative: $\{x \mid x < 0\}$. $P'(x) = 3x^2 - 12x + 9 = 3(x-1)(x-3)$. Positive: $\{x \mid x < 1 \text{ or } x > 3\}$; zero: $\{1, 3\}$; negative: $\{x \mid 1 < x < 3\}$. $P''(x) = 6x - 12 = 6(x-2)$. Positive: $\{x \mid x > 2\}$; zero: $\{2\}$; negative: $\{x \mid x < 2\}$.

4.8 EXERCISES (page 205)

1. $P(x) = (x-3)(3x^2 + 5x + 17) + 36$.   **3.** $P(x) = (x - \frac{1}{2})(4x^2 - x + 4\frac{1}{2}) + \frac{1}{4}$.   **5.** $k = -4$.
7. $P(x) = (x-5)(x+4)(x+3); \{-4, -3, 5\}$ **11.** $P(x) = (x+1)(2x-5)(x+4); \{-4, -1, \frac{5}{2}\}$
13. a) False. If $P(x) = x^n + c^n$, then $P(c) = c^n + c^n = 2c^n$ and $2c^n \neq 0$ whenever $c \neq 0$. Hence, $x - c$ is not a factor of $x^n + c^n$. **b)** True. If $P(x) = x^n - c^n$ and n is even, then $P(-c) = (-c)^n - c^n = c^n - c^n = 0$, so $(x - (-c)) = (x + c)$ is a factor of $P(x)$.

5.1 EXERCISES (page 209)

1. $5i; -6$ **3.** $2 + 11i; -24 + 16i$ **5.** $1 + i; -66 + 43i$ **9.** $5 - 5i; -13i$
13. $-\frac{4}{3} + \frac{11}{18}i; -\frac{11}{18} - \frac{43}{54}i$ **15.** $3\sqrt{2} + 4i; 1 + 7\sqrt{2}\,i$ **17.** $-1 - 4i$ **21.** $6 + 4i$
23. $x = 4, y = 3$ **25. a)** $i^4 = 1; i^5 = i; i^6 = -1; i^7 = -i; i^8 = 1$ **b)** $i^{17} = i; i^{100} = 1; i^{103} = -i$

5.2 EXERCISES (page 212)

1. $-i$ **3.** $-\frac{1}{10} + \frac{1}{5}i$ **5.** $\frac{1}{15} + \frac{2}{15}i$ **7.** $-1i$ **9.** $-2.6 + 0.8i$ **11.** $\{-5i, 5i\}$
13. $\{-25i, 25i\}$ **15.** $\{-8i, 8i\}$

5.3 EXERCISES (page 215)

1. $\{3 - i, 3 + i\}$ **5.** $\{6 - \sqrt{33}, 6 + \sqrt{33}\}$ **13.** False   **15.** True

5.4 EXERCISES (page 221)

1. $P(x) = (x - 2i)(x^2 + 2ix)$ **3.** $P(x) = (x - i)(x^2 + (1+i)x + i)$
5. $P(x) = (x+3)(x - 4i)(x + 4i); \{-3, 4i, -4i\}$ **7.** $P(x) = (x - i)(x + i)(x - 3); \{i, -i, 3\}$
11. $P(x) = x^3 + 2x^2 + 9x + 18$ **13.** $P(x) = x^3 - 4x^2 + 6x - 4$
17. $P(x) = (x - 1)\left(x + \frac{1}{2} - \frac{\sqrt{3}}{2}i\right)\left(x + \frac{1}{2} + \frac{\sqrt{3}}{2}i\right); \left\{1, -\frac{1}{2} + \frac{\sqrt{3}}{2}i, -\frac{1}{2} - \frac{\sqrt{3}}{2}i\right\}$

5.5 EXERCISES (page 226)

1. Negative: $\{x \mid x < 3\}$; zero: $\{3\}$; positive: $\{x \mid x > 3\}$.   **3.** Negative: $\{x \mid x < 2\}$; zero: $\{2\}$; positive: $\{x \mid x > 2\}$.   **5.** Negative: $\{x \mid x > -5 \text{ and } x \neq 0\}$; zero: $\{-5, 0\}$; positive: $\{x \mid x < -5\}$.
7. $a = -7; b = -9$.   **9.** $P(x)$ negative: $\{x \mid |x| < \sqrt{7}\}$; $P(x)$ zero: $\{-\sqrt{7}, \sqrt{7}\}$; $P(x)$ positive: $\{x \mid |x| > \sqrt{7}\}$. $P'(x)$ negative: $\{x \mid x < -\sqrt{3} \text{ or } 0 < x < \sqrt{3}\}$; $P'(x)$ zero: $\{-\sqrt{3}, 0, \sqrt{3}\}$;

$P'(x)$ positive: $\{x \mid -\sqrt{3} < x < 0 \text{ or } x > \sqrt{3}\}$. $P''(x)$ negative: $\{x \mid |x| < 1\}$; $P''(x)$ zero: $\{-1, 1\}$; $P''(x)$ positive $\{x \mid |x| > 1\}$.

15. 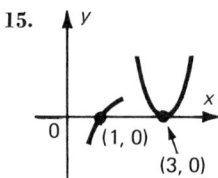   17.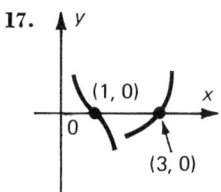

5.7a EXERCISES (page 232)

1. Possible: $\{-1, 1\}$; actual: $\{-1, 1\}$ **3.** Possible; $\{\pm 6, \pm 3, \pm 2, \pm 1\}$; actual: $\{-1, 6\}$
5. Possible: $\{\pm 14, \pm 7, \pm 2, \pm 1\}$; actual: $\{-2, 1\}$ **7.** Possible: $\{-1, -\frac{1}{2}, \frac{1}{2}, 1\}$; actual: $\{-\frac{1}{2}\}$
9. Possible: $\{\pm 1, \pm \frac{1}{2},\}$; actual: $\{-1, \frac{1}{2}, 1\}$

5.7b EXERCISES (page 233)

1. The possible rational zeros are $\{\pm 4, \pm 2, \pm 1\}$. By Theorem 5.7c, the possible rational zeros are $\{4, 2, 1\}$. There are no actual rational zeros.   **3.** Possible rational zeros: $\{\pm 6, \pm 3, \pm 2, \pm 1\}$; by Theorem 5.7b: $\{-6, -3, -2, -1\}$; actual rational zeros: $\{-3, -2, -1\}$.   **7.** Possible rational zeros: $\{\pm 3, \pm \frac{3}{2}, \pm 1, \pm \frac{3}{4}, \pm \frac{1}{2}, \pm \frac{1}{4}\}$; by Theorem 5.7b: $\{-3, -\frac{3}{2}, -1, -\frac{3}{4}, -\frac{1}{2}, -\frac{1}{4}\}$; actual rational zeros: $\{-\frac{1}{2}\}$.

5.8 EXERCISES (page 236)

3. Between 1 and 2; irrational   **5.** Between 2 and 3; rational, $\frac{5}{2}$ **9.** $P(x)$ has a zero between 0 and 1, which is closer to 1 than to 0.

6.1a EXERCISES (page 241)

1. $\frac{\pi}{6}$; first quadrant   **5.** $\frac{-4\pi}{15}$; fourth quadrant   **7.** 15°; first quadrant   **11.** 50°; first quadrant   **13.** $\left(\frac{\sqrt{3}}{2}, \frac{1}{2}\right)$ **15.** $\left(-\frac{\sqrt{2}}{2}, \frac{\sqrt{2}}{2}\right)$ **17.** $\left(-\frac{\sqrt{3}}{2}, \frac{1}{2}\right)$

6.1b EXERCISES (page 244)

1. $\left(\frac{1}{2}, \frac{\sqrt{3}}{2}\right)$ **3.** $\left(-\frac{\sqrt{3}}{2}, \frac{1}{2}\right)$ **5.** $(-1, 0)$ **13.** $\left(-\frac{1}{2}, -\frac{\sqrt{3}}{2}\right)$ **15.** $\left(\frac{1}{2}, -\frac{\sqrt{3}}{2}\right)$
17. x-axis   **19.** y-axis   **25.** y-axis   **27.** y-axis

6.2 EXERCISES (page 247)

1. $\sin t = 0$; $\cos t = 1$ **5.** $\sin t = \frac{\sqrt{2}}{2}$; $\cos t = -\frac{\sqrt{2}}{2}$ **7.** $\sin t = -\frac{\sqrt{3}}{2}$; $\cos t = -\frac{1}{2}$
11. $\sin t = \frac{1}{2}$; $\cos t = -\frac{\sqrt{3}}{2}$ **13.** $\frac{3}{5}$ **15.** $-\frac{8}{17}$ **17.** $-\frac{4}{5}$ **19.** $\sin t$: decreases from 1 to 0; $\cos t$: decreases from 0 to -1 **21.** $\sin t$: increases from -1 to 0; $\cos t$: increases from 0 to 1

6.3 EXERCISES (page 254)

1.

Function	Period	Greatest ordinate
sine	2π	1
g	2π	2

3.

Function	Period	Greatest ordinate
sine	2π	1
h	2π	1

5.

Function	Period	Greatest ordinate
sine	2π	1
h	2π	1

7.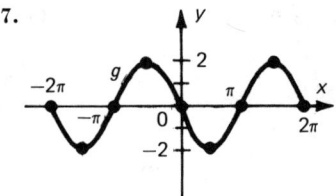

6.4 EXERCISES (page 260)

3.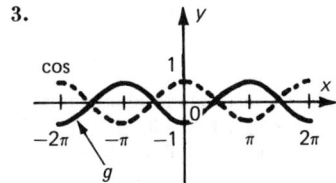

5.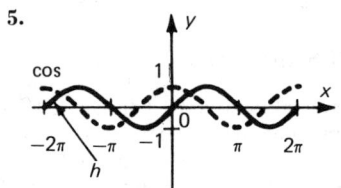

6.5a EXERCISES (page 263)

1. $\cos t$ **3.** $\sin t$ **5.** $\sin t$ **7.** $\frac{63}{65}; \frac{56}{65}; \frac{120}{169}; \frac{7}{25}$ **9.** $\frac{\sqrt{2}+\sqrt{6}}{4}; \frac{\sqrt{6}+\sqrt{2}}{4}; \frac{\sqrt{3}}{2}; 0$

6.5b EXERCISES (page 266)

11. $\left\{ t \mid t = \frac{7\pi}{6} + 2n\pi \text{ or } t = \frac{11\pi}{6} + 2n\pi, n \in I \right\}$ **13.** $\left\{ t \mid t = \frac{\pi}{3} + n\pi \text{ or } t = \frac{2\pi}{3} + n\pi, n \in I \right\}$

15. $\left\{ t \mid t = n\pi \text{ or } t = \frac{\pi}{3} + 2n\pi \text{ or } t = \frac{2\pi}{3} + 2n\pi, n \in I \right\}$ **21.** $\left\{ t \mid t = \frac{n\pi}{2}, n \in I \right\}$

23. $\{t \mid t = n\pi, n \in I\}$

7.1a EXERCISES (page 273)

1. $\frac{\sqrt{3}}{3}$ **3.** 1 **5.** $-\frac{\sqrt{3}}{3}$ **7.** 1 **9.** $\sqrt{3}$ **11.** -1

7.1b EXERCISES (page 277)

1.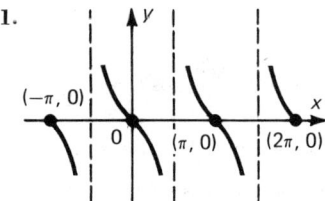

5.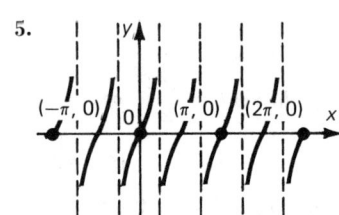

7. $y = 2x + \left(1 - \dfrac{\pi}{2}\right)$. 9. $y = \tfrac{4}{3}x + \left(\dfrac{\sqrt{3}}{3} - \dfrac{2\pi}{9}\right)$. 11. $\tfrac{4}{3}$ 13. $-\tfrac{7}{24}$ 15. b) Yes. Function f is one-to-one and, hence, has an inverse. d) $y = x$; $y = \tfrac{1}{2}x + \left(\dfrac{\pi}{4} - \dfrac{1}{2}\right)$; $y = \tfrac{1}{4}x + \left(\dfrac{\sqrt{3}}{4} - \dfrac{\pi}{3}\right)$.

7.2 EXERCISES (page 279)

1. $\left\{\dfrac{\pi}{4}\right\}$ 3. $\left\{\dfrac{\pi}{3}\right\}$ 5. $\left\{-\dfrac{\pi}{4}\right\}$ 7. $\left\{\dfrac{\pi}{6}, -\dfrac{\pi}{6}\right\}$ 9. $\left\{0, \dfrac{\pi}{6}\right\}$ 11. $\left\{\dfrac{\pi}{4} + n\pi,\ 1.25 + n\pi\right\}$
13. $\{-1.11 + n\pi,\ 1.43 + n\pi\}$ 15. $\{.32 + n\pi,\ 1.37 + n\pi\}$

7.3 EXERCISES (page 282)

1. $\dfrac{\sqrt{3}}{3}$ 3. $-\sqrt{3}$ 5. $-\dfrac{\sqrt{3}}{3}$ 7. 1 9. $\dfrac{\sqrt{3}}{3}$ 11. $\sqrt{3}$

13. 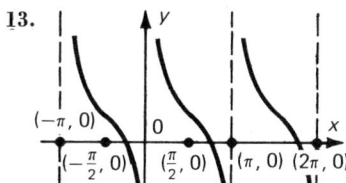   15.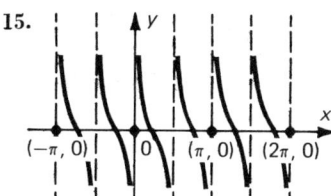

21. $\left\{\dfrac{\pi}{4}\right\}$ 23. $\left\{\dfrac{5\pi}{6}\right\}$ 25. $\left\{\dfrac{\pi}{4}, \dfrac{\pi}{2}\right\}$

7.4a EXERCISES (page 286)

1. $2\dfrac{\sqrt{3}}{3}$ 3. -2 5. $\sqrt{2}$

7. 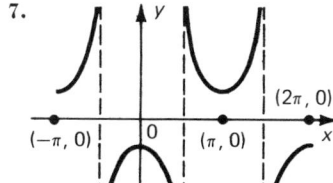   9.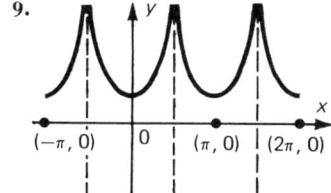

15. $\left\{x \mid 0 \leq x < \dfrac{\pi}{3} \text{ or } \dfrac{\pi}{2} < x < \dfrac{3\pi}{2} \text{ or } \dfrac{5\pi}{3} < x \leq 2\pi\right\}$

17. $\left\{x \mid \dfrac{\pi}{6} \leq x < \dfrac{\pi}{2} \text{ or } \dfrac{\pi}{2} < x \leq \dfrac{5\pi}{6} \text{ or } \dfrac{7\pi}{6} \leq x < \dfrac{3\pi}{2} \text{ or } \dfrac{3\pi}{2} < x \leq \dfrac{11\pi}{6}\right\}$ 19. $\left\{\dfrac{\pi}{3}, \dfrac{5\pi}{3}\right\}$ 21. $\sin x$

23. $\dfrac{\sin^2 x}{\cos^2 x}$

7.4b EXERCISES (page 289)

1. $\dfrac{2\sqrt{3}}{3}$ 3. -2 5. 1 17. $\left\{x \mid \dfrac{\pi}{6} < x < \dfrac{5\pi}{6} \text{ or } \dfrac{7\pi}{6} < x < \dfrac{11\pi}{6}\right\}$

19. $\{0 < x < \pi \text{ or } \pi < x < 2\pi\}$ 21. $\left(\dfrac{7\pi}{6}, \dfrac{11\pi}{6}\right)$ 25. $\cos^2 x$ 27. $\dfrac{1}{\cos x} + \dfrac{1}{\sin x}$

8.1 EXERCISES (page 299)

1. 100 milligrams; 25 milligrams; 6.25 milligrams   3. 6400 milligrams   5. $\tfrac{1}{4}W_0$; $\tfrac{1}{16}W_0$; $\tfrac{1}{64}W_0$
7. 8 times as great   9. $\tfrac{1}{2}$ hour   11. 1 hour   13. $N(t) = N_0(2)^{t/12}$ 15. $n = 10$
17. a) $1040   b) $1040.40   c) $1040.60

8.2a EXERCISES (page 303)

1. 1 3. $\tfrac{16}{9}$ 5. 64 7. 4 9. 0.00000001 13. $\dfrac{5}{x^3}$ 15. $\dfrac{1}{125x^3}$ 17. $5x^2$
21. $\{\tfrac{1}{3}, -\tfrac{1}{3}\}$ 23. $\{x \mid x > \tfrac{1}{5} \text{ or } x < 0\}$ 25. 7.53×10^1 27. 1.46×10^{-1} 29. 7.12×10^5

8.2b EXERCISES (page 307)

1. 5 3. 2 5. $\tfrac{1}{2}$ 7. $\tfrac{1}{5}$ 19. $\{\tfrac{1}{4}\}$ 21. $\{x \mid x < 0 \text{ or } x > \tfrac{1}{32}\}$

8.3 EXERCISES (page 310)

1. 125 3. 512 5. 0.01024 7. 0.001 9. 32 11. 625

8.4 EXERCISES (page 313)

1. a) 81 b) 16 c) $1{,}000{,}000$ d) 3 5. True. The function $\exp_{0.5}$ is strictly decreasing and $\pi < \tfrac{22}{7}$.
7. $\{x \mid x < \tfrac{1}{2}\}$ 9. $\{x \mid x > 1\}$ 11. $\{x \mid x > \tfrac{2}{5}\}$

8.5a EXERCISES (page 316)

1. 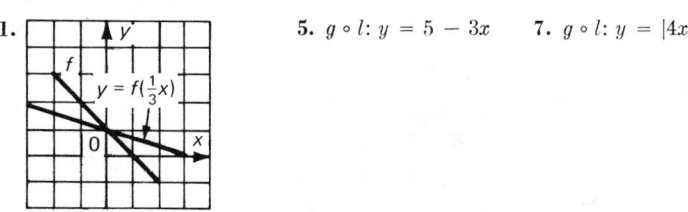   5. $g \circ l: y = 5 - 3x$ 7. $g \circ l: y = |4x|$

9. $g \circ l: y = \sin(2x)$ 11. $g \circ l: y = 2\cos(-2x)$

8.5b EXERCISES (page 318)

3. [graph showing points $(-1, \frac{10}{11})$, $(0, 1)$, $(1, 1.1)$, $(2, 1.21)$]

7. $\{3\}$ 9. $\{-3\}$ 11. $\{3\}$ 13. $\{-1\}$ 15. $\{9\}$
17. $\{\frac{1}{16}\}$ 19. $\{127\}$ 21. $\{14.5\}$

8.6 EXERCISES (page 322)

1. $\frac{1}{4}k \approx 0.173$ 3. $\frac{\sqrt{2}}{2}k \approx 0.490$ 5. $\sqrt{2}\,k \approx 0.980$ 7. $2\sqrt{2}\,k \approx 1.960$
9. $4\sqrt{2}\,k \approx 3.920$ 13. $g'(x) = -k(\frac{1}{2})^x$ 15. $G'(x) = -2k(\frac{1}{4})^x$ 17. $f'(x) = 5k \cdot 2^x$
19. $F'(x) = \frac{k}{3} \cdot 2^x$

8.7 EXERCISES (page 324)

1. $f'(x) = 20 \cdot e^{2x}$ 3. $F'(x) = -40e^{10x}$

8.8 EXERCISES (page 327)

1. $\log_2 64 = 6$ 3. $\log_2 \sqrt{2} = \frac{1}{2}$ 5. 5 7. -7 9. 8 11. $\frac{1}{16}$ 13. 10 19. 10

8.9 EXERCISES (page 330)

1. 1 3. $\frac{1}{2}$ 5. 5 7. 125 9. 8 11. 10 13. 9 15. 16 17. 5 19. 1.2
21. **a)** True. $\log_{10} 10{,}000 = 4$, $\log_{10} 1000 = 3$, and $\log_{10} 10 = 1$.

8.10 EXERCISES (page 334)

1. 1024 3. 4096 5. 7 9. 7

8.11 EXERCISES (page 337)

1. $r + s$ 3. $r + 2s$ 5. $\frac{r+s}{2}$ 7. 2.51 9. 13.2 11. 3.90 13. 0.0546
15. 0.776 (Answers are given to the nearest hundredth.)   17. 1.18 19. 1.17 21. 0.310
23. e^π is greater.

8.12 EXERCISES (page 340)

1. 2.48 3. 13.3 5. 7.74 sec.   7. **a)** \$2208   **b)** 17.5 years

8.13 EXERCISES (page 343)

1. 5.248 3. -0.811 5. 4.847 7. -1.198 9. 2.41 11. 3.16

9.1 EXERCISES (page 346)

1. $\{1, -1\}$ **3.** $\{0\}$ **5.** $\{0\}$ **7.** $\{-1\}$ **9.** Improper; $f(x) = x + \dfrac{1}{x}$

11. Improper; $h(x) = x^2 + 2x + 4 + \dfrac{-1}{x-2}$ **13.** Improper; $F(x) = 3x - 2 + \dfrac{-9}{2x+3}$

15. Improper; $H(x) = \tfrac{1}{2} + \dfrac{9 - 24x}{8x^2 + 4x - 6}$ **17.** $\dfrac{3x-1}{x^2-x}; \dfrac{-x-1}{x^2-x}; \dfrac{2}{x^2-x}; \dfrac{x-1}{2x}$

19. $\dfrac{2x^2 + 18}{x^2 - 9}; \dfrac{12x}{x^2 - 9}; 1; \dfrac{x^2 + 6x + 9}{x^2 - 6x + 9}$

9.2 EXERCISES (page 350)

1. $\{1\}$ **3.** $\varnothing$ **5.** $\{-\tfrac{2}{3}\}$. **7.** $\varnothing$ **9.** $\{x \mid x < 0 \text{ or } x > 1\}$ **11.** $\{x \mid -1 < x < 1\}$
13. $\{x \mid -4 < x \leq 1 \text{ or } x \geq 3\}$ **15.** $\{x \mid -5 < x \leq -\tfrac{7}{2} \text{ or } 0 \leq x < \tfrac{1}{3}\}$
17. **19.** **25.**

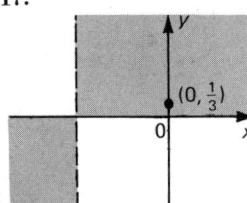
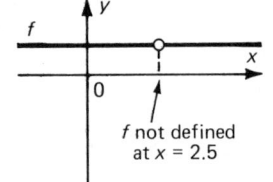

9.3 EXERCISES (page 354)

1. $f'(x) = 42x + 44$ **3.** $f'(x) = 30x^4 - 12x$ **5.** $f'(x) = 14x - 5$ **7.** $f'(x) = -\dfrac{1}{x^2}$

9. $f'(x) = -\dfrac{2}{(2x+3)^2}$ **11.** $f'(x) = \dfrac{5}{(3x+1)^2}$ **13.** $f'(x) = \dfrac{-32x}{(x^2-16)^2}$

15. $h'(x) = 64x^3 + 288x^2 + 432x + 216$ **17.** $h'(x) = 36x^3 - 72x^2 + 32x$

19. $h'(x) = 4x^3 + 6x^2 + 4x + 2$ **21.** $h'(x) = (136x^4 + 90x^2 + 5)(2x^2 + 5)^6$

23. $h'(x) = \dfrac{-24}{(2x+.1)^5}$ **25.** $h'(x) = \dfrac{-60x}{(x^2-1)^4}$

9.4 EXERCISES (page 358)

1. Positive: $r(x)$: R; $r'(x)$: $\{x \mid x < 0\}$; $r''(x)$: $\left\{x \,\middle|\, |x| > \dfrac{\sqrt{3}}{3}\right\}$; zero: ϕ; $\{0\}$; $\left\{-\dfrac{\sqrt{3}}{3}, \dfrac{\sqrt{3}}{3}\right\}$;
negative: $\varnothing$; $\{x \mid x > 0\}$; $\left\{x \,\middle|\, |x| < \dfrac{\sqrt{3}}{3}\right\}$.

5. **9.** **12.**

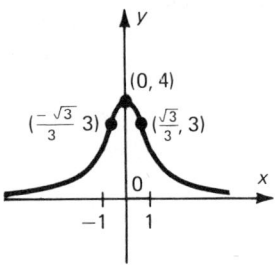
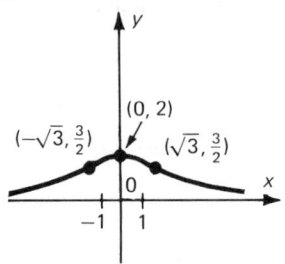
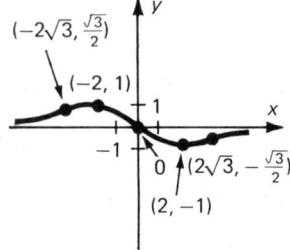

9.5 EXERCISES (page 364)

1. $y = 2$ **3.** As $x \to \infty$, $r(x) \to \infty$; as $x \to -\infty$, $r(x) \to -\infty$ **5.** $y = \frac{1}{4}$ **7.** $y = \frac{1}{3}$
9. a) 0 **b)** The x-axis is a horizontal asymptote for the graph of f.

9.6a EXERCISES (page 370)

1. Yes **3.** Yes **5.** Yes **9.** No

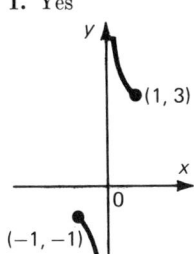 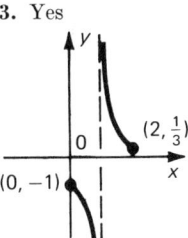 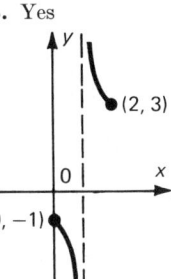 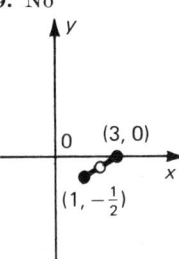

13. 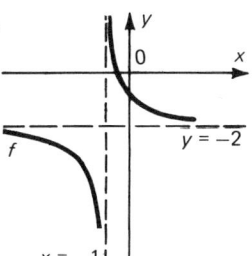 **15.**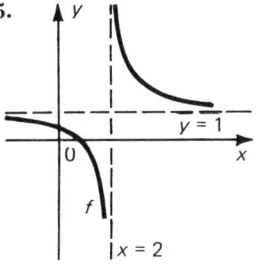

9.6b EXERCISES (page 373)

1. 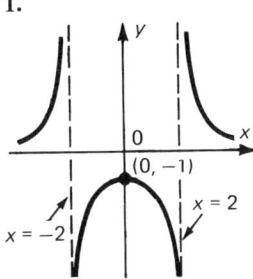 **3.** 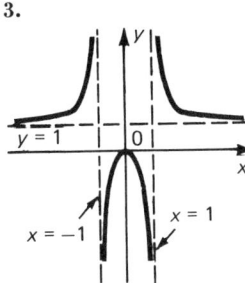 **5.**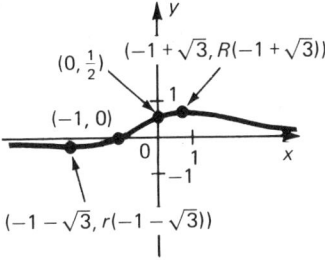

10.1 EXERCISES (page 375)

1. $\left(3, -\frac{7\pi}{4}\right), \left(3, \frac{9\pi}{4}\right)$ **3.** $\left(\frac{1}{2}, -\frac{4\pi}{3}\right), \left(\frac{1}{2}, \frac{8\pi}{3}\right)$ **5.** $\left(1, \frac{5\pi}{6}\right), \left(1, -\frac{19\pi}{6}\right)$ **7.** (b)
9. (a) **11.** (c) **13.** (b) **15.** (c)

17.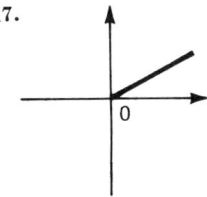

10.2 EXERCISES (page 378)

1. $(5\sqrt{3}, -5)$ **3.** $(-7, 7\sqrt{3})$ **5.** $(-3\sqrt{2}, -3\sqrt{2})$ **7.** $\left(10, -\dfrac{\pi}{6}\right)$ **9.** $(8, 0)$

11. $\left(10\sqrt{2}, \dfrac{\pi}{4}\right)$

15. 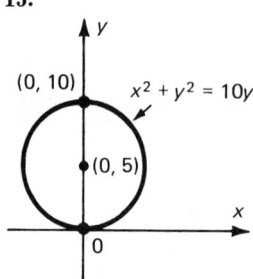   **17.** 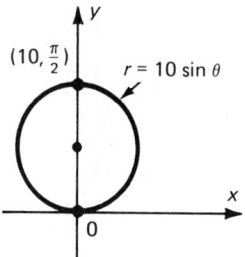   **19.**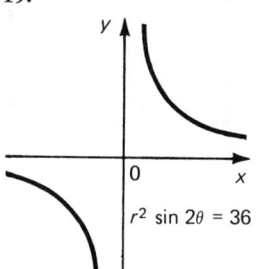

10.3 EXERCISES (page 381)

1. $r = \dfrac{6}{1 - \frac{1}{2}\cos\theta}; \; e = \frac{1}{2}; \; b = 12$ **3.** $r = \dfrac{4}{1 - \frac{3}{4}\cos\theta}; \; e = \frac{3}{4}; \; b = \frac{16}{3}$

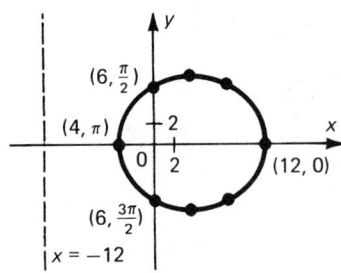

5. $r = \dfrac{8}{1 - \cos\theta}; \; e = 1; \; b = 8$ **9.**

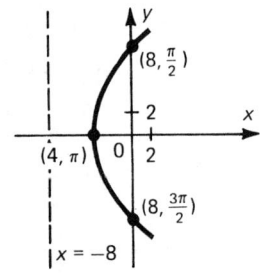 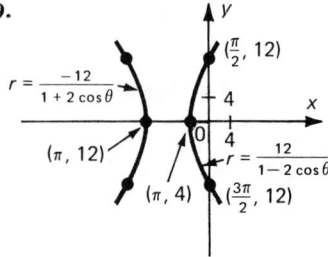

13. $r = \dfrac{eb}{1 - e\sin\theta}$ **15.** $r = \dfrac{eb}{1 - e\sin\theta}, \; r = \dfrac{-eb}{1 + e\sin\theta}$

10.4 EXERCISES (page 383)

1. 3. 5. 7. 9.

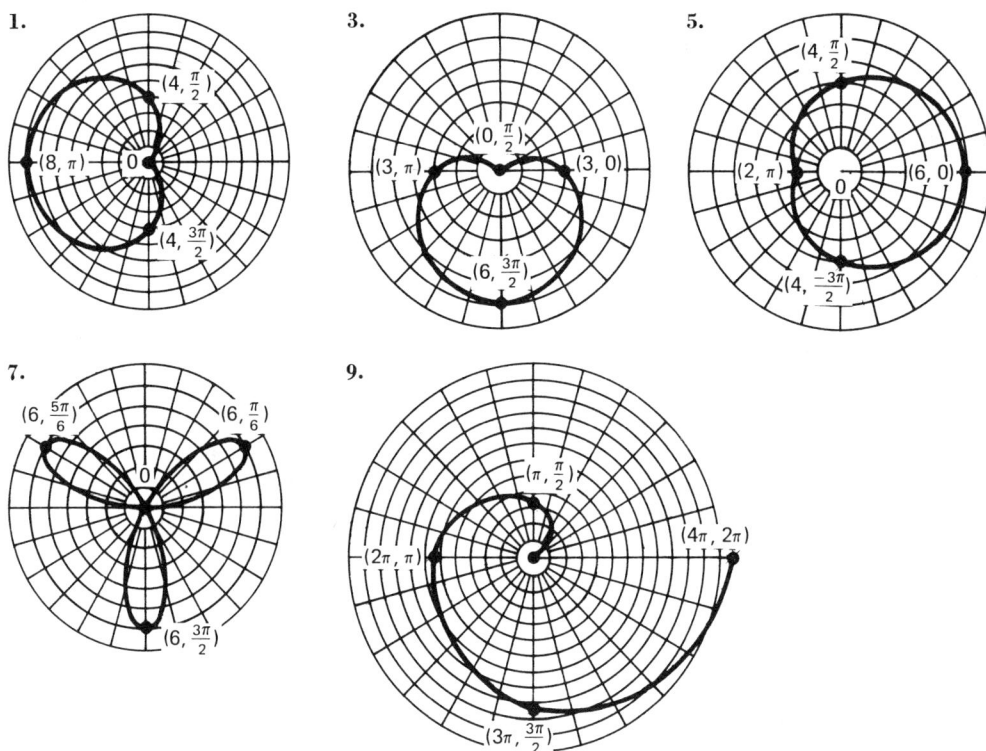

Index

Abscissa, 44
Absolute value
  and distance, 7-8
  conditions involving, 10
  definition, 8
  properties of, 8-9
Absolute value function, 79-80
Addition formulas for trigonometric functions, 263, 292
Angle(s)
  cosine of, 247
  degree measure of, 238
  directed, 238
  initial side of, 238
  of intersection of two lines, 67
  radian measure of, 239-241
  sine of, 247
  standard position, 239
  terminal side of, 238
Archimedean Law, 19
Arithmetic sequence, 2
Asymptotes, 277, 282, 285, 286, 360-372

Base of a power, 301
Binary logarithms, 326-327, 331-334
Bounds
  greatest lower, 19
  least upper, 18
  of sequences, 14-15
  of sets, 14-15

Cartesian plane, 44
Characteristic of a logarithm, 335
Circle, unit, 243
Closed interval, 10
Coefficients of a polynomial, 159
Common logarithms, *see* Logarithms
Completeness Axiom, 18
Completing the square, 137-138
Complex numbers
  conjugates, 210-211
  definition, 208

Complex numbers (*Continued*)
  difference of, 209
  imaginary part, 208
  product of, 209
  quotient of, 211-212
  real part, 208
  sum of, 208-209
Complex roots of quadratic equations, 213-214
Complex zeros of polynomials, 214
Composite function, 131, 132
Composition of functions, 129-133, 313-316
Concavity, 112-113, 139, 190-195
Conic sections, 381-383
Conjugates, complex, 210-211
Constant functions, 62, 70, 159
Constant sequences, 4
Continuous functions, 82-85, 169-170
Converging sequences, 13, 16-18, 22-24, 29-31, 33-34
Coordinate systems
  polar, 376-377
  rectangular, 43-46
Cosecant function
  definition, 288
  graph of, 288-290
  properties of, 288-290
Cosine
  of an angle, 247
  of a real number, 246
Cosine function
  addition formula for, 263
  definition, 246
  difference formula for, 262
  double angle formula for, 263
  graph of, 257-259
  half-angle formula for, 264
  properties of, 256-259
  range of, 258
  slope function of, 268-269
Cotangent function
  definition, 280
  graph of, 281
  properties of, 282

Decreasing functions, 52, 60, 184
Decreasing sequences, 4, 21
Degree of polynomial, 159
Dependent variable, 41
Derivative function, *see also* Slope Function, 154-156, 352-355
Difference formulas for trigonometric functions, 262-263
Difference of functions, 90-91
Difference of sequences, 26, 29-31
Directrix of parabola, 116, 381
Discriminant of a quadratic equation, 145-147
Distance
  from point to line, 77-78
  on a number line, 5, 7-10
Diverging sequences, 13
Domain of function, 36, 47-48
Double angle formulas for trigonometric functions, 263, 292

e, 321-322
Ellipse, 382-383
Even function, 100, 102
Exponential functions
  definition, 299
  graph of, 313-316, 323-324
  inverse function of, 325-327, 328
  properties of, 295, 298, 310, 313-316, 317
  slope function of, 318-322, 323-324
  with integral domain, 295, 296, 301-302
  with rational domain, 298, 305, 308, 309-310
  with real domain, 299, 311-313
Exponents
  integral, 301-303
  properties of, 301, 309, 312
  rational, 305-309
  real, 311-312
Extension of function, 40

Factor Theorem, 203
Fibonacci sequence, 2
Focus of parabola, 116, 381
Functions
  absolute value, 79–80
  composite, 131–132
  composition of, 129–133, 313–316
  constant, 62, 70
  continuous, 82–85
  as correspondence, 35–38
  decreasing, 52, 60, 184
  definition, 36
  dependent variable of, 41
  derivative, 154–156, 352–355
  difference of, 90–91
  domain, 36, 47–48
  even, 100, 102
  exponential, 294–326, 328
  extension of, 40
  graphs of, 42–49
  increasing, 51, 60, 184
  independent variable of, 41
  inverse, 57–61, 105–108, 325–327, 328
  limit of, 87–88, 92–94, 361–362
  linear, 70–72
  logarithmic, 326–343
  as mapping, 43
  maximal domain, 42
  maximum of, 141
  minimum of, 141
  monotone, 51–55
  multiplication by real number, 118–121
  nondecreasing, 51
  nonincreasing, 52
  notation for, 39–41, 43
  odd, 100, 102
  one-to-one, 55, 60
  periodic, 250
  polynomial, 158–236
  product of, 90
  quadratic, 110–156
  quotient of, 90–91
  range, 36, 48
  rational, 344–375
  real valued of single real variable, 46
  reciprocal, 91
  restriction of, 40, 53
  slope, 154–156, 175–182, 352–355
  sum of, 90, 128
  symmetric, 100–103
  trigonometric, 237, 246–269, 270–293
  wrapping, 243–244
  zeros of, 160
Fundamental Theorem of Algebra, 217

Geometric sequence, 2, 14, 16–18
Greatest lower bound, 19–20

Half-angle formulas for trigonometric functions, 264
Half-life of radioactive substance, 295
Hyperbola, 382–383

i, 208
Identities, see Trigonometric Identities
Inclination of a line, 68
Increasing function(s), 51, 60, 184
Increasing sequence(s), 4, 21
Independent variable, 41
Inflection, point of, 195
Intercepts of a line, 69
Intermediate Value Theorem, 171
Intervals, 10, 12
Inverse function, 57–61, 105–108, 325–328
Irreducible quadratic expression, 220–225

Least upper bound, 18, 20
Limit
  of function, 87–88, 92–94, 361–362
  of sequence, 4–6, 11–12, 13, 16–18, 22–24, 29–31, 33–34
Linear factors, 216
Linear functions, 70–72
Linear interpolation, 332
Line(s)
  angle of intersection of, 67
  distance from point to, 77–78
  inclination of, 68
  intercepts of, 69
  parallel, 74
  perpendicular, 74–75
  slope of, 64–66
  of symmetry, 96, 139
  tangent, 150–153
Logarithmic functions
  base a, 328–330
  base e, 234
  base 2, 326–327, 331–334
  base 10, 334
  definition, 328
  graph of, 329–330
  properties of, 330, 334, 336–337
Logarithms
  binary (base 2), 326–327, 331–334
  characteristic of, 335
  common (base 10), 334, 335, 338–339

Logarithms (*Continued*)
  computing with, 338–339
  converting to different bases, 340–345
  natural (base e), 334
  properties of, 334, 336–337

Mantissa of common logarithm, 335
Mappings, 43, 129–130
Maximal domain, 42
Maximum value of function, 141, 187–188
Minimum value of function, 141, 188
Monotone functions, 51–55
Monotonic sequences, 22

Natural logarithms, 334
Nondecreasing functions, 51
Nondecreasing sequences, 21–23
Nonincreasing functions, 52
Nonincreasing sequences, 22–23

Odd functions, 100, 102
One-to-one functions, 55, 60
Open interval, 10, 12
Ordered pair, 36
Ordinate, 44
Origin of rectangular coordinate system, 43

Parabola(s), 114–116, 125–126, 140, 381–382
Parallel lines, 74
Periodic function, 250
Perpendicular lines, 74–75
Point of inflection, 195
Polar coordinates, 376–386
  conic sections in, 381–383
  definition, 376
  graphs of curves in, 379–386
  initial ray of, 376
  pole of, 376
  relationship to rectangular coordinates, 378–379
Polynomial functions
  concavity of graphs of, 190–195
  constant, 159
  continuity of, 169–170
  definition, 159
  graphs of, 163–166, 196–201, 222–225
  points of inflection of graphs, 195
  real polynomial function, 159
  relative maximum and minimum points of, 187–188

Polynomial functions (*Continued*)
  of single variable, 159
  slope functions of, 172–182
  zeros of, 159–161, 203–205, 213–236
  zero polynomial function, 159
Polynomials
  coefficients of, 159
  definition, 158
  degree of, 159
  irreducible, 220–225
  Factor Theorem for, 203
  Remainder Theorem for, 202
  zeros of, 160–161, 203–205, 213–236
Principal root of a number, 306
Product
  of functions, 90
  of sequences, 25–27, 29–31
Pythagorean identity for trigonometric functions, 261, 291–292

Quadratic equations, 145–148, 213–215
  complex roots of, 213–215
  discriminant of, 145–147
  real roots of, 145–148
Quadratic expressions, irreducible, 220–225
Quadratic formula, 145
Quadratic functions
  concavity of graph, 139
  definition, 110
  of form $q(x) = x^2$, 110–116
  of form $q(x) = ax^2$, 123–126
  of form $q(x) = ax^2 + c$, 128
  of form $q(x) = ax^2 + bx + c$, 134–141
  line of symmetry of graph, 139
  slope function of, 155
  tangents to graph of, 150–153
  vertex of graph, 140–141
  zeros of, 160
Quadrants, 45
Quotient
  of functions, 90–91
  of sequences, 26, 29–31, 33–34

Range of function, 36, 48
Rational functions
  definition, 345
  graphs of, 347–350, 356–359, 373–375
  horizontal asymptotes of graph, 360–365
  improper, 345
  proper, 345
  slope function of, 352–355

Rational functions (*Continued*)
  vertical asymptotes of graph, 366–372
  zeros of, 348
Real valued function of single real variable, 46
Reciprocal function, 91
Recursive sequences, 2
Reflection(s)
  of a point in a line, 96
  of a point in a point, 97
Relative maximum, minimum points, 187–188
Remainder Theorem, 202
Restriction of a function, 40, 53

Scientific notation, 304
Secant function
  definition, 284
  graph of, 285
  properties of, 284–286
Sequences
  arithmetic, 2
  bounded, 14–15
  constant, 4
  converging, 13, 16–18, 22–24, 29–31, 33–34
  decreasing, 4, 21
  definition, 1
  difference of, 26, 29–31
  diverging, 13
  Fibonacci, 2
  as functions, 37–38, 49
  geometric, 2, 14, 16–18
  increasing, 4, 21
  limit of, 4–6, 11–12, 13, 16–18, 22–24, 29–31, 33–34
  monotonic, 22
  nondecreasing, 21–23
  nonincreasing, 22–23
  product of, 25–27, 29–31
  quotient of, 26, 29–31, 33–34
  recursive, 2
  sum of, 25–27, 29–31
  unbounded, 14–15
Sine
  of an angle, 247
  of a real number, 246
Sine function
  addition formula for, 263
  definition, 246
  difference formula for, 263
  double angle formula for, 263
  graph of, 252–254
  half-angle formula for, 264
  properties of, 249–254
  range of, 251
  slope function of, 267, 269
Slope
  of a line, 64–66
  of a segment, 63–64

Slope function(s), 154–156, 172–182, 267–269, 318–324, 352–355
Sum
  of functions, 90, 128
  of sequences, 25–27, 29–31
Symmetric functions, 100–103
Symmetry
  of graphs of function and its inverse, 105–108
  with respect to a line, 96
  with respect to the origin, 97–98
  with respect to a point, 97
  with respect to y-axis, 97

Tangent function
  definition, 270
  graph of, 271–273, 274–277
  properties of, 270–273
  slope function of, 274–275, 278–279
Tangent lines, 150–153
Translates of graph, 132, 136
Translation, 136
Trigonometric functions, 237, 246–269, 270–293
Trigonometric identities
  addition formula for cosine function, 263
  addition formula for sine function, 263
  addition formula for tangent function, 292
  difference formula for cosine function, 262
  difference formula for sine function, 263
  double angle formula for cosine function, 263
  double angle formula for sine function, 263
  double angle formula for tangent function, 292
  half-angle formula for cosine function, 264
  half-angle formula for sine function, 264
  Pythagorean identities, 261, 291–292

Unbounded sequences, 14–15
Unit circle, 243

Vertex of parabola, 140–141

Wrapping function, 243–244

Zero of multiplicity k, 219
Zeros of function, 159–161, 203–205, 213–236, 348

INDEX **410**